Artificial Intelligence and Machine Learning Applications in Civil, Mechanical, and Industrial Engineering

Gebrail Bekdaş
Istanbul University–Cerrahpaşa, Turkey

Sinan Melih Nigdeli
Istanbul University–Cerrahpaşa, Turkey

Melda Yücel
Istanbul University–Cerrahpaşa, Turkey

A volume in the Advances in Computational
Intelligence and Robotics (ACIR) Book Series

Published in the United States of America by
    IGI Global
    Engineering Science Reference (an imprint of IGI Global)
    701 E. Chocolate Avenue
    Hershey PA, USA 17033
    Tel: 717-533-8845
    Fax: 717-533-8661
    E-mail: cust@igi-global.com
    Web site: http://www.igi-global.com

Library of Congress Cataloging-in-Publication Data

Names: Bekdas, Gebrail, 1980- editor. | Nigdeli, Sinan Melih, 1982- editor.
  | Yucel, Melda, 1995- editor.
Title: Artificial Intelligence and machine learning applications in civil,
  mechanical, and industrial engineering / Gebrail Bekdas, Sinan Melih
  Nigdeli, and Melda Yucel, editors.
Description: Hershey, PA : Engineering Science Reference, [2019] | Includes
  bibliographical references. | Summary: "This book examines the
  application of artificial intelligence and machine learning civil,
  mechanical, and industrial engineering"-- Provided by publisher.
Identifiers: LCCN 2019021269 | ISBN 9781799803010 (h/c) | ISBN
  9781799803027 (s/c) | ISBN 9781799803034 (eISBN)
Subjects: LCSH: Artificial intelligence. | Civil engineering--Data
  processing. | Machine learning. | Mechanical engineering--Data
  processing. | Industrial engineering--Data processing.
Classification: LCC TA347.A78 A79 2019 | DDC 620.00285/63--dc23
LC record available at https://lccn.loc.gov/2019021269

This book is published in the IGI Global book series Advances in Computational Intelligence and Robotics (ACIR) (ISSN: 2327-0411; eISSN: 2327-042X)

British Cataloguing in Publication Data
A Cataloguing in Publication record for this book is available from the British Library.

For electronic access to this publication, please contact: eresources@igi-global.com.

Advances in Computational Intelligence and Robotics (ACIR) Book Series

Ivan Giannoccaro
University of Salento, Italy

ISSN:2327-0411
EISSN:2327-042X

MISSION

While intelligence is traditionally a term applied to humans and human cognition, technology has progressed in such a way to allow for the development of intelligent systems able to simulate many human traits. With this new era of simulated and artificial intelligence, much research is needed in order to continue to advance the field and also to evaluate the ethical and societal concerns of the existence of artificial life and machine learning.

The **Advances in Computational Intelligence and Robotics (ACIR) Book Series** encourages scholarly discourse on all topics pertaining to evolutionary computing, artificial life, computational intelligence, machine learning, and robotics. ACIR presents the latest research being conducted on diverse topics in intelligence technologies with the goal of advancing knowledge and applications in this rapidly evolving field.

COVERAGE

- Fuzzy Systems
- Algorithmic Learning
- Brain Simulation
- Computational Logic
- Synthetic Emotions
- Heuristics
- Artificial Life
- Computer Vision
- Artificial Intelligence
- Pattern Recognition

IGI Global is currently accepting manuscripts for publication within this series. To submit a proposal for a volume in this series, please contact our Acquisition Editors at Acquisitions@igi-global.com or visit: http://www.igi-global.com/publish/.

Titles in this Series

For a list of additional titles in this series, please visit:
https://www.igi-global.com/book-series/advances-computational-intelligence-robotics/73674

Virtual and Augmented Reality in Education, Art, and Museums
Giuliana Guazzaroni (Università Politecnica delle Marche, Italy) and Anitha S. Pillai (Hindustan Institute of Technology and Science, India)
Engineering Science Reference • © 2020 • 385pp • H/C (ISBN: 9781799817963) • US $225.00

Advanced Robotics and Intelligent Automation in Manufacturing
Maki K. Habib (The American University in Cairo, Egypt)
Engineering Science Reference • © 2020 • 325pp • H/C (ISBN: 9781799813828) • US $245.00

Language and Speech Recognition for Human-Computer Interaction
Amitoj Singh (MRS Punjab Technical University, India) Munish Kumar (MRS Punjab Technical University, India) and Virender Kadyan (Chitkara University, India)
Engineering Science Reference • © 2020 • 225pp • H/C (ISBN: 9781799813897) • US $185.00

Control and Signal Processing Applications for Mobile and Aerial Robotic Systems
Oleg Sergiyenko (Universidad Autónoma de Baja California, Mexico) Moises Rivas-Lopez (Universidad Autónoma de Baja California, Mexico) Wendy Flores-Fuentes (Universidad Autónoma de Baja California, Mexico) Julio Cesar Rodríguez-Quiñonez (Universidad Autónoma de Baja California, Mexico) and Lars Lindner (Universidad Autónoma de Baja California, Mexico)
Engineering Science Reference • © 2020 • 340pp • H/C (ISBN: 9781522599241) • US $225.00

Examining Fractal Image Processing and Analysis
Soumya Ranjan Nayak (Chitkara University, India) and Jibitesh Mishra (College of Engineering and Technology, India)
Engineering Science Reference • © 2020 • 305pp • H/C (ISBN: 9781799800668) • US $245.00

AI and Big Data's Potential for Disruptive Innovation
Moses Strydom (Emeritus, France) and Sheryl Buckley (University of South Africa, South Africa)
Engineering Science Reference • © 2020 • 405pp • H/C (ISBN: 9781522596875) • US $225.00

Handbook of Research on the Internet of Things Applications in Robotics and Automation
Rajesh Singh (Lovely Professional University, India) Anita Gehlot (Lovely Professional University, India) Vishal Jain (Bharati Vidyapeeth's Institute of Computer Applications and Management (BVICAM), New Delhi, India) and Praveen Kumar Malik (Lovely Professional University, India)
Engineering Science Reference • © 2020 • 433pp • H/C (ISBN: 9781522595748) • US $295.00

701 East Chocolate Avenue, Hershey, PA 17033, USA
Tel: 717-533-8845 x100 • Fax: 717-533-8661
E-Mail: cust@igi-global.com • www.igi-global.com

Table of Contents

Detailed Table of Contents

This chapter presents a summary review of the development of artificial intelligence (AI). The chapter provides definitions of AI, together with its basic features, and illustrates the development process of AI and machine learning. Further, the authors outline the developments of applications from the past to today and the use of AI in different categories. Finally, they describe prediction applications using artificial neural networks for engineering applications. The outcomes of this review show that the usage of AI methods to predict optimum results is the current trend and will be more important in the future.

This chapter reveals the advantages of artificial neural networks (ANNs) by means of prediction success and effects on solutions for various problems. With this aim, first, the authors explain multilayer ANNs and their structural properties. Then, they present feed-forward ANNs and a type of training algorithm called back-propagation, which used this type of networks. Different structural design problems in civil engineering are optimized and handled for obtaining prediction results thanks to the usage of ANNs.

Steel wire ropes are frequently subjected to dynamic reciprocal bending movement over sheaves or drums in cranes, elevators, mine hoists, and aerial ropeways. This kind of movement initiates fatigue damage on the ropes. It is a quite significant case to know bending cycles to failure of rope in service, which is also known as bending over sheave fatigue lifetime. It helps to take precautions in the plant in advance and eliminate catastrophic accidents due to the usage of rope when allowable bending cycles are exceeded. To determine the bending fatigue lifetime of ropes, experimental studies are conducted. However, bending over sheave fatigue testing in laboratory environments require high initial preparation cost and a long time to finalize the experiments. Due to those reasons, this chapter focuses on a novel prediction perspective to the bending over sheave fatigue lifetime of steel wire ropes by means of artificial neural networks.

Chapter 4

Ivo Bukovsky, Czech Technical University in Prague, Czech Republic
Peter M. Benes, Czech Technical University in Prague, Czech Republic
Martin Vesely, Czech Technical University in Prague, Czech Republic

This chapter recalls the nonlinear polynomial neurons and their incremental and batch learning algorithms for both plant identification and neuro-controller adaptation. Authors explain and demonstrate the use of feed-forward as well as recurrent polynomial neurons for system approximation and control via fundamental, though for-practice, efficient machine learning algorithms such as Ridge Regression, Levenberg-Marquardt, and Conjugate Gradients; authors also discuss the use of novel optimizers such as ADAM and BFGS. Incremental gradient descent and RLS algorithms for plant identification and control are explained and demonstrated. Also, novel BIBS stability for recurrent HONUs and for closed control loops with linear plant and nonlinear (HONU) controller is discussed and demonstrated.

Chapter 5

Melda Yucel, Istanbul University-Cerrahpaşa, Turkey
Aylin Ece Kayabekir, Istanbul University-Cerrahpaşa, Turkey
Sinan Melih Nigdeli, Istanbul University-Cerrahpaşa, Turkey
Gebrail Bekdaş, Istanbul University-Cerrahpaşa, Turkey

In this chapter, an application for demonstrating the predictive success and error performance of ensemble methods combined via various machine learning and artificial intelligence algorithms and techniques was performed. For this reason, two single methods were selected, and combination models with a Bagging ensemble were constructed and operated with the goal of optimally designing concrete beams covering with carbon-fiber-reinforced polymers (CFRP) by ensuring the determination of the design variables. The first part was an optimization problem and method composing an advanced bio-inspired metaheuristic called the Jaya algorithm. Machine learning prediction methods and their operation logics were detailed. Performance evaluations and error indicators were represented for the prediction models. In the last part, performed prediction applications and created models were introduced. Also, the obtained predictive success of the main model, as generated with optimization results, was utilized to determine the optimal predictions of the test models.

Computer vision methods are widespread techniques mostly used for detecting cracks on structural components, extracting information from traffic flows, and analyzing safety in construction processes. In recent years, with the increasing usage of machine learning techniques, computer vision applications have been supported by machine learning approaches. So, several studies have been conducted which apply machine learning techniques to image processing. As a result, this chapter offers a scientometric analysis for investigating the current literature of image processing studies for the civil engineering field in order to track the scientometric relationship between machine learning and image processing techniques.

Chapter 7

In this chapter, the authors realized prediction applications of concrete compressive strength values via generation of various hybrid models, which are based on decision trees as main a prediction method. This was completed by using different artificial intelligence and machine learning techniques. In respect to this aim, the authors presented a literature review. The authors explained the machine learning methods that they used as well as with their developments and structural features. Next, the authors performed various applications to predict concrete compressive strength. Then, the feature selection was applied to a prediction model in order to determine parameters that were primarily important for the compressive strength prediction model. The authors evaluated the success of both models with respect to correctness and precision prediction of values with different error metrics and calculations.

Innovation and technology are trending in the industry 4.0 revolution, and dealing with environmental issues is no exception. The articulation of artificial intelligence (AI) and its application to the green economy, climate change, and sustainable development are becoming mainstream. Water is a resource that has direct and indirect interconnectedness with climate change, development, and sustainability goals. In recent decades, several national and international studies revealed the application of AI and algorithm-based studies for integrated water management resources and decision-making systems. This chapter identifies major approaches used for water conservation and management. On the basis of a literature review, the authors will outline types of approaches implemented through the years and offer instances of the ways different approaches selected for water conservation and management studies are relevant to the context.

Urbanization, industrialization, and increase in population lead to depletion of groundwater and also deteriorate its quality. Madurai city is one of the oldest cities in India. In this study, the authors assessed the quality of groundwater using various statistical techniques. The researchers collected groundwater samples from 11 bore wells and 5 dug wells in the post-monsoon season, in 2002, and analyzed the samples for physicochemical characterization in the laboratory. They analyzed around 17 physicochemical parameters for all the samples. The aim of the descriptive statistical analysis was understanding the correlation between each parameter. Then, the authors carried out cluster analysis to identify the most affected bore well and dug well in Madurai city.

Chapter 10

Smart transportation is a framework that leverages the power of Information and Communication Technology for acquisition, management, and mining of traffic-related data sources. This chapter categorizes them into probe people and vehicles based on Global Positioning Systems, mobile phone cellular networks, and Bluetooth, location-based social networks, and transit data with the focus on smart cards. For each data source, the operational mechanism of the technology for capturing the data is succinctly demonstrated. Secondly, as the most salient feature of this study, the transport-domain applications of each data source that have been conducted by the previous studies are reviewed and classified into the main groups. Possible research directions are provided for all types of data sources. Finally, authors briefly mention challenges and their corresponding solutions in smart transportation.

Chapter 11

The rapid growth in the number of drivers and vehicles in the population and the need for easy transportation of people increases the importance of public transportation. Traffic becomes a growing problem in Istanbul which is Turkey's greatest urban settlement area. Decisions on investments and projections for the public transportation should be well planned by considering the total number of passengers and the variations in the demand on the different regions. The success of this planning is directly related to the accurate passenger demand forecasting. In this study, machine learning algorithms are tested in a real world demand forecasting problem where hourly passenger demands collected from two transfer stations of a public transportation system. The machine learning techniques are run in the WEKA software and the performance of methods are compared by MAE and RMSE statistical measures. The results show that the bagging based decision tree methods and rules methods have the best performance.

Chapter 12

Sinem Büyüksaatçı Kiriş, Istanbul University-Cerrahpasa, Turkey
Tuncay Özcan, Istanbul University-Cerrahpasa, Turkey

Vehicle routing problem (VRP) is a complex problem in the Operations Research topic. School bus routing (SBR) is one of the application areas of VRP. It is also possible to examine the employee bus routing problem in the direction of SBR problem. This chapter presents a case study for capacitated employee bus routing problem with data taken from a retail company in Turkey. A mathematical model was developed based on minimizing the total bus route distance. The number and location of bus stops were determined using k-means and fuzzy c-means clustering algorithms. LINGO optimization software was utilized to solve the mathematical model. Then, due to NP-Hard nature of the bus routing problem, simulated annealing (SA) and genetic algorithm (GA)-based approaches were proposed to solve the real-world problem. Finally, the performances of the proposed approaches were evaluated by comparing with classical heuristics such as saving algorithm and nearest neighbor algorithm. The numerical results showed that the proposed GA-based approach with k-means performed better than other approaches.

Chapter 13

Ramazan Ünlü, Gumushane University, Turkey

Manual detection of abnormality in control data is an annoying work which requires a specialized person. Automatic detection might be simpler and effective. Various methodologies such as ANN, SVM, Fuzzy Logic, etc. have been implemented into the control chart patterns to detect abnormal patterns in real time. In general, control chart data is imbalanced, meaning the rate of minority class (abnormal pattern) is much lower than the rate of normal class (normal pattern). To take this fact into consideration, authors implemented a weighting strategy in conjunction with ANN and investigated the performance of weighted ANN for several abnormal patterns, then compared its performance with regular ANN. This comparison is also made under different conditions, for example, abnormal and normal patterns are separable, partially separable, inseparable and the length of data is fixed as being 10,20, and 30 for each. Based on numerical results, weighting policy can better predict in some of the cases in terms of classifying samples belonging to minority class to the correct class.

Chapter 14

Didem Filiz, Ankara University, Turkey
Ömer Özgür Tanrıöver, Ankara University, Turkey

In this chapter, the authors explore keystroke dynamics as behavioral biometrics and the effectiveness of state-of-the-art machine learning algorithms for identifying and authenticating users based on keystroke data. One of the motivations of this study is to explore the use of classifiers to the field of keystroke dynamics. In different settings, recent machine learning models have been relatively inexpensive and effective with limited data. Therefore, the authors conducted experiments with two different keystroke dynamics datasets with limited data. They demonstrated the effectiveness of the models using a dataset obtained from touch screen devices (mobile phones) and also on normal keyboards. Although there are similar recent studies that explore different classification algorithms, their main aim has been anomaly detection. However, the authors experimented with classification methods for binary and multiclass user identification and authentication using two different keystroke datasets from touchscreens and keyboards.

Preface

Various optimization techniques have been utilized from the past to the present in order to solve engineering problems and obtain optimized designs. These techniques consist of classical mathematical methods, and these methods remain incapable of optimally solving real-life problems that have a complex structure with more than one requirement at the same time or of showing non-linear behavior. This situation shows the truth of the need for new methods.

Nowadays, optimization techniques—which are advanced and more flexible in structure—have emerged with the help of developing technology and computers. These are known as heuristic methods. They involve an initial situation. Metaheuristic methods are one step ahead of heuristic methods. They also involve different algorithms. However, although they avail much in terms of their ability to provide solutions to many (multiple) problems, they require a very long time to process those problems and determine the resulting solutions to those problems.

In order to solve these problems, and in order to obtain fast and effective solutions to a range of different problems, a different approach and one that is effective is the use of artificial intelligence (AI) and machine learning (ML) methods. With these methods, it is not necessary to have continuous optimization. Moreover, the optimum values of multi models can be determined simultaneously and quickly. In this way, it is possible to determine the optimum value for multiple designs without the need for new and repeated optimization processes via the learner model. At the same time, the optimal values for various new models that have been constructed with the hybrid models enable users to estimate results quickly and accurately.

Factors such as economic objectives (such as the goal of keeping costs to a minimum), the ensuring the feasibility of achieving an intended purpose and a correct design; esthetic objectives, etc.; have played a role in addition to the objective of providing all required safety features, among the principal aims of engineering designs. In the process of deciding on the appearance of the designs, the designer may decide to provide solutions all such aims at the same time. Such a decision often depends on the experience of the engineers. Some conditions are optional. However, certain conditions must be carried out. For instance, these include carefully and correctly calculating various variables that comprise the design; and correctly incorporating controlled material usage in the framework of a safe and economic design.

In this respect, optimization methods become useful for solving various engineering problems most appropriately. Especially nowadays, metaheuristics—which are derived from advanced optimization methods—avail much in terms of various problems. Such problems include determining the best tool; minimizing structural/material costs, optimizing structural weight; and managing projects. On the other hand, the use of artificial intelligence and machine learning techniques (ones derived from today's technologies) is gradually increasing in multiple areas of engineering. Many problems can be solved rapidly, accurately, and effectively, thanks to these methods. These methods provide for the realization of various activities such as forecasting, inference, and feature identification. At the same time, these methods—when executed—can provide ease and efficiency in terms of time, cost, and effort.

With this goal, meeting the required/preferred conditions is ensured by resolving various engineering problems through optimizing operations. This is being realized by using a variety of different methods and by using the most appropriate values of the parameters of the problem. The ability of hybrid models to predict outcomes is improved both in terms of speed and of correctness. These hybrid models are created by using various AI and ML methods. This is a result of this process.

In recent years, machine learning and artificial intelligence applications have been gaining greater and greater importance. Especially, they can provide effective solutions for different problems in engineering applications. The main purpose of this book is to introduce concepts of artificial intelligence and machine learning; to explain methods of machine learning; and to introduce various engineering applications that are performed with these methods. In this respect, various examples of solutions which have been realized with AI and ML methods will be reflected on. This book is intended as a guide for various fundamental problems from the fields of civil engineering, mechanical engineering, and industrial engineering, along with other engineering fields. Thus, the book can be a guide for students, researchers, and academics in various fields of engineering.

Furthermore, some of the engineering problems in the book are related to examples that involve optimum values obtained by optimization, and that are associated with training with machine learning methods, and which thus minimize the time and effort required to generate new designs. Other problems reflect efficiency and speed in solving of problem with using one of these methods. The book provides new approaches in terms of ML methods and forms of usage.

The applications of artificial intelligence and machine learning—which have become widespread with advancing technology—are no longer limited to certain areas. In many countries, in the field of education—such as in universities; research centers; etc.—such applications are conveyed in various lectures or courses, being sources of such information. In particular, it is thought that different applications can be studied together by undergraduate and graduate students alike.

In addition, the book tackled artificial intelligence and machine learning concepts which are derived from the most recent and technological methods and which represent practices that are performed in various branches of engineering. They are also suitable for use by researchers.

Artificial Intelligence (AI) applications are the most interesting and fruitful subject of the new era. The most trendy applications of AI are related to technological devices such as human-like robots; electronic devices; autonomous vehicles; drones; etc. There is no doubt that these technologies will be a part of future life. But now, we use AI in our everyday life at present. We cannot realize that AI applications are responsible from a process standpoint.

The main scope of the book is to express information concerning AI applications depending on a prediction process via machine learning applications. Similar applications from everyday life using AI and machine learning are also given. Engineering applications, especially in the fields of civil, mechanical, and industrial engineering—may critically need AI techniques for organization, planning, and prediction.

The structure of the book is as follows:

In Chapter 1, a review of AI and ML methods are presented. The development of methods is given by the explanation of basic features. The development of applications from past to present can be found. Prediction applications using artificial neural networks (ANNs) are also mentioned at the end of the chapter.

The main aim second chapter is to present ANN in more detail—including the presentation of multilayer ANNS, feed-forward ANNs and back-propagation. Two prediction problems for the application of structural mechanics are presented. The first problem involves the prediction of the optimal design of tubular columns. In this application, the optimal results of several cases of design were used in training; and a metaheuristic algorithm called Teaching Learning-Based Optimization (TLBO) is used for optimization. For the second application, the optimal design of an I-beam is predicted for vertical deflection minimization. For optimal results used in machine learning, a Flower Pollination Algorithm (FPA) is used. At the end of Chapter 2, a code for the optimal design of the tubular column problem is also given.

In the third chapter of the book, a novel prediction proposal is presented for predicting the lifetime fatigue of bent sheaves of steel wire ropes. In the prediction, ANN is employed. For providing a prediction regarding the lifetime fatigue of bent sheaves of steel wire ropes used in cranes, elevators, mine hoisting, and aerial ropeways, experimental methods are generally used. However, the fatigue testing of bent sheaves requires high cost and time. For that reason, for this application, there are great advantages to using ANN methods.

Chapter 4 is related to nonlinear polynomial neurons and batch learning algorithms for applications such as plant identification and neuro-controller adaptations. This application employs Ridge Regression, Levenberg-Marquardt, and Conjugate Gradients for ML algorithms. A novel optimizer was also used. Comparison results were discussed.

In Chapter 5, a civil engineering application is presented. In the retrofit applications presented, the shear force capacity of the reinforced concrete (RC) beam is increased by carbon-fiber-reinforced polymer (CFRP) wrapping. In the application presented, a methodology for the optimization and prediction of optimal results is presented in the application.

Chapter 6 includes computer vision methods. In engineering, computer vision methods are used for detecting cracks in structural members; extracting information from traffic flows; performing safety analyses in construction processes; etc. Computer vision applications are supported via ML for image processing.

The discussion of predictive applications in civil engineering continues in Chapter 7. In this chapter, the prediction of the compressive strength of high-performance concrete (HPC) is presented. This prediction is realized via the generation of several hybrid models, based on decision trees.

Chapter 8 is related to integrated water management resources and decision-making systems. The use of AI method and soft computing algorithms for water management is important, because water resources have direct and indirect inter-connectedness with climate change, development, and sustainability.

In Chapter 9, a brief case study is presented regarding the analysis of ground water quality using statistical techniques. The case study is done for the city of Madurai in India.

Chapter 10 is related to smart transportation. Applications regarding probing people-based and vehicle-based data sources are presented. By using technologies, smart transportation is a formwork providing a means for the acquisition, management, and mining of traffic-related data sources.

In Chapter 11, passenger demand prediction via ML methods is presented. The rapid growth of vehicles and the need for the easy transportation of people increases the demand for public transportation. In Chapter 11, a case study of Istanbul's public transportation system is also presented.

Chapter 12 includes a clustering-based bus stop selection for bus routing problems. As an operational research topic, the problem of vehicle routing is a complex and important one. In chapter 12, it is possible to find discussion of methods such as K-Means, fuzzy C-Means, simulated annealing, genetic algorithm, saving algorithm, and nearest neighborhood algorithm.

Chapter 13 presents an ANN application for chart pattern recognition in unbalanced control data. Comparatively, the automatic detection of abnormality in control data is more effective than manual detection is. A weighting strategy implemented via the use of ANN was investigated.

Chapter 14 is related to biometric identification based on Keystroke dynamics. Several ML methods are compared and discussed.

Chapter 1
Review and Applications of Machine Learning and Artificial Intelligence in Engineering:
Overview for Machine Learning and AI

Melda Yucel
Istanbul University-Cerrahpaşa, Turkey

Gebrail Bekdaş
Istanbul University-Cerrahpaşa, Turkey

Sinan Melih Nigdeli
Istanbul University-Cerrahpaşa, Turkey

ABSTRACT

This chapter presents a summary review of the development of artificial intelligence (AI). The chapter provides definitions of AI, together with its basic features, and illustrates the development process of AI and machine learning. Further, the authors outline the developments of applications from the past to today and the use of AI in different categories. Finally, they describe prediction applications using artificial neural networks for engineering applications. The outcomes of this review show that the usage of AI methods to predict optimum results is the current trend and will be more important in the future.

ARTIFICIAL INTELLIGENCE AND MACHINE LEARNING

Alongside generally related to computer sciences and engineering disciplines, artificial intelligence (AI) pertains to science and technology, and benefits many fields, such as biology and genetics, psychology, language learning and comprehension, and mathematics.

As AI is accepted as a branch of a science, being considered as a field of research and technology at the same time, its various definitions are from many different sources. Some of these are as follows:

DOI: 10.4018/978-1-7998-0301-0.ch001

1. This approach, which is human-oriented, should be an experimental science that includes hypothesis and experiences within (Russell & Norvig, 1995).
2. AI can be defined as an effort for developing computer operations, which will ensure to find out the similarities of this structure, via an understanding of the human thinking structure (Uygunoğlu & Yurtçu, 2006).
3. McCarthy (2007) remarked that of AI is a science and engineering discipline, which is generated by intelligent machines, especially computer programming, too.
4. Luger (2009) defined AI as a computer science branch that is related to automation of intelligent behaviors.
5. AI is an expert system to understand intelligent beings, and establish and make the process of decision-making productive, rapid, and simple (Patil et al., 2017).

Based on these different definitions, AI is a technology, which is made of computers, which function similarly to humans' intelligent structure and thinking behaviors, intelligently thinking software or computer-controlled robots.

However, although this technology is a case, which is related to using computers in order to understand human intelligence, it should not be limited to methods that are measurable only according to biological factors (McCarthy, 2007). On the other hand, in recent times, the main reason of importantly advancing of AI technology consists in developments in computer functions, which are integrated with human intelligence, such as reasoning, nominated ability of discernment, learning, and solving probles.

Machine learning, which is seen as a subfield of AI, is a technology related to designing and developing algorithms and techniques, which ensure devices to learn, such as computers (Olivas, Guerrero, Sober, Benedito, & López, 2009). In this respect, the concept of machine learning expresses the generated changes in systems by tasks, which include the actions which are realized with AI, such as recognition, robot-controlling, detection/identification, and prediction. Also, with occurring changes, previously generated systems are developed or new systems are synthesized. Figure 1 shows the structure of a typical AI tool, which illustrates this case more clearly.

This tool senses information coming from its surrounding and realizes a suitable modelling. Actions are calculated based on the possible effects from the models being predicted. Conversely, changes, which can occur in any of AI components Figure 1 shows, may be regarded as learning. Also, different learning mechanisms may be run, based on the change of subsystems.

The definition of learning in the dictionary is gain of knowledge and success through study, understanding or experience, become skillful and changes that occurred via experience in behavioral tendencies (Nilsson, 1998). Consequently, learning happens in a long and iterative process, and generates alterations, which ensure certain attainments and experiences as a result of this process, too. The development of many technological devices (e.g., computer) and becoming like humans are unavoidable when various activities are carried out. In this case, machine learning is considered as gain of experience or information of machines, such as computers, as a result of a variety of events (e.g., developing various decision-making mechanisms and foreknowing similar states) which may be lived in the future.

As the authors remarked above, in order to have a successful learning process in devices, firstly actualized learning actions of biologic beings, namely people and animals, and the features of processes should be understood properly. However, this information is insufficient alone and in terms of engineering, too; some vital information and calculations are needed (Nilsson, 1998). Some of these are as follows:

Figure 1. Artificial intelligence system
(Nilsson, 1998)

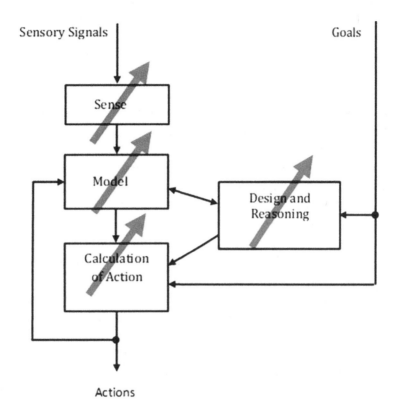

1. The use of metalevel knowledge makes the control of solving-problem strategies more advanced. Although this is a very hard problem, investigation of some current systems can be relatively considerable as an area of research (Luger, 2009).
2. A program can only learn behaviors and factors, and which formation (mathematical or logical definition) of itself may be represented. However, unfortunately, almost all of learned systems depend on very limited abilities about representing the knowledge (McCarthy, 2007).
3. Some tasks are not described well without a sample. Therefore, it is possible to determine input-output couples for tasks. Yet, a clear relation may not occur between inputs and desired outputs.
4. The current amount of information of specific duties can be too much, because open coding can be performed by people.

DEVELOPMENT PROCESS OF ARTIFICIAL INTELLIGENCE AND MACHINE LEARNING

Machine learning technology is a branch of AI that focuses on independent computers and machines, and realizes the conversion of these to smart entities. Various features which belong to humans (e.g., physical, genetic, and intelligence) benefit from this conversion.

Since the discovery and usage of computers and various technological calculators, hand in hand with the continuous development of technology, the development of more innovative and differentiated machines has occurred. Nevertheless, only in the 20th century these machines were able to think, experience, and learn lessons from mistakes, by making various comparisons like humans.

In this respect, one of the first studies in the literature is cybernetics, meaning communication science and automatic control systems between livings and machines, which Arturo Rosenblueth, Norbert Wiener, and Julian Bigelow developed in 1943. In 1948, Weiner's famous book was published with this name, too. Between 1948 and 1949, Grey Walter performed an experiment in Bristol, based on the opinion that little count brain cells can cause complex behaviors with autonomous robots, which are turtles called Elsie and Elmer. Also, in 1949, Donald Hebb proved a simple and current rule for altering connection weights between neurons which ensure to realize the learning (Buchanan, 2006).

The real important developments about AI in terms of machine learning occurred in 1950 and after. For example, in 1950, Alan Turing published the Turing test, which is required for measuring intelligent behaviors and which he included in the book called *Computing Machinery and Intelligence*. In this book, the author discussed the required conditions to be able to think that a machine is intelligent. According to Turing, it must be thought that the machine is certainly smart in case a machine shams a human successfully, compared to a smart observer (McCarthy, 2007).

The application in Figure 2 is a method which Turing generated and named as a simulation game. In this test, a machine and a human are brought to face opposite, and in a separate section another person, called querier, is present. The querier cannot see directly any entities within the test and talk with them. However, the querier actually does not know which entity is a machine or a human, and can communicate with them using a device, such as a terminal, which can write only text. From the querier's perspective, the distinction of a human from a computer is based on given answers by entities existed in test, to ask questions, which are asked with just the device. When the machine or the human cannot be distinguished, it may be assumed the machine is intelligent, according to Turing.

Nevertheless, the querier is free to ask any question, regardless how deceptive or indirect, to reveal the identity of the computer. For example, the querier can want arithmetic calculation from both, by assuming a high probability that the computer gives more accurate results than the human. Contrary to this strategy, the computer may need to know when it may be unsuccessful about finding correct answers to a set of problems as these, to resemble the human. The querier can want to ask both entities to discuss a poem or artwork, to find out their identity based upon human sensitive structure. Also, a strategy is required to allow the computer for information related to the sensual formation of mankind (Luger, 2009).

In this respect, the Turing test is a significant application to measure similarity between a machine and a human. The results showed that machines have similarities with humans at high rate, but these results cannot be considered as right exactly. The clearest reason stands in some features, which come from human creation; machines do not have such sensual and humanized features. Therefore, machines cannot carry out everything as humanoid and, consequently, the fact that machines are more successful than or superior to humans may not be deduced through all kinds of applications.

On the other hand, in the 1950s, the use of the k-means algorithm started. It is one of the commonly attracted clustering methods in machine learning and works to estimate the values of a complex and multidimensional dataset with the help of its several average points (Olivas et al., 2009).

In the same period, another study consisted in chess programs, which Claude Shannon (1950) and Alan Turing (1953) wrote for von-Neumann-style traditional computers. Then, in 1951, Marvin Minsky and Dean Edmonds, who were graduate students at the Mathematics Department of Princeton University,

Figure 2. Application of Turing test
(Luger, 2009)

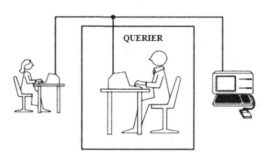

set up the first neural network computer. Between the years 1952 and 1962, Arthur Samuel wrote the first program which can play games, with the aim of having the upper hand in a world champion by gaining adequate skill in checkers. In this period, which started in 1952, the idea that computers can carry out only determined actions by themselves was refuted, because this written program learned rapidly playing a game better than its designer did. Logic Theorist, which was written by Allen Newell, J. C. Shaw, and Herbert Simon, was a demonstration of the first functioning AI program (Russell & Norvig, 1995).

In addition, in 1958, Rosenblatt proposed a simple neural network called "perceptron," which was seen as a milestone in machine learning history (Han & Kamber, 2006). Again, in 1958, McCarthy, who became an important name in the field of AI, discovered the LISP programming language, which is the second oldest programming language still used today. At the beginning of 1960s, Margaret Masterman and friends designed semantic networks for machine translation. In 1964, the thesis of Danny Bobrow at the MIT sustained that computers were good at understanding natural language as well as solving algebraic problems, too. Joel Moses demonstrated the power of symbolic reasoning for integral problems through the Macsyma program (a computer-aided algebra system) in his doctoral thesis, which the MIT published in 1967. This program is the first successful knowledge-based program in the field of math. Another event in this year was the design of a chess program named MacHack, which is based on knowledge, and as good as being able of getting a degree of C-class in a tournament. Its author was Richard Greenblatt, from the MIT (Buchanan, 2006).

In conjunction with an increased adoption of the approach of knowledge-based systems, the intelligibility problem of a molecular structure was solved through the information which was acquired with the mass spectrometer Ed Feigenbaum, Bruce Buchanan, and Joshua Lederberg had developed in 1969. The program input was formed from the base (beginning) formula of the molecule, and, when the created molecule was bombarded by cathode (electron) rays, the mass spectrometer put forth masses of various parts of the molecule. For instance, the mass spectrometer can include a density at m=15 corresponding to the mass of a methyl part. By producing all possible structures, which were interconnected with the formula, with a simple version of the program, the effect on each part of the mass spectrometer was compared with an original spectrometer by being predicted (Russell & Norvig, 1995).

In 1970 and after, with the rising of machine learning capabilities, more developments were experienced in the fields of industry and neural networks. Moreover, with machines and robots designed for space studies, the aim to use them to simulate the human became widespread gradually.

Coming to 1980, the first expert systems and commercial applications became more and more developed. For example, Lee Erman, Rick Hayes-Roth, Victor Lesser, and Raj Reddy developed the Hearsay-II speech-understanding system, which has a blackboard model as a skeleton structure. With respect to the middle of the 1980s, Werbos defined how to work neural networks by using the back-propagation algorithm, for the first time in 1974. Pattern recognition, which is another technology, is an issue closely related to machine learning; Bishop (1995) and Ripley (1996) explained the usage of neural networks for it. On the other hand, Mitchell (1997) wrote a book covering many machine learning techniques, also including genetic algorithms and reinforcement learning. Again, in 1997, the program called The Deep Blue defeated Garry Kasparov, who was then world chess champion. At the end of the 20th century, Rod Brooks improved the Cog Project at the MIT, which allowed a major progress on building a humanoid robot. With these advances, it is sighted that any machines and systems have abilities on human level and even come to a grade that will rival the human. For example, Cynthia Breazeal's thesis *Social Machines* described Kismet, which is a robot whose face can express its feelings. Also, in 2005, Robot Nomad explored Antarctica's far areas, to search for meteorite samples (Buchanan, 2006; Witten, Hall, & Frank, 2011).

ARTIFICIAL INTELLIGENCE AND MACHINE LEARNING

Machine learning improved with AI technology can be used for many different applications in various areas. Performed applications and developments in these areas are explained as follows:

1. **Expert Systems:** These systems generate solutions to existing problems in a similar way to solve-style of any problem by an expert. When viewed from this aspect, expert systems were created through conversion to a format that computers can apply for solving similar problems by gaining information from a human expert.

Dendral, which is one of the oldest expert systems and is benefiting from domain knowledge about solving problems, was developed at the end of the 1960s, at Stanford. Generally, the structure of organic molecules is too big; this increases the possibility of the probable number of structures for molecules is higher. Therefore, this system was developed to determine the structure of organic molecules, with mass spectrometer information related to chemical bonds in molecules, and chemical formulas. Another system that has a feature to be first is Mycin, which was developed in 1974. Mycin was improved by using expert medical information, in a way to present cure advices by diagnosing spinal meningitis and bacterial infections in the blood (Luger, 2009).

As another example, in 1991 Bell Atlantic developed a system which allows to assign a company's suitable technician based on the problem the customer reports on the phone. The system, which Bell Atlantic developed with machine learning methods, has features for becoming an expert system, which provides savings for more than ten billion dollars per year through carrying out this decision of improvement in 1999 (Witten et al., 2011). The following systems were also developed as machine learning methods: *Internist*, to diagnose internal diseases; *Prospector*, to determine the kind of mineral deposits and possible locations, according to geological information about a site; *Dipmeter Advisor*, to comment the results of oil well drilling (Luger, 2009).

2. **Robotics:** Besides, AI is generated in order to perform the tasks and actions, which are carried out by people, with a better and quick performance; numbers and usage areas of robots, which are one of the practices developing with AI technology for fulfilling these tasks, increase day by day.

Indeed, AI technology and machine learning, which is a branch of it, are closely related to robotics technology. The reason is that, generally, artificial neural networks (ANNs) or other classification systems are used for various tasks, which are required to control robots (Onwubolu & Babu, 2004). However, nowadays, structures with a simple form (e.g., algorithms, all sort software, Internet robots, and apps), which are AI tools, together with structures with an advanced form (e.g., robots, driverless vehicles, smart watches, and the other mechanical devices) exist. On the other hand, Floridi (2017) predicted that generated digital technologies and automations would take place of employees in fields such as agricultural and production sectors, while in service industry robots would come instead of humans. Even so, the human performs as an interface between a vehicle and a GPS, documents in different languages, a finished meal and its ingredients, or in cases such as diagnosing a disease which matched with symptoms. Thus, robots replacing humans are posing a danger.

3. **Gaming:** Gaming is one of the oldest fields of work of AI. In 1950, as soon as computers became programmable, Claude Shannon and Alan Turing wrote the first chess program. Since that time, a resolute progression has been at stake in game standards, reaching up to the point that today's systems can challenge across to a world champion (Russell & Norvig, 1995).

The largest part of the first researches in this field was performed by using common board games, such as checkers, chess, and fifteen crosswords. As board games have specific features, they represent an ideal topic for AI research. Many games are played by using a well-described set of rules. This condition facilitates the definition of the field of research and saves the researcher from most of the complexities and uncertainness, which is inherent to less structured problems (Luger, 2009). Less abstract games, such as football or cricket, did not attract AI researchers' attention much. As increasing number of competitors in games makes more complex and affects the operating of the installed functions.

Together with the progress in technology, researchers have included other kinds of games (e.g., football, tennis, and volleyball), which require human talent, physical capacity or humanoid responses such as reflex, alongside intelligent games (e.g., chess and checkers), which are preferred as a beginning-phase study field. Researchers obtained real-like values, besides observing the usage AI technology in these kinds of game.

4. **Speech Recognition with Natural Language Process**: Natural language processing is a research and development field, which deals with written and talked (digitized and recorded) language and its data. Natural language processing is known as the science and technology of features and usage of human language, and is mentioned as NLP technology. NLP comes from the first letters of the words "Natural Language Process," and emphasizes the origins of the languages people talk and their role in information processing. The substantial execution areas of theory, tool, and techniques developed in the NLP contain the topics of machine translation, text summarization, text mining, and information extraction.

Further, Olivas et al. (2009) sustained that NLP methods are able to process a manageable amount of data, like people do, and analyze big amounts of data easily. For example, NPL's combination of speech parts automatically is close to faultless performance, which belongs to an adequate human.

One of the first designs is the Lunar system, which William Woods developed in 1973. This system was the first natural language program, which allowed geologists for asking English questions about rock/stone samples from the Apollo Moon Mission, and was used by other people practically (Russell & Norvig, 1995). Besides, examples of modern-day developments of this technology are especially virtual assistants as smart phones applications that transcribe speech, and systems which automatically correct errors in messages and e-mails (Biswas, 2018).

5. **Finance, Banking, and Insurance Applications**: Expenditures, banking actions, and various financial activities reflect the various features of people, and finance, bank, and insurance institutions benefit from new technologies for increasing the number of customers who earn and for meeting these customers' expectations (i.e., customer satisfaction).

An example of application of these technologies can be determining which persons a loan corporation will give how much credit. In this case, loan companies apply a decision procedure by calculating a numeric parameter according to the information they obtain from surveys to applicants. If this parameter is above the defined threshold, the application is accepted, while it is refused if the parameter is below the threshold (Witten et al., 2011). Another example of application of NPL technologies is that, in various sectors, it is possible to plan sale timing by predicting the next purchasing time of a product/service, buying operations for a travel, flight, car or house, the customer's profile, when the operation may be realized, and its rate of possibility of occurrence (Maimon & Rokach, 2010).

6. **Medicine and Health**: Medicine and health science is the primary areas area in which AI and, especially, machine learning applications are used most. They are used for diagnosing and treating of many diseases; moreover, machine learning practices are adopted for recognizing genetic and biological features and improving new technologies.

For instance, determining the most proper medicine dose for a person in treatment of chronic diseases generally occurs by trial and error. Machine learning methods are used in order to speed this process, and identify the suitable medicine treatment for each person (Olivas et al., 2009). Also, these methods help deduce phenotype information from the data of the gene expression, determine how the sequencing of protein and DNA will occur, examine pattern regulations on biological gene sequences and with regard to the generation of different gene patterns, and find out recurrence frequency (Han & Gao, 2008; Maimon & Rokach, 2010; Olivas et al. 2009).

7. **Vision Systems and Image Recognition Technology**: Systems, which are created by benefitting from sight sense and characters of beings within nature, are vision systems. These systems can perform position determination, image selection and combination, pattern and photo analogy, debugging, and many similar processes with image detection actions, mainly by using properties and advantages of the sight sense.

The vision system based on optical flow is an example, and is used for controlling of responses of airplanes during landing. In addition, these systems are intended for aims, such as performing the movement of mobile robots in an environment by detecting location of existing barriers and free spaces, monitoring many astronomical bodies in space, determining varying land cover because of abrupt changes in green zones by identifying damages, which are natural or man-made within ecosystems (Han & Gao, 2008; Russell & Norvig, 1995).

Except from these applications, vision systems are frequently preferred in various areas with many purposes, especially geographical positioning, face recognition, determining vehicle route, detecting faulty products, which can occur during production activities, and diagnosing diseased/healthy cells.

8. **Publication, Web, and Social Media Applications**: Increase of the spending time on the Web and social media, and varying of carried out actions cause to create of a big data stack. Also, the improvement of the new technologies is becoming possible via using the data from people's use of social media accounts and various Internet sites in many scopes.

Particularly, one of these is the sequencing of Web pages by search engines. This problem can be solved via machine learning, depending on a training set creating the data, which is obtained in line with the decisions persons make previously. By analyzing training data, a learning algorithm can predict the relation level of query, thanks to its features, such as URL address content for a new query or whether it takes place at a header label. On the other hand, machine learning applications uses the search terms to select adverts (e.g., choices of other users about advices to similar products on shopping a book or a film/music CD) (Witten et al., 2011).

PREDICTION APPLICATIONS VIA ARTIFICIAL NEURAL NETWORKS

In various study fields, ANNs benefit many topics, such as feature extraction, clustering, classification, notably prediction and forecasting. In engineering science, which is the primary of these fields, ANNs can represent quite successful results, especially about predicting the values of design parameters in handled problems.

Also, in civil engineering, some studies were performed to rapidly predict parameter values belong to a problem, for designing a model. One of these is a study that developed of a model which can predict salinity rate of Murray River in South Australia from 14 days ago, by used ANN and GA methods by Bowden, Maier, and Dandy (2002).

Atici (2011) used ANNs, besides multiple regression analyses, to predict the compressive strength in different curing periods of concrete mixtures consisting of various amounts of blast furnace slag and fly ash. These amounts depend on values and features of additives which are obtained with ultrasonic pulse velocity and nondistructive testing. Momeni, Armaghani, and Hajihassani (2015) used ANNs combined with the particle swarm optimization method, with the aim of increasing network performance, to predict the unconfined compressive strength of granite and limestone rock samples from an area in Malaysia.

In 2015, Aichouri et al. (2015) developed an ANN with the aim of modeling the rain-runoff relation in a basin, which has a semiarid Mediterranean climate in Algeria. Veintimilla-Reyes, Cisneros, and Vanegas (2016) created a model based on ANN, which allowed to predict the water flow in Tomebamba River in real time and for a specific day of the year. In addition, model inputs were precipitation and flow information, which occurred in determined stations. The reseaarchers used real data from a system which was placed in the stations.

On the other hand, Chatterjee et al. (2017) determined the structural damages of multi-storey reinforced buildings via ANN, which they used to train the particle swarm optimization algorithm. Also, Cascardi, Micelli, and Aiello (2017) devised a model based on ANN for predicting the compression strength of reinforced circular columns, which are covered unremittingly by a wrap with fiber reinforced polymer. Shebani and Iwnicki (2018) used ANNs to predict the wearing rate of wheel and ray according to different wheel-rail contact cases, such as under dry, wet, and oiled conditions and after sanding.

At the same time, ANNs together with metaheuristic algorithms work well, and become hybrid, too. Besides, they are quite suitable in prediction issues, and enable various applications that determined by predicting optimum parameters for design problems under investigation.

For example, Ormsbee and Reddy (1995) realized optimum control of pumping systems to supply water. Their study aimed at handling the working times of a pump as decision variable, and at minimizing its operating costs. The researchers trained with a neural network the prediction data they had obtained from the simulation model they used for the purpose, and then provided the optimization of the system control by using the trained data in a genetic algorithm. Further, they compared the performance with genetic algorithm and ANN methods, by developing a multi-stages prediction model that included the firefly algorithm that is used to determine optimum values of the parameters in model, which is used in prediction duration, with the support of a vector machine for predetermining the daily lake level according to three different horizon lines, in Kişi et al.'s (2015) study. Sebaaly, Varma, and Maina (2018) developed a model which combined ANNs with a genetic algorithm, by using a data set including information belonging to numerous asphalt mixtures.

The above-mentioned examples show that the use of AI methods, mainly ANNs, are the increasing trend in science and engineering when the problem solving is too complex, impossible or the process needs too much time.

REFERENCES

Aichouri, I., Hani, A., Bougherira, N., Djabri, L., Chaffai, H., & Lallahem, S. (2015). River flow model using artificial neural networks. *Energy Procedia*, *74*, 1007–1014. doi:10.1016/j.egypro.2015.07.832

Atici, U. (2011). Prediction of the strength of mineral admixture concrete using multivariable regression analysis and an artificial neural network. *Expert Systems with Applications*, *38*(8), 9609–9618. doi:10.1016/j.eswa.2011.01.156

Biswas, J. (2018). *9 Complex machine learning applications that even a beginner can build today* Retrieved from https://analyticsindiamag.com/machine-learning-applications-beginners/

Bowden, G. J., Maier, H. R., & Dandy, G. C. (2002). Optimal division of data for neural network models in water resources applications. *Water Resources Research*, *38*(2), 2–1. doi:10.1029/2001WR000266

Buchanan, B. G. (2006). *Brief history.* Retrieved from https://aitopics.org/misc/brief-history

Cascardi, A., Micelli, F., & Aiello, M. A. (2017). An Artificial Neural Networks model for the prediction of the compressive strength of FRP-confined concrete circular columns. *Engineering Structures, 140*, 199–208. doi:10.1016/j.engstruct.2017.02.047

Chatterjee, S., Sarkar, S., Hore, S., Dey, N., Ashour, A. S., & Balas, V. E. (2017). Particle swarm optimization trained neural network for structural failure prediction of multistoried RC buildings. *Neural Computing & Applications, 28*(8), 2005–2016. doi:10.100700521-016-2190-2

Floridi, L. (2017). *If AI is the future, what does it mean for you?* Retrieved from https://www.weforum.org/agenda/2017/01/the-future-ofai-and-the-implications-for-you

Han, J., & Gao, J. (2008). Data mining in e-science and engineering. In H. Kargupta, J. Han, S. Y. Philip, R. Motwani, & V. Kumar (Eds.), *Next generation of data mining* (pp. 1–114). Boca Raton, FL: CRC Press. doi:10.1201/9781420085877.pt1

Han, J. & Kamber, M. (2006). Data mining: concepts and techniques (2nd ed.). CL: Morgan Kaufmann.

Kisi, O., Shiri, J., Karimi, S., Shamshirband, S., Motamedi, S., Petković, D., & Hashim, R. (2015). A survey of water level fluctuation predicting in Urmia Lake using support vector machine with firefly algorithm. *Applied Mathematics and Computation, 270*, 731–743. doi:10.1016/j.amc.2015.08.085

Luger, G. F. (2009). *Artificial intelligence: structures and strategies for complex problem solving* (6th ed.). Boston, MA: Pearson Education.

Maimon, O., & Rokach, L. (2010). *Data mining and knowledge discovery handbook* (2nd ed.). New York, NY: Springer. doi:10.1007/978-0-387-09823-4

McCarthy, J. (2007). *What is artificial intelligence?* Retrieved from http://www.formal.stanford.edu/jmc/whatisai.html

Momeni, E., Armaghani, D. J., & Hajihassani, M. (2015). Prediction of uniaxial compressive strength of rock samples using hybrid particle swarm optimization-based artificial neural networks. *Measurement, 60*, 50–63. doi:10.1016/j.measurement.2014.09.075

Nilsson, N. J. (1998). *Introduction to machine learning an early draft of a proposed textbook.* Retrieved from http://ai.stanford.edu/~nilsson/MLBOOK.pdf

Olivas, E. S., Guerrero, J. D. M., Sober, M. M., Benedito, J. R. M., & López, A. J. S. (Eds.). (2009). *Handbook of research on machine learning applications and trends: algorithms, methods, and techniques.* Hershey, PA: IGI Global.

Onwubolu, G. C., & Babu, B. V. (2004). *New optimization techniques in engineering* (Vol. 141). Heidelberg, Germany: Springer-Verlag. doi:10.1007/978-3-540-39930-8

Ormsbee, L. E. & Reddy, S. L. (1995). Pumping system control using genetic optimization and neural networks. *IFAC Proceedings Volumes, 28*(10), 685-690.

Patil, A., Patted, L., Tenagi, M., Jahagirdar, V., Patil, M., & Gautam, R. (2017). Artificial intelligence as a tool in civil engineering – a review. In *Proceedings of National conference on advances in computational biology, communication, and data analytics (ACBCDA 2017)*. India: IOSR Journal of Computer Engineering.

Russell, S. J., & Norvig, P. (1995). *Artificial intelligence: a modern approach*. Englewood Cliffs, NJ: Prentice Hall.

Sebaaly, H., Varma, S., & Maina, J. W. (2018). Optimizing asphalt mix design process using artificial neural network and genetic algorithm. *Construction & Building Materials, 168*, 660–670. doi:10.1016/j.conbuildmat.2018.02.118

Shebani, A., & Iwnicki, S. (2018). Prediction of wheel and rail wear under different contact conditions using artificial neural networks. *Wear, 406*, 173–184. doi:10.1016/j.wear.2018.01.007

Uygunoğlu, T., & Yurtçu, Ş. (2006). Yapay zekâ tekniklerinin inşaat mühendisliği problemlerinde kullanımı. *Yapı Teknolojileri Elektronik Dergisi, 1*, 61–70.

Veintimilla-Reyes, J., Cisneros, F., & Vanegas, P. (2016). Artificial neural networks applied to flow prediction: A use case for the Tomebamba river. *Procedia Engineering, 162*, 153–161. doi:10.1016/j.proeng.2016.11.031

Witten, I. H., Hall, M. A., & Frank, E. (2011). *Data mining: practical machine learning tools and techniques*. USA: Morgan Kaufmann Series.

Chapter 2
Artificial Neural Networks (ANNs) and Solution of Civil Engineering Problems:
ANNs and Prediction Applications

Melda Yucel
Istanbul University-Cerrahpaşa, Turkey

Sinan Melih Nigdeli
Istanbul University-Cerrahpaşa, Turkey

Gebrail Bekdaş
Istanbul University-Cerrahpaşa, Turkey

ABSTRACT

This chapter reveals the advantages of artificial neural networks (ANNs) by means of prediction success and effects on solutions for various problems. With this aim, first, the authors explain multilayer ANNs and their structural properties. Then, they present feed-forward ANNs and a type of training algorithm called back-propagation, which used this type of networks. Different structural design problems in civil engineering are optimized and handled for obtaining prediction results thanks to the usage of ANNs.

MULTILAYER ARTIFICIAL NEURAL NETWORKS

Artificial neural networks (ANNs) are computer systems that realize the learning function, which is one of the principal features of human brain. They carry out the learning activity by the help of samples.

These systems are used in various applications, such as prediction, classification, and problem control in practice. Besides, they are a computational stencil composing from a group of artificial neurons (i.e., nodes, namely input, hidden, and output neurons), which are connected to each other and have the weight value; they are based on biologic neural networks. In this regard, ANNs are an adaptive system, which

DOI: 10.4018/978-1-7998-0301-0.ch002

can change its structure by internal or external information flowing along the network, along the learning process. This learning or training is a process whose weights are determined (Olivas et al., 2009). In addition, the weight value of every connection between nodes reflects the effect of each neuron on output value, and nodes are processed according to these weight values. Therefore, information that will be occurred takes form with respect of these weights, too.

A layer name is given to the section that is made altogether of more than one neuron cell. The main structure of ANNs includes only input and output layers. However, multilayer ANNs contain a layer called "hidden layer," too. The input layer from these three layers, which compose the network structure, represents the training data, and these data are processed according to the model continuously.

Similarly, the second layer is a hidden layer. In this layer, neurons are trained by the data, and they are not open to a direct intervention from the external environment. Although generally only one hidden layer is used in practice, the number of layers is at discretion, and can be between 0 and a high number. Output values arranged with weight values produced by the last hidden layer are transferred to units comprised of an output layer, so the network spreads the prediction values generated for training data (Han & Kamber, 2006; Veintimilla-Reyes et al., 2016).

In this way, how compatible the obtained outputs are with actuals is observed by comparing the ultimate results obtained in the output layer and the real results, and error rate (variation) is investigated. It is possible to say that a correct training of the network, in case of variation between the actual output and output, which is obtained from the neural network, is in an allowable limit (Ormsbee & Reddy, 1995). If this rate is not within an admissible range, the network structure is arranged again by making the required improvements. If required, the weights of the connections can be modified, too. Operations are iteratively continued, and the network structure is completed when the error achieves the minimum acceptable level. The following list reports some positive and negative properties of ANNs:

- They can produce solutions properly to samples whose values are not known or were not applied previously by generalization, through learning the linear or nonlinear relationship between input and output data about any problem from current samples. Networks have the ability to work rapidly due to their learning ability, adaptability to different problems, and the structure of their components, which can be worked simultaneously. At the same time, they require less information, and their implementation is easy (Uygunoğlu & Yurtçu, 2006).
- They can process external information, based on their previous experiences, and simplify the complex and time-consuming problems, owing to their mapping ability (Gholizadeh, 2015).
- Neural networks include long training times, thus they are more suitable for applications that allow long training. In a network structure, generally, many parameters, which are determined in the best way experimentally, are needed (Han & Kamber, 2006).
- They can solve problems related to uncertain models or data that have variables, which include much missing and noisy information. This error tolerance feature addresses data mining problems, because real data are generally dirty and follow the open-possibility structures statistical models typically desire (Maimon & Rokach, 2010).
- When an ANN structure is very small, the desired function cannot be carried out. When it is the network is very big, that is it learns all created samples from a large search area, it does not generalize inputs of which it does not know the value or which it has not seen before, substantially. From this respect, neural networks show extreme/peak behavior, in case numerous parameters are present in the model (Russell & Norvig, 1995)

Feed-Forward Artificial Neural Networks

Multilayer feed-forward ANNs are comprised of one input layer, one or more hidden layers, and one output layer; also in this form of ANN, which is named as "perceptron," all nodes and layers are arranged in a way that allows a flow to go forward (Han & Kamber, 2006; Olivas et al., 2009). Therefore, the connections in a feed-forward ANN are in a single direction, and cycle does not consist. In these networks, every node can establish contact only with the nodes in the next layer. Nodes have not connection with other/neighbor nodes in the layers, which are taken place of themselves, or nodes in the previous layer (Russell & Norvig, 1995).

Every neuron (node) in the input layer transforms into outputs in the same way as it processes given information, thus it takes on an information-processing task. Knowledge that will be generated thanks to the linking paths between these neurons, can be produced and stored as connection weights, which are related to the power of the relationship between different nodes. Although each neuron applies its function slowly and faultily, the main structure of a neural network can carry out various tasks effectively and allow to obtain remarkable results (Maimon & Rokach, 2010).

Figure 1 shows the structure of a feed-forward neural network composed of neurons that are arranged as three layers. Here, the current input layer by i, the related input neuron by j, and the output neuron in next layer by k are represented respectively, and input neuron by X_i, weight values of connections by w_{ijk}, and the bias value by b_{ij} are expressed respectively as $i, j, k = 1, 2, ..., n$.

Dependent variables expressed via neurons in the output layer are predicted by means of neurons in the input layer in this network structure and corresponding to independent variables. Each input value is multiplied via interconnection weights related to the output node. The output value is obtained via processing all obtained multiplication results by means of an activation function (Gandomi et al., 2013). As in this process, the hidden layer and its neurons are included with the aim of evaluating the information according to various factors, learn it, and rearrange it by experiencing.

Figure 1. Multilayer feed-forward neural network
(Chou et al., 2015)

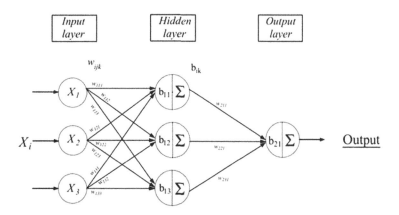

The Back-Propagation Algorithm

Similarly to any artificial intelligence model, ANNs have learning ability, too. The most common and efficient learning algorithm, which is used to train a multilayer ANN, is known as back-propagation algorithm (Chou et al., 2015). On the other hand, in the learning process, errors, which will be occurred between consecutive layers in the structure of ANNs, can be spread to a general network differently. The most frequently preferred way is the back-propagation algorithm, which is based on rearranging the error by spreading it backward.

Generally, the back propagation algorithm applied to multilayer feed-forward neural networks uses the weight changes of connections to minimize the output error generated by the function of the network.

The name "back-propagation" is due to the errors in the output node spreading backward from the output layer, in the same way as towards the hidden layer; these errors are used in an equation, which ensures the weights of the hidden layer are updated. These errors propagating backward are determined as the difference between real and predicted values, in the output layer (Chong & Zak, 2001; Luger, 2009).

Each neuron in the layers obtains a new output value/prediction (feed-forward process) (Figure 2). A nonlinear logistic or sigmoid function operates the net value as activated via weighted summation of the values which result from the multiplied output from the nodes connected to this output in the previous layer. These nodes have connection weights between each other, together with the threshold value (named as "bias") as a net input value in each node. In this process, the difference between the real output and the output value the network forecasts is the error value, which is transferred towards the input layer by the back-propagation algorithm. Outputs (inputs) are obtained from the previous layer, and connection weights are expressed as error value y_n and w_{nj}, respectively (Chou et al., 2015; Han & Kamber, 2006).

The value, which is expressed as activation level (net), too, is obtained by summing up the weighted summation (Σ) result and the bias value (θ_j); the output prediction is carried out by processing this value via the activation (threshold) function, which is indicated as f. These operation steps are calculated by means of the equations below:

$$net = \sum y_n w_{nj} + \theta_j \tag{1}$$

Figure 2. Output prediction via activation of the back-propagation algorithm for the hidden or output layer node
(Han & Kamber, 2006)

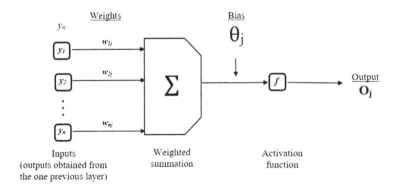

$$f\left(net\right)=O_j=\frac{1}{\left(1+e^{-net}\right)} \tag{2}$$

The error value presenting the difference between the real output (T_j) and the value (O_j), which is generated for any node of the output layer following a calculation process as in Equation (2) by the activation function, is δ_j. Also, the error value which is obtained for a node in the hidden layer is expressed as δ_k. These values are known as error propagating backward on the network, and are calculated via the following equations:

$$\delta_j=O_j\left(1-O_j\right)\left(T_j-O_j\right) \tag{3}$$

$$\delta_k=O_k\left(1-O_k\right)\sum_j\delta_jw_{kj} \tag{4}$$

Obtained errors are spread to weights on layer connections orderly. This operation is named as "training of the network;" in each of the performed t cycles (iteration), the updating of the weights is carried out according to Equation (5), for the hidden-output node, and Equation (7), for the input-hidden node.

For the hidden-output node:

$$w_{kj}\left(t\right)=w_{kj}\left(t-1\right)+\Delta w_{kj}\left(t\right) \tag{5}$$

$$\Delta w_{kj}\left(t\right)=\eta\delta_jO_k+\alpha\Delta w_{kj}\left(t-1\right) \tag{6}$$

For the input-hidden node:

$$w_{nk}\left(t\right)=w_{nk}\left(t-1\right)+\Delta w_{nk}\left(t\right)\triangleleft \tag{7}$$

$$\Delta w_{nk}\left(t\right)=\eta\delta_ky_n+\alpha\Delta w_{nk}\left(t-1\right) \tag{8}$$

where η is the learning speed (rate) of the network, α is the parameter of momentum, and Δw_{nk} is the weight change occurred in a cycle.

ARTIFICIAL NEURAL NETWORKS PREDICTION APPLICATIONS FOR OPTIMUM DESIGNS

Tubular Column Cost Optimization via Artificial Neural Networks

The Cost Optimization Problem

This subsection tackles the cost optimization problem of a tubular column under compressive load (Figure 3). The aim is finding out the optimal center diameter (d, in cm) and thickness (t, in cm) of the section, which ensure the minimization of the total cost including material and construction.

As Figure 3 illustrates, the axial compressive load (P) is applied to the center of the tubular column at the A-A cross-section. Also, the length of the column is indicated with l (cm).

The other components belonging to the problem are design constants that are yield stress (σ_y) taken as 500 kgf/cm^2, density (ρ) taken as 0.0025 kgf/cm^3, and modulus of elasticity (E) taken as 850000 kgf/cm^2. Besides the design variables, there are the center diameter and thickness, which will be optimized in order to minimize the cost. Therefore, the objective function of the expressed optimization problem can be written as in Equation (9). Also, constraint functions are six for the design problem, but only two are related to the design. The other four constraints are related to minimum and maximum limit values of the design variables. On the other hand, constraints functions are g_1 and g_2, which are expressed as axial compressive load capacity of the column and Euler buckling limit, respectively. All of them are formulated as follows:

$$f(d,t)=9.8dt+2d \tag{9}$$

$$g_1 = \frac{P}{\pi dt\sigma_y} - 1 \leq 0 \tag{10}$$

$$g_2 = \frac{8Pl^2}{\pi^3 Edt\left(d^2 + t^2\right)} - 1 \leq 0 \tag{11}$$

Equations (12) and (13). Show the ranges of the design variables which are subjected to optimization.

$$2 \leq d \leq 14 \tag{12}$$

$$0.2 \leq t \leq 0.9 \tag{13}$$

Figure 3. Tubular column with A-A cross-section design parameters
(Hsu & Liu, 2007)

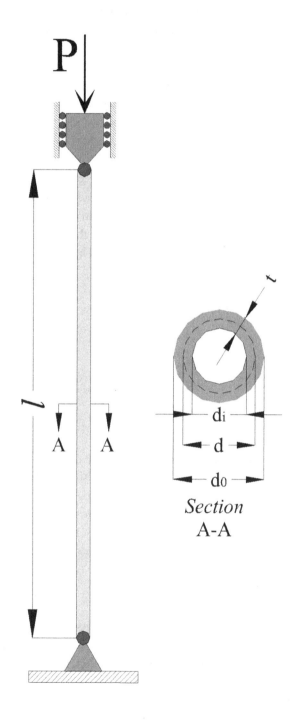

Teaching-Learning-Based Optimization

During the optimization process, the researchers used the teaching-learning-based optimization (TLBO) method (Rao et al., 2011), in order to obtain optimum values of the diameter and thickness of the center by ensuring the cost optimization for a large number of P-l data (the written Matlab, 2018, computer code for the optimization process via TLBO is represented in the Appendix). This algorithm was generated based on the teaching-learning process between a teacher and his/her students in a classroom. Also, the algorithm considers the interactions of the students with each other and the effects of the teacher's interaction with the students, with the aim of improving knowledge of both each student and the whole class towards an iterative process. Therefore, the teacher, who is seen as the most experienced person, impacts on the students' achievement of a predetermined target, and is expected to increase the knowledge level of the whole class, depending on his/her ability (Temür & Bekdaş, 2016; Toğan, 2012).

This process is composed of two separate phases (i.e., teacher and learner phases) with the TLBO method. The first process is the teacher phase, which takes form according to the teacher's education and skills to increase the students' grades. The second process is the student or learner phase and consists in increasing knowledge as a result of the students sharing. These phases are formulated as Equations (14) and (15), respectively:

$$X_{i,new} = X_{i,j} + rand(0,1)\left(X_{i,best} - (TF)X_{i,mean}\right) \tag{14}$$

$$X_{i,new} = \begin{cases} \text{if} & \begin{array}{l} f(X_a) < f(X_b), X_{i,j} + rand(0,1)\left(X_{i,a} - X_{i,b}\right) \\ f(X_a) > f(X_b), X_{i,j} + rand(0,1)\left(X_{i,b} - X_{i,a}\right) \end{array} \end{cases} \tag{15}$$

In these formulas, $X_{i,j}$ is the old value of the related design variable within the initial matrix, and $X_{i,new}$ is its new value. Also, $X_{i,best}$ and $X_{i,mean}$ are the best values with regards to the objective function (i.e., teacher solution) and average all-candidate solutions of this design variable, respectively. The value TF expresses the teacher's teaching force, namely the teaching factor taken as 1 or 2, randomly. Furthermore, solutions $X_{i,a}$ and $X_{i,b}$ in the equation of the learner phase are selected randomly from the initial matrix, and $f(X_a)$ and $f(X_b)$ are the values of their objective function, too.

Artificial Neural Network Model and Its Training

In this chapter, the authors investigate prediction applications performed by using ANN models for different problems. One of these is cost optimization of a tubular column. Thus, the authors carried out optimization via the TLBO method and then operated the obtained results in the prediction process they executed via ANN.

In this problem, the ANN is an effective, rapid, and well-convergent tool to predict values. Thus, the researchers adopted feed-forward networks trained with the back-propagation algorithm.

Numerical Examples

In numerical applications, firstly, the optimization process produced 10000 data (composed form P-l random values) and the optimum design variables for these data. The researchers realized this process by using the axial force (*P*) ranging in 100-5000 kgf and the column length (*l*) ranging in 100-800 cm, as random. These means allowed to create a dataset including different P-l design couples, together with optimum design variables (Yucel, Bekdaş, Nigdeli, & Sevgen, 2018).

Following the generation of the dataset, the training is done via ANNs. In this way, the researchers determined two input nodes (*P* and *l* values) and three output nodes (*d*, *t*, and the objective function *Min f(d,t)*), in order to construct an ANN structure. This ANN model was trained by the Matlab neural net fitting application available in the machine learning toolbox. For this reason, the nodes of the hidden layer are 10, which this tool usually assigns as a constant.

Table 1, shows prediction error rates of the trained main model, in comparison with optimum values, which were obtained via the TLBO method.

Moreover, by virtue of the prediction success of the main model, a test model was generated by using the same ranges of *P* and *l* outputs. Table 2 shows optimization values for this model composed of 10 data.

Tables 3-5 report the obtained optimum design variable predictions, including Min f(d,t). In addition, these Tables show error metrics from mean absolute error (MAE), mean absolute percent error (MAPE), and mean-squared error (MSE) for design variables and objective function, by comparing TLBO-ANN.

Table 1. Error rates and evaluation of success for the main model

Optimum Parameters	Error Metrics		
	Mean Absolute Error (MAE)	Mean Absolute Percentage Error (MAPE)	Mean Square Error (MSE)
d	0.127	1.815	0.122
t	0.021	6.531	0.002
Min f (d,t)	0.398	0.931	2.262

Table 2. Optimum results obtained via the TLBO method for the test model

P (kgf)	*l* (cm)	d (cm)	t (cm)	g_1	g_2	Min f (d,t)
171	551	4.2840	0.200	-0.8729	0	16.9646
1373	306	5.7978	0.200	-0.2462	-1.11×10^{-16}	22.9593
958	110	2.3883	0.255	-1.11×10^{-16}	-1.11×10^{-16}	10.7535
749	411	5.7669	0.200	-0.5866	-2.22×10^{-16}	22.8371
2043	701	11.5059	0.200	-0.4348	0	45.5635
433	568	5.9607	0.200	-0.7688	-2.22×10^{-16}	23.6043
3786	270	5.8814	0.409	-1.11×10^{-16}	-2.22×10^{-16}	35.3832
2702	217	4.7244	0.364	0	-5.55×10^{-16}	26.3062
2810	259	5.6466	0.316	-1.11×10^{-16}	0	28.8245
870	686	8.5318	0.200	-0.6754	-1.11×10^{-16}	33.7859

The analysis of the error results highlights that these rates are very small and acceptable in comparison with determining the values of design variables by using the main model. The reason is the biggest error value was realized for thickness indicated as t and is 8.776% in terms of MAPE.

Therefore, for the test model, the calculation of the constraint functions g_1 and g_2 is possible because prediction results are reasonable and proper. Table 6 shows the calculation of constraint functions; importantly, only one value exceeds the limitation.

Table 3. d predictions for new samples via artificial neural network

P (kgf)	l (cm)	ANN (10 neuron)	Error Values for TLBO		
		d (cm)	*MAE*	*MAPE*	*MSE*
171	551	4.3687	0.0847	1.9780	0.0072
1373	306	5.7872	0.0105	0.1819	0.0001
958	110	2.4948	0.1064	4.4569	0.0113
749	411	5.7955	0.0286	0.4954	0.0008
2043	701	11.4767	0.0293	0.2542	0.0009
433	568	5.8880	0.0727	1.2198	0.0053
3786	270	5.7367	0.1447	2.4605	0.0209
2702	217	4.7064	0.0180	0.3805	0.0003
2810	259	5.6852	0.0386	0.6840	0.0015
870	686	8.5527	0.0209	0.2452	0.0004
		Average	0.0554	1.2356	0.0049

Table 4. t predictions for new samples via artificial neural network

P (kgf)	l (cm)	ANN (10 neuron)	Error Values for TLBO		
		t (cm)	*MAE*	*MAPE*	*MSE*
171	551	0.197	0.0026	1.3181	0.0000
1373	306	0.228	0.0288	14.3849	0.0008
958	110	0.265	0.0098	3.8375	0.0001
749	411	0.219	0.0198	9.8751	0.0004
2043	701	0.187	0.0128	6.3985	0.0002
433	568	0.209	0.0092	4.5756	0.0001
3786	270	0.479	0.0693	16.9100	0.0048
2702	217	0.408	0.0441	12.1022	0.0019
2810	259	0.373	0.0564	17.8151	0.0032
870	686	0.198	0.0011	0.5428	0.0000
		Average	0.0254	8.7760	0.0011

Table 5. Min f (d,t) predictions for new samples via artificial neural network

P (kgf)	l (cm)	ANN (10 neuron)	Error Values for TLBO		
		Min f(d,t)	MAE	MAPE	MSE
171	551	17.2501	0.2855	1.6829	0.0815
1373	306	22.9529	0.0064	0.0279	0.0000
958	110	11.2577	0.5042	4.6883	0.2542
749	411	22.9484	0.1114	0.4878	0.0124
2043	701	45.7778	0.2143	0.4703	0.0459
433	568	23.3401	0.2642	1.1193	0.0698
3786	270	35.4033	0.0200	0.0566	0.0004
2702	217	26.3915	0.0854	0.3245	0.0073
2810	259	28.8227	0.0018	0.0061	0.0000
870	686	33.9598	0.1739	0.5147	0.0302
		Average	0.1667	0.9378	0.0502

Conclusions

The analysis of error values shows that the prediction model was improved by using the ANN method, which is one of the most used and intelligent machine learning algorithms, that is an efficient tool for determining optimum values close to real/target data. Thus, the ANN model can be accepted as successful in optimizing design variables rapidly, effectively, and easily. For this reason, the researchers realized a prediction application to generate the test model; according to the obtained predictions, these values are very close to real optimum data. Thus, differences, in other words, errors did not reach very big values.

According to error values, the MSE metric is very small for all design variables and for the objective function. Also, the MAE shows the same situation as the MSE; indeed, their values are in the range of only 0.0254-0.1667, and the biggest one belongs to *Min f(d,t)* predictions. Percentage errors, namely

Table 6. The calculations for constraint functions for the new samples via artificial neural network

P (kgf)	l (cm)	g_1	g_2
171	551	-0.8737	-0.0443
1373	306	-0.3398	-0.1213
958	110	-0.0780	-0.1549
749	411	-0.6256	-0.1035
2043	701	-0.3946	0.0766
433	568	-0.7762	-0.0080
3786	270	-0.1231	-0.0802
2702	217	-0.1045	-0.0991
2810	259	-0.1570	-0.1694
870	686	-0.6744	-0.0019

the MAPE, are in the range 0.9378-8.7760%. The value 8.7760% of the MAPE is related to thickness (*t*). The reason is that the values of the variable *t* are more precise, so predictions should be very close and similar to real values, in order to ensure a decrease in errors.

Even so, in order to calculate the constraint functions belonging to the test model, design couples can be validated by the success of the ANN prediction model. Therefore, the researchers calculated constraints (i.e., g_1 and g_2) by using ANN predictions of the new test samples. The results showed that the prediction model is significantly successful because design constraints are proper to limitations. One design model exceeded the limits for g_2. In this regard, error values of design variables can be neglected as independent (single).

This tool can be useful and efficient for determining design data for various usage fields, such as engineering, medicine, and science. Also, it ensures the optimization of cost, time especially effort. In the future, ANN models that aim at preventing waste of time and effort with the other residuals can be useful.

I-Beam Section Optimization by Using Artificial Neural Network Optimum Predictions

Vertical Deflection Minimization of I-Section Beam

The second problem is related to the minimization of vertical deflection, which occurs in an I-section beam.

This beam is acted from two loads composed by a vertical load (*P*) and a horizontal load (*Q*), which are applied to the midpoint of the central axis of a beam span and to the middle axis point of a beam web, respectively. From these loads, Q was dealt with as constant 50 kN. However, the vertical design load *P* was normally constant like Q in optimization, but in this study the authors determined P values in different ranges, randomly. This range is 100-750 kN. Also, the beam span length (*L*) was determined between 100-350 cm. For the aim of this application, different optimum design models benefit from the training of an ANN model (Yucel, Nigdeli, & Bekdaş, 2019).

In this problem, design variables are composed of section properties, which are beam height (*h*), beam width, beam flange thickness (t_f), and beam web thickness (t_w), and all of them are denominated

Figure 4. Design variables and loads of an I-section beam
(Yang et al., 2016)

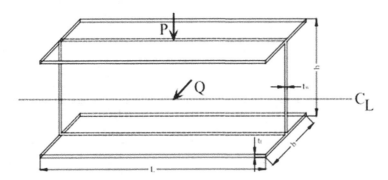

in centimeters. Figure 4 shows the design loads (inputs for the ANN model) and design variables of the beam section (outputs for the ANN model)(Yang et al., 2016):

While the researchers determined the optimum design variables in the optimization process they realized by the flower pollination algorithm (FPA), the targeted objective function is as in Equation (16). This function is concerned with the optimization of section properties by ensuring the minimization of vertical deflection. In this equation, the elasticity modulus E is 20000 kN/cm² and the moment of inertia (I) of the I-beam is formulated with Equation (17):

$$f(x) = \frac{PL^3}{48EI} \tag{16}$$

$$I = \frac{t_w(h-2t_f)^3}{12} + \frac{bt_f^3}{6} + 2bt_f\left(\frac{h-t_f}{2}\right)^2 \tag{17}$$

Constraint functions are g_1 and g_2, which are related to the beam section area limitation (this cannot exceed 300 cm²) and allowable moment stress limitation (this cannot be bigger than 6 kN/cm²), respectively. Equations (18) and (19) represent formulations of constraints:

$$g_1 = 2bt_f + t_w(h-2t_f) \le 300 \tag{18}$$

$$g_2 = \frac{1.5PLh}{t_w(h-2t_f)^3 + 2bt_w\left(4t_f^2 + 3h(h-2t_f)\right)} + \frac{1.5QLb}{t_w^3(h-2t_f) + 2t_wb^3} \le 6 \tag{19}$$

Also, the researchers handled design variables within specific ranges intended for desired design conditions. For this purpose, the limits of design variables are determined in Equations (20)-(23):

$$10 \le h \le 100 \tag{20}$$

$$10 \le b \le 60 \tag{21}$$

$$0.9 \le t_w \le 6 \tag{22}$$

$$0.9 \le t_f \le 6 \tag{23}$$

The Metaheuristic Method: The Flower Pollination Algorithm

The flower pollination algorithm (FPA) is known as a metaheuristic method and, nowadays, is preferred to a large extent for solving various optimization problems.

This method was proposed by Xin-She Yang in 2012 (Yang, 2012) and, in the development process of the algorithm, flowery plants are the inspiration for determining optimization stages (Bekdaş, Nigdeli, & Yang, 2015). In this respect, the FPA method is composed of two separate stages containing cross-pollination, which is realized between flowers of different plants coming from the same type, and self-pollination, which occurs between self-flowers of the same plant, or if a flower is one by one. Also, these stages are simulated as global search and local search in optimization.

During cross-pollination, pollinators move from one flower to another with one type of search flight, which is known as Lévy distribution rule, as in Equation (25). Also, this flight formulation is as follows:

$$Lévy = \left(\frac{1}{\sqrt{2\pi}} \right) \int^{-1.5} e^{\left(-\frac{1}{2\int} \right)} \tag{24}$$

$$X_{i,new} = X_{i,j} + Lévy \left(X_{i,best} - X_{i,j} \right) \tag{25}$$

In these equations, Lévy and ϵ expressions are Lévy distribution and a value ranging 0 and 1 determined randomly, respectively. $X_{i,j}$ is the old value of ith design variable belonging to j^{th} candidate solution within the initial matrix. Also, $X_{i,new}$ is the new solution value for concerned design variable, and this updates iteratively during optimization. $X_{i,best}$ is the best solution, which ensures the minimum/maximum objective function. The self-pollination process performs with the following equation:

$$X_{i,new} = X_{i,j} + \int \left(X_{i,m} - X_{i,k} \right) \tag{26}$$

Here, $X_{i,m}$ and $X_{i,k}$ are two different candidate solutions as m and k, which are determined randomly, ranging 0 and 1 for the ith design variable.

In this study, too, the FPA was benefited with the aim of optimizing the design variables belonging to the I-section beam by ensuring the minimization of the cost. In this regard, the researchers could determine optimum values for the design variables, which are beam height and width, together with flange and web thickness of beam.

Artificial Neural Networks Model

As in the authors mentioned before, ANNs are smart methods coming from machine learning and artificial intelligence. As they think, analyze, and exhibit behavior like a human, it is thought that ANNs are intelligent. In this direction, the researchers adopted a prediction application for determining optimum values for the I-section beam design by exploiting these advantages of ANNs. Thus, they developed an effective and rapid prediction tool via ANNs.

Firstly, the researchers used optimum values of the design variables for generating a dataset as training process. They realized the training process with the Matlab R2018a program by using the neural net fitting application. Besides, they generated the 750 P-L design model in the optimization process for this aim. They handled these as outputs for the ANN model and determined them randomly during optimization. Moreover, they selected optimum values of the design variables as inputs in the dataset. For this, they could create the main ANN training model with the vertical load (P) and the span length of beam (L), together with section values of the variables (h, b, t_w, and t_f).

The aim of this process was predicting optimum designs for a new model. Therefore, the researchers proposed a new test model containing ten different design combinations, and they could realize the prediction operations owing to the obtained ANN prediction tool.

Determination of Artificial Neural Network Predictions Belonging to the Test Model

Table 7 shows optimum values for the test model containing 10 different designs. Also, the optimum values determined via the main ANN prediction model are also in Tables 8-11.

Following the prediction of design variables, the researchers carried out calculations of minimum objective functions. Table 12 shows objective functions the researchers determined via predictions with their error metrics.

CONCLUSION

According to tables including the optimum results for design variables within test model, error values are very small, so predictions are close to real optimum values of the variables h and b, especially h. However, design variables h and b have the same or very similar values for test model designs. As a result, all prediction process should be analyzed to understand and demonstrate the success of the ANN model.

Thus, the analysis of the variables t_w and t_f show that the error measurements are a bit considerable comparing to other variables. Nevertheless, not so big errors occurred in these variables in the test

Table 7. Optimum values belonging to designs within the test model

L (cm)	P (kN)	h	b	t_w	t_f	Min f (x)
120	652	100.0000	60.0000	0.9000	1.7766	0.002018
350	520	100.0000	60.0000	1.6303	1.1733	0.049381
285	743	100.0000	60.0000	1.6833	1.1289	0.038774
150	200	100.0000	60.0000	0.9000	1.7766	0.001209
345	264	100.0000	60.0000	1.1171	1.5989	0.020572
100	690	100.0000	60.0000	0.9000	1.7766	0.001236
250	442	100.0000	59.9999	1.0635	1.6428	0.012915
310	675	100.0000	60.0000	1.9552	0.9000	0.049937
270	482	100.0000	60.0000	1.2077	1.5244	0.018465
220	355	100.0000	60.0000	0.9000	1.7766	0.006771

Table 8. h predictions of designs within the test model by using an artificial neural network

Predictions via ANNs	-	Error Metrics Comparing to FPA		
h (cm)	Error	Absolute Error	Error %	Squared Error
99.9989	0.0011	0.0011	0.0011	0.0000
100.0121	-0.0121	0.0121	0.0121	0.0001
100.0283	-0.0283	0.0283	0.0283	0.0008
100.0045	-0.0045	0.0045	0.0045	0.0000
100.0061	-0.0061	0.0061	0.0061	0.0000
99.9984	0.0016	0.0016	0.0016	0.0000
99.9970	0.0030	0.0030	0.0030	0.0000
100.0279	-0.0279	0.0279	0.0279	0.0008
100.0029	-0.0029	0.0029	0.0029	0.0000
99.9942	0.0058	0.0058	0.0058	0.0000
-	-	MAE	MAPE	MSE
Average	-	0.0093	0.0093	0.0002
RMSE	-	0.0141	-	-

Table 9. b predictions of designs within the test model by using an artificial neural network

Predictions via ANNs	-	Error Metrics Comparing to FPA		
b (cm)	Error	Absolute Error	Error %	Squared Error
59.9968	0.0032	0.0032	0.0053	0.0000
60.0180	-0.0180	0.0180	0.0300	0.0003
60.0295	-0.0295	0.0295	0.0492	0.0009
59.9652	0.0348	0.0348	0.0580	0.0012
60.0047	-0.0048	0.0048	0.0080	0.0000
59.9933	0.0067	0.0067	0.0112	0.0000
59.9290	0.0710	0.0710	0.1183	0.0050
60.0200	-0.0200	0.0200	0.0333	0.0004
59.9336	0.0664	0.0664	0.1106	0.0044
59.9266	0.0734	0.0734	0.1223	0.0054
-	-	MAE	MAPE	MSE
Average	-	0.0328	0.0546	0.0018
RMSE	-	0.0424	-	-

Table 10. t_w predictions of designs within the test model by using an artificial neural network

Predictions via ANNs	-	Error Metrics Comparing to FPA		
t_w (cm)	Error	Absolute Error	Error %	Squared Error
0.8851	0.0149	0.0149	1.652	0.0002
1.5850	0.0454	0.0454	2.782	0.0021
1.7029	-0.0196	0.0196	1.163	0.0004
0.9401	-0.0401	0.0401	4.457	0.0016
1.0728	0.0443	0.0443	3.965	0.0020
0.9265	-0.0265	0.0265	2.948	0.0007
1.0968	-0.0333	0.0333	3.133	0.0011
1.7537	-0.0420	0.0420	2.451	0.0018
1.2371	-0.0295	0.0295	2.447	0.0009
0.9422	-0.0422	0.0422	4.684	0.0018
-	-	MAE	MAPE	MSE
Average	-	0.0338	2.9681	0.0012
RMSE	-	0.0346	-	-

Table 11. t_f predictions of designs within the test model by using an artificial neural network

Predictions via ANNs	-	Error Metrics Comparing to FPA		
t_f (cm)	Error	Absolute Error	Error %	Squared Error
1.7867	-0.0101	0.0101	0.5669	0.0000
1.2128	-0.0395	0.0395	3.3666	0.0000
1.1206	0.0083	0.0083	0.7344	0.0000
1.7410	0.0356	0.0356	2.0065	0.0000
1.6336	-0.0347	0.0347	2.1718	0.0000
1.7486	0.0280	0.0280	1.5764	0.0000
1.6088	0.0340	0.0340	2.0714	0.0000
1.0812	0.0238	0.0238	2.1558	0.0000
1.4951	0.0293	0.0293	1.9223	0.0000
1.7341	0.0426	0.0426	2.3957	0.0000
-	-	MAE	MAPE	MSE
Average	-	0.0286	1.8968	0.0000
RMSE	-	0.0000	-	-

Table 12. Calculations of Min f (x) values via artificial neural networks predictions for the test model samples

Objective Function Values Calculated via Predictions	-	Error Metrics Comparing to FPA		
Min f (x)	Error	Absolute Error	Error %	Squared Error
0.0020	0.0000	0.0000	0.277	0.0000000
0.0486	0.0008	0.0008	1.613	0.0000006
0.0388	0.0000	0.0000	0.070	0.0000000
0.0012	0.0000	0.0000	1.225	0.0000000
0.0203	0.0002	0.0002	1.143	0.0000001
0.0012	0.0000	0.0000	1.006	0.0000000
0.0131	-0.0002	0.0002	1.357	0.0000000
0.0461	-0.0003	0.0003	0.653	0.0000001
0.0187	-0.0002	0.0002	1.191	0.0000000
0.0069	-0.0001	0.0001	1.614	0.0000000
-	-	MAE	MAPE	MSE
Average	-	0.0002	1.015	0.0000001
RMSE	-	0.0003	-	-

model. MAPE is 2.9681% for t_w and 1.8968% for t_f, both with extremely small values of RMSE, which is closely related to the success of the ANN.

In addition, the analysis of the obtained results of objective functions indicates that error values are extremely small and differ only 1.614% from real optimum results. As a result, the ANN prediction model is an effective tool and suitable to approach real observations.

REFERENCES

Bekdaş, G., Nigdeli, S. M., & Yang, X. S. (2015). Sizing optimization of truss structures using flower pollination algorithm. *Applied Soft Computing*, *37*, 322–331. doi:10.1016/j.asoc.2015.08.037

Chong, E. K. P., & Zak, S. H. (2001). *An introduction to optimization* (2nd ed.). New York, NY: John Wiley & Sons.

Chou, J. S., Lin, C. W., Pham, A. D., & Shao, J. Y. (2015). Optimized artificial intelligence models for predicting project award price. *Automation in Construction*, *54*, 106–115. doi:10.1016/j.autcon.2015.02.006

Gandomi, A. H., Yang, X. S., Talatahari, S., & Alavi, A. H. (2013). Metaheuristic algorithms in modeling and optimization. In A. H. Gandomi, X. S. Yang, S. Talatahari, & A. H. Alavi (Eds.), *Metaheuristic applications in structures and infrastructures* (pp. 1–24). Elsevier. doi:10.1016/B978-0-12-398364-0.00001-2

Gholizadeh, S. (2015). Performance-based optimum seismic design of steel structures by a modified firefly algorithm and a new neural network. *Advances in Engineering Software, 81*, 50–65. doi:10.1016/j. advengsoft.2014.11.003

Han, J., & Kamber, M. (2006). *Data mining: concepts and techniques* (2nd ed.). San Francisco, CA: Morgan Kaufmann.

Hsu, Y. L., & Liu, T. C. (2007). Developing a fuzzy proportional derivative controller optimization engine for engineering design optimization problems. *Engineering Optimization, 39*(6), 679–700. doi:10.1080/03052150701252664

Luger, G. F. (2009). *Artificial intelligence: structures and strategies for complex problem solving* (6th ed.). Boston, MA: Pearson Education.

Maimon, O., & Rokach, L. (Eds.). (2010). *Data mining and knowledge discovery handbook* (2nd ed.). New York, NY: Springer. doi:10.1007/978-0-387-09823-4

Mathworks. (2018). MATLAB R2018a. The MathWorks Inc., Natick, MA.

Olivas, E. S., Guerrero, J. D. M., Sober, M. M., Benedito, J. R. M., & López, A. J. S. (Eds.). (2009). *Handbook of research on machine learning applications and trends: Algorithms, methods, and techniques*. Hershey, PA: IGI Global.

Ormsbee, L. E. & Reddy, S. L. (1995). Pumping system control using genetic optimization and neural networks. *IFAC Proceedings Volumes, 28*(10), 685-690.

Russell, S. J., & Norvig, P. (1995). *Artificial intelligence: A modern approach*. Englewood Cliffs, NJ: Prentice Hall.

Temür, R., & Bekdaş, G. (2016). Teaching learning-based optimization for design of cantilever retaining walls. *Structural Engineering and Mechanics, 57*(4), 763–783. doi:10.12989em.2016.57.4.763

Toğan, V. (2012). Design of planar steel frames using teaching–learning based optimization. *Engineering Structures, 34*, 225–232. doi:10.1016/j.engstruct.2011.08.035

Uygunoğlu, T., & Yurtçu, Ş. (2006). Yapay zekâ tekniklerinin inşaat mühendisliği problemlerinde kullanımı. *Yapı Teknolojileri Elektronik Dergisi, 1*, 61–70.

Veintimilla-Reyes, J., Cisneros, F., & Vanegas, P. (2016). Artificial neural networks applied to flow prediction: A use case for the Tomebamba River. *Procedia Engineering, 162*, 153–161. doi:10.1016/j. proeng.2016.11.031

Yang, X. S. (2012). Flower pollination algorithm for global optimization. In Jérôme Durand- Lose, & Nataša Jonoska (Eds.), *International conference on unconventional computation and natural computation* (pp. 240-249). Berlin, Germany: Springer. 10.1007/978-3-642-32894-7_27

Yang, X. S., Bekdaş, G., & Nigdeli, S. M. (Eds.). (2016). *Metaheuristics and optimization in civil engineering*. Springer International Publishing. doi:10.1007/978-3-319-26245-1

Yucel, M., Bekdaş, G., Nigdeli, S. M., & Sevgen, S. (2018). Artificial neural network model for optimum design of tubular columns. In *Proceedings of 7th International conference on applied and computational mathematics (ICACM '18)* (pp. 82-86). Rome, Italy: International Journal of Theoretical and Applied Mechanics.

Yucel, M., Nigdeli, S. M., & Bekdaş, G. (2019, Sept. 22-25). *Generation of an artifical neural network model for optimum design of I-beam with minimum vertical deflection.* Paper presented at 12th HSTAM International Congress on Mechanics. Thessaloniki, Greece.

APPENDIX

```
%App. (2.1): Cost Optimization of Tubular Column Under Compressive Load via
TLBO Algorithm

clear all

% DESIGN CONSTANTS AND VALUES
% yield: yield strength (kgf/cm^2)
% E: elasticity modulus (kgf/cm^2)
% ro: density (kgf/cm^3)
% P: axial compressive load (kgf)
% l: column length (cm)

yield=500;  E=0.85E6;  ro=0.0025;

% DESIGN VARIABLES OF PROBLEM
% d: average diameter of column crossection (cm)
% t: thickness of column crosssection (cm)

% LIMITATIONS FOR DESIGN VARIABLES
% mind: lower limit for average diameter
% maksd: upper limit for average diameter
% mint: lower limit for thickness
% makst: upper limit for thickness

mind=2; maksd=14; mint=0.2; makst=0.9;

% PARAMETERS OF TLBO ALGORITHM
% stop_criteria: stopping criteria of algorithm (maximum iteration)
% pn: population size (learner number)

stop_criteria=20000; pn=15;

tic

for ii=1:10000
    P(ii)=round(100+4900*rand());
    l(ii)=round(100+700*rand());
```

```
OPT_P_l_comb(ii,1)=P(ii);        %recording of P random values
OPT_P_l_comb(ii,2)=l(ii);        %recording of l random values

% GENERATION OF INITIAL MATRIX

for i=1:pn

% Assigning of random values for design variables between determined limits
        d=mind+(maksd-mind)*rand;
        t=mint+(makst-mint)*rand;

        % OBJECTIVE FUNCTION OF PROBLEM
        Fd_t=9.8*d*t+2*d;

        % CONSTRAINT FUNCTIONS
        % g1: axial load capacity (compressive strength should not exceed the
yield strength in design.)
        % g2: buckling limit for column (Euler buckling load)

        g1=(P(ii)/(pi*d*t*yield))-1;
        g2=8*P(ii)*(l(ii)^2)/((pi^3)*E*d*t*(d^2+t^2))-1;

        % OPT: Initial solution matrix
        OPT(1,i)=d;
        OPT(2,i)=t;
        OPT(3,i)=g1;
        OPT(4,i)=g2;
        OPT(5,i)=Fd_t;

        if OPT(3,i)>0
            OPT(5,i)=10^6;
        end
        if OPT(4,i)>0
            OPT(5,i)=10^6;
        end
    end

    % ITERATION PROCESS
    for dongu=1:stop_criteria

% GENERATION OF NEW SOLUTION MATRIX ACCORDING TO TLBO ALGORITHM RULES
```

```
        % Teacher phase
        for i=1:pn

% Obtaining of the best solution (teacher ensuring that objective function is
minimum in initial solution matrix, and design variables belonging to this so-
lution
                [p,r]=min(OPT(5,:));
                best_d=OPT(1,r);
                best_t=OPT(2,r);

% Determining again of design variables according to TLBO rules

                % TF: teaching factor
                TF=round(1+rand);
                d=OPT(1,i)+rand*(best_d-TF*mean(OPT(1,:)));
                t=OPT(2,i)+rand*(best_t-TF*mean(OPT(2,:)));

% Controlling of lower and upper limits of new generated design variables

                if d>maksd
                    d=maksd;
                end
                if d<mind
                    d=mind;
                end
                if t>makst
                    t=makst;
                end
                if t<mint
                    t=mint;
                end

                % Objective function of problem
                Fd_t=9.8*d*t+2*d;

                % Inequality constraints of problem
                g1=(P(ii)/(pi*d*t*yield))-1;
                g2=8*P(ii)*(l(ii)^2)/(pi^3*E*d*t*(d^2+t^2))-1;

                %OPT_Y: New solution matrix
                OPT_Y(1,i)=d;
                OPT_Y(2,i)=t;
                OPT_Y(3,i)=g1;
```

```
            OPT_Y(4,i)=g2;
            OPT_Y(5,i)=Fd_t;
```

% Penalized of solutions not providing required constraints in respect to design, with a big value (10^6)

```
            if OPT_Y(3,i)>0
                OPT_Y(5,i)=10^6;
            end
            if OPT_Y(4,i)>0
                OPT_Y(5,i)=10^6;
            end
        end
```

% UPDATING OF MATRIX WITH VALUE OF VARIABLES BELONGING TO BETTER SOLUTION, BY COMPARING OF CURRENT VALUES WITH NEWLY PRODUCED VALUES IN TERMS OF MINIMIZATION OF OBJECTIVE FUNCTION

```
        for i=1:pn
            if OPT(5,i)>OPT_Y(5,i)
                OPT(:,i)=OPT_Y(:,i);
            end
        end

        % Learner phase
        for i=1:pn
```

% Determining again of design variables according to learner phase rules

% Selecting randomly of required two solutions (students) as different from each other

```
            xi=ceil(rand*pn);
            xj=ceil(rand*pn);

            while xi==xj
                xi=ceil(rand*pn);
                xj=ceil(rand*pn);
            end

            if OPT(5,xi)<OPT(5,xj)
                d=OPT(1,i)+rand*(OPT(1,xi)-OPT(1,xj));
                t=OPT(2,i)+rand*(OPT(2,xi)-OPT(2,xj));
```

```
        else
            d=OPT(1,i)+rand*(OPT(1,xj)-OPT(1,xi));
            t=OPT(2,i)+rand*(OPT(2,xj)-OPT(2,xi));
        end

% Controlling of lower and upper limits of new generated design variables
        if d>maksd
            d=maksd;
        end
        if d<mind
            d=mind;
        end
        if t>makst
            t=makst;
        end
        if t<mint
            t=mint;
        end

        % Objective function of problem
        Fd_t=9.8*d*t+2*d;

        % Inequality constraints of problem
        g1=(P(ii)/(pi*d*t*yield))-1;
        g2=8*P(ii)*(l(ii)^2)/(pi^3*E*d*t*(d^2+t^2))-1;

        %OPT_Y: New solution matrix
        OPT_Y(1,i)=d;
        OPT_Y(2,i)=t;
        OPT_Y(3,i)=g1;
        OPT_Y(4,i)=g2;
        OPT_Y(5,i)=Fd_t;

% Penalized of solutions not providing required constraints in respect to de-
sign, with a big value (10^6)
        if OPT_Y(3,i)>0
            OPT_Y(5,i)=10^6;
        end
        if OPT_Y(4,i)>0
            OPT_Y(5,i)=10^6;
        end
    end
```

% UPDATING OF MATRIX WITH VALUE OF VARIABLES BELONGING TO BETTER SOLUTION, BY COMPARING OF CURRENT VALUES WITH NEWLY PRODUCED VALUES IN TERMS OF MINIMIZATION OF OBJECTIVE FUNCTION

```
        for i=1:pn
            if OPT(5,i)>OPT_Y(5,i)
                OPT(:,i)=OPT_Y(:,i);
            end
        end

    end  %end of the iteration

    OPT_cycle=zeros(5,1);
    [min_value,rank]=min(OPT(5,:));
    OPT_cycle(1,1)=OPT(1,rank);       %d
    OPT_cycle(2,1)=OPT(2,rank);       %t
    OPT_cycle(3,1)=OPT(3,rank);       %g1
    OPT_cycle(4,1)=OPT(4,rank);       %g2
    OPT_cycle(5,1)=OPT(5,rank);       %min Fd_t
```

% OPT_cycle vector is that recording of result ensuring the minimum conditions for each compressive load (that in end of iteration)

% OPT_CEV is result storage matrix containing thickness, center diameter with objective function values of vector that ensuring minimization of objective function in updated optimization matrix generating for each compressive load

```
    OPT_CEV(:,ii)=OPT_cycle(:,:);
end % end of any compressive load cycle
toc
```

Chapter 3
A Novel Prediction Perspective to the Bending Over Sheave Fatigue Lifetime of Steel Wire Ropes by Means of Artificial Neural Networks

Tuğba Özge Onur
Zonguldak Bulent Ecevit University, Turkey

Yusuf Aytaç Onur
Zonguldak Bulent Ecevit University, Turkey

ABSTRACT

Steel wire ropes are frequently subjected to dynamic reciprocal bending movement over sheaves or drums in cranes, elevators, mine hoists, and aerial ropeways. This kind of movement initiates fatigue damage on the ropes. It is a quite significant case to know bending cycles to failure of rope in service, which is also known as bending over sheave fatigue lifetime. It helps to take precautions in the plant in advance and eliminate catastrophic accidents due to the usage of rope when allowable bending cycles are exceeded. To determine the bending fatigue lifetime of ropes, experimental studies are conducted. However, bending over sheave fatigue testing in laboratory environments require high initial preparation cost and a long time to finalize the experiments. Due to those reasons, this chapter focuses on a novel prediction perspective to the bending over sheave fatigue lifetime of steel wire ropes by means of artificial neural networks.

DOI: 10.4018/978-1-7998-0301-0.ch003

INTRODUCTION

Wire ropes are frequently used in elevators, cranes, bridges, aerial ropeways, mine hoisting, and so on. Personnel, materials, cargo, etc. are lifted by steel wire ropes in a vast variety of material handling systems. Carbon steel rods are drawn to make steel wires with different shapes and sizes. The very high strength of the rope wires allows wire ropes to endure large tensile loads and to run over sheaves with relative small diameters. One or several layers of steel wires laid helically around a center wire form a strand. Traditional stranded steel wire ropes have six or eight strands wound around a core. Rotation resistant ropes have higher number of strands in order to resist rotation (Feyrer, 2015). Cars and counter-weights are suspended by steel wire ropes in traction elevators. That is, steel wire ropes are used to lift personnel and freight in the car of the elevator (Janovsky, 1999). In cranes, ropes are used to lift, convey, and discharge heavy goods from one location to another location within a specific area. Crane ropes are selected to maintain a certain lifetime period in service (Suner, 1988). A Koepe (friction) system is often used in mine hoisting to lift heavy loads from deep shafts by means of steel wire ropes with large diameters (Onur, 2012). Wire ropes deteriorate gradually as a result of normal running or misuse while operating. Those deteriorations exhibit themselves in different ways after a certain period of time. Mostly, degradations that occur on steel wire ropes are due to fatigue. Furthermore, under almost all operational conditions, wire ropes are subjected to fatigue due to alternate bending and longitudinal movements. The fatigue in ropes can be divided into two main categories in general. One of the dominant fatigue types in rope applications is tension-tension fatigue in which ropes are subjected to alternate tensile load with time, such as with suspension bridges. Another type is known as bending over sheave (BoS) fatigue in which ropes are subjected to dynamic repetitive bending and straightening travel due to the winding of wire rope on a drum or sheaves, such as with cranes.

Fatigue causes degradation on the rope and reduces the lifetime of steel wire ropes. Knowing the lifetime of the rope is an important issue in terms of occupational safety. Rope manufacturers are also eager to extend their rope's lifetime. Therefore, it is an important research topic to investigate the fatigue lifetime of steel wire ropes.

This study addresses the BoS fatigue of steel wire ropes. Numerous studies have been conducted to shed light on the effect of BoS fatigue on the lifetime of the steel wire ropes.

Gibson et al. (Gibson et al., 1974) performed bending fatigue tests by using 6x36 Warrington Seale rope with a steel core, 6x24 Warrington Seale rope with a fiber core, and 6x26 Warrington Seale rope with a steel core. Each rope has right regular lay. The 6x24 Warrington Seale rope is made of galvanized high carbon steel, and the other two ropes are made of bright high carbon steel. Samples with diameters of 12.7 millimeters (1/2 inch) and 19.05 millimeters (3/4 inch) were used for bending fatigue tests. According to the results, the 6x36 Warrington Seale rope and 6x26 Warrington Seale rope had almost the same fatigue performance, while the 6x24 Warrington Seale rope had lower values. The authors explained that this rope may be used in applications where a low modulus of elasticity is desired. Therefore, it should not be used in applications where better fatigue performance is expected. The authors also measured temperature fluctuations at 45.72 m/min (150 feet/min) and declared that a low diameter ratio and high tensile load would lead to a rapid increase in temperatures occuring on the rope (Onur, 2010).

Bartels et al. (Bartels, McKewan, &Miscoe, 1992) examined the factors that affect the life of wire rope. Two 50.8 mm (2 inch) diameter 6x25 Filler ropes with fiber cores were degraded on a bending fatigue machine. The authors determined the number of wire breaks, residual breaking loads, and percent elongations at break due to the number of bending cycles. The test results indicated that once a wire rope nears the end of its service life, both deterioration and the consequent loss of rope strength begin to increase at an accelerated rate (Onur, 2010). Verreet (Verreet, 1998) used Feyrer's theoretical rope breaking lifetime and discarding lifetime formulas and prepared a computer program by taking into consideration the parameters such as rope type, rope diameter, sheave diameter, tensile load, wire strength, and the length of the most stressed rope section in his research report. He found generalized rope breaking lifetime and discarding lifetime results related to drive group and load collective. Suh and Chang (Suh & Chang, 2000) conducted experimental studies to investigate the effects of the stress variation, mean stress, and rope lay length on the lifetime of wire ropes subjected to tensile-tensile fatigue. Authors observed that outstanding temperature variation was in the range of 2.5–4 Hz frequency. Other authors (Chaplin, Ridge, &Zheng, 1999; Ridge, Chaplin, &Zheng, 2001) assessed effects of wire breaks, abrasive wear, corrosion, plastic wear, torsional imbalance, and slack wires and strands on the BoS fatigue lifetime. Therefore, the similar types of damages that are specified were created on the rope samples, which were 6x25 Filler rope and rotation resistant rope with 34 compacted strands before the test. Furthermore, the Feyrer's theoretical rope fatigue lifetime equations were used to compare experimental results gathered. In Gibson's handbook (Gibson, 2001), an attempt was made to correlate rope contact length on sheave with the rope fatigue lifetime. It is found that there was a reduction in the rope fatigue lifetime until contact length is equal to the pitch length and there is no change in larger contact lengths. Torkar and Arzensek (Torkar & Arzensek, 2002) performed a failure analysis of a 6x19 Seale steel rope broken while running on a crane. The authors concluded that the main reasons for the failure of the wire rope were fatigue and poor inspection. Feyrer's book (Feyrer, 2015) presented recent developments in instrumentations and experimental measurements in order to exhibit static and dynamic behaviors of rope structures. The author revealed the effects of tensile load and diameter ratio, rope wire strength, rope diameter, rope bending length on the sheave or drum, rope core type, zinc coating, lubrication, groove geometry and its material, and reverse bending on the BoS fatigue lifetime. A regression analysis was performed by using the obtained experimental results. Gorbatov et al. (Gorbatov et al., 2007) investigated the variation of the rope BoS fatigue lifetime with the type of rope core and type of rope core lubricant. A 6x36 Warrington-Seale rope with 16 mm in diameter was used. The authors revealed that the steel wire rope, which has jute core and is lubricated by ASKM-1A, has a superior BoS fatigue lifetime than a steel wire rope, which has a hemp core and is lubricated by E-1. In the technical bulletin published by Wire Rope Works Inc. (Wirerope Works Inc., 2008), the rope lifetime curve developed by the wire rope industry was explained in order to extract the effect of sheave diameter accurately to the performance of wire ropes subjected to the BoS fatigue by using rope samples that were 12.7 mm (1/2 inch) in diameter. The report revealed that increasing the diameter ratio provides longer BoS fatigue lifetime. In the case of the diameter ratio becoming 55 instead of 40, the BoS fatigue lifetime increased 91%. Pal et al. (Pal, Mukhopadhyay, Sharma, & Bhattacharya, 2018) investigated why a 6x36 steel wire rope failed after nine months of service and tried to estimate bending fatigue life as per the Niemann equation. The authors recommended changing the schedule of the investigated wire rope from nine to six months. Zhang et al. (Zhang, Feng, Chen, Wang, &Ni, 2017) used a custom-made bending fatigue test machine to judge the bending fatigue behavior of a 6x19 steel wire rope that had pre-broken wires before the test. The effect of different distributions of pre-broken wires on bending fatigue behavior was determined. The authors concluded that the broken wires changed the stress state of the inner wire strands and this led to a concentration of severe wear, which accelerated the density of broken wires. Kim et al. (Sung-Ho, Sung-Hoon, & Jae-Do, 2014) studied an 8x19 Seale elevator rope. Rope samples were corroded by a salt-water test chamber and a bending fatigue test apparatus was used to evaluate the effect of corrosion fatigue on life expectancy. The authors concluded that a greater amount of corrosion, an increase in the tensile load, and repeated bending cycles produced a rapid increase in the number of broken wires.

However, in addition to the investigations summarized above, although the usage of optimization methods like artificial neural networks (ANN) and a genetic algorithm (GA) for the prediction of bending fatigue lifetime of steel wire ropes is not sufficiently discussed in the present literature, Dou and Wang (Dou & Wang, 2012) used ANN to analyze the characteristics of wire ropes. According to their idea, because of the recent advances in computer technology, the use of ANN become more feasible. Also, ANN has the potential to solve certain types of complex problems that have not been satisfactorily handled by more traditional methods.

BoS fatigue testing of steel wire ropes requires high initial preparation cost and significant time to finalize the experiments and performance experienced by investigating numerous parameters in order to determine the rope lifetime in the related literature. In addition, mathematical equations must be found in order to evaluate the results obtained from the performed tests. Furthermore, it is impossible to handle all kinds of parameters and conditions in the tests performed in order to determine the rope lifetime. For example, there are limited studies on how the rope BoS fatigue lifetime will change as the rotation speed increases and that determine the effects of parameters such as insufficient lubrication and twisting on the rope on BoS fatigue lifetime. Due to those reasons, the use of computer systems is accepted as an alternative method to minimize the high cost and performance performed during the tests. For this purpose, techniques such as artificial neural networks (ANN), fuzzy logic, and a genetic algorithm can be used as an alternative to traditional methods in determining rope life with developing computer technology. ANN also have the ability to generate solutions quickly by creating an algorithm. ANN, which are basically an algorithm based on the human nervous system, are used more frequently than genetic algorithms because of their faster solutions for the problems that are large scaled, nonlinear, and discontinuous/missing.

In this chapter, authors focus on a novel prediction perspective to the bending over sheave fatigue lifetime of steel wire ropes by means of artificial neural networks. On the other hand, in order to research the robustness and accuracy of the designed ANN model, theoretical calculations are done by using Feyrer equations (Feyrer, 2015). Furthermore, the results obtained from ANN are compared with the data presented in the literature (Onur & İmrak, 2012).

ROPES

The ropes generally consist of several strands laid helically in one or more layers around a core of fiber or steel. A strand has two or more wires wrapped in a peculiar sequence around its core. Ropes are used extensively as traction component of systems for lifting and transporting due to their flexible structure and convenient use (Cürgül, 1995). Ropes can be thought as a machine that has many moving parts and should be manufactured properly. It should be selected according to the place it is to be used. Furthermore, ropes are manufactured in various compositions, like machines, when their application areas are taken into consideration. An identical rope may be suitable in one system, while it may not suitable for another system (Cookes Limited, 2007). Ropes are divided into two groups, fiber ropes and wire ropes, according to the material they are made of. The ropes made by steel wires are the highest stressed and the most important component of transport machines. Different operating conditions are necessary to manufacture various types of ropes. The characteristic properties of wire ropes are little known in practice.

If the operating personnel lack the necessary information about the maintenance, control, and the use of wire ropes, a proper and safe operation cannot be carried out. Apart from the designer, the user must also have the necessary information about wire ropes (Demirsoy, 1991). As a result of the improvements in wire ropes, chains, which were used previously in cranes, are not used in crane manufacturing today. The superiorities of wire ropes over the chains can be listed as follows:

- Due to their light weight, there is little mass effect on the lifting machines working at elevated speeds.
- Their operating safety is high and operation controls are easy.
- They resist light impacts since they are more flexible than chains.
- They operate silently at high speeds.
- The unit prices are cheaper than the chains, conditionally.

Wire ropes don't break suddenly like chains and, with taking care of initiation of broken wires, safety precautions can be taken. The above qualities outline the reasons to choose wire ropes over chains. Furthermore, wire ropes have a high lifting capacity. They don't lose much of their strength due to fatigue and moisture effects while operating. Operation safety is quite high due to the fact that the load is distributed over a large number of wires in the rope. Wire ropes can be operated at high operation speeds. There is a convenient rate between their weights and lifting capacity, and they have a large amount of elastic elongation. It is possible to visually control them easily during operation. The lifting capacity and operation properties of wire ropes don't change with low temperatures. However, ropes with fiber cores should not be operated at temperatures above 100˚C, and ropes with steel cores should not be operated at temperatures above 250˚C. The rope end connection should be checked very carefully at low and high temperatures (Demirsoy, 1991). A wire rope structure is depicted in Figure 1.

Wire ropes can also be classified considering core material. Core material is either steel or fiber; wire ropes can also be classified as parallel lay rope and cross lay rope, according to their strand arrangements (Adıvar çelik halat, 2010) and can also be classified as special ropes (Çelik Halat ve Tel Sanayi A.Ş., 1999; Cürgül, 1995; Demirsoy, 1991; Türk Standartları Enstitüsü, 2005). Wire ropes are widely used in mining, oil wells, the transportation of heavy loads, elevators, trams, naval applications, fishery, forestry, ships and yachts, aerial ropeways, and in general engineering applications (Çelik Halat ve Tel Sanayi A.Ş., 1999).

Figure 1. Wire rope structure

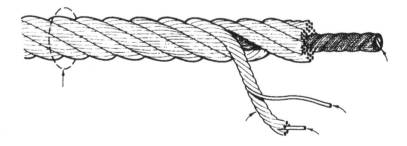

ARTIFICIAL NEURAL NETWORKS

The nervous system consists of nerves and specialized cells known as neurons. There are approximately 10^{11} neurons in the human body (Kohonen, 1988). The brain is composed of the combination of the neurons, which exist not only in the brain, but also in the whole body in the nervous system. Neurons can be called a network since they work in groups. Each neuron in this network has input (dendrite), output (axon), connections (synapse), and a cell nucleus. A neuron structure is depicted in Figure 2.

Connections provide the communication between the cells. A neuron takes the signals incoming from other neurons to the inlet part via the connections. Incoming signals are weakened or amplified by the connections, then these signals are transmitted to the cell nucleus. If the signals incoming to the cell nucleus exceed a certain threshold value as a result of their interaction with each other, a signal is sent to the outlet and the nerve becomes active. If threshold value is not be exceeded, the signal cannot be sent. In view of this situation, artificial neural networks models have been developed. The first study on artificial neural networks was made in 1943 by a neurophysicist, McCulloch, and a mathematician, Pitts. In this study, a cell model was developed according to a definition of an artificial nerve (McCulloch & Pitts, 1943). In 1949, a study was conducted giving an idea about how a neural network can perform the learning, and the "Hebbian Rule," which is the basis of most of the learning rules used today, has emerged from this study (Hebb, 1949). A model that reacts to the stimuli and is able to be adaptable was created for the first time in 1954 (Farley & Clark, 1954). The perceptron was developed by Rosenblatt and was the most important step after the emergence of the Hebbian Rule in 1958. This development also forms the basis for the learning algorithms of today's machines (Şen, 2004). The ADALINE (Adaptive Linear Combiner) learning rule was developed in 1960. This learning rule has passed the literature as Widrow-Hoff learning rule. Its most important feature is to aim to minimize the error that occurred during the training of the model (Office of Naval Research Contract, 1960). In 1969, the book entitled "Perceptron" was published and indicated that single-layer perceptrons could not solve complex problems, with the example of an XOR problem. After the beginning of the more effective use of computers in the 1980s, remarkable improvements emerged about ANN. It can be said that the modern era started in ANN with the study of "Neural Networks and Physical Properties" in 1982. Nonlinear networks were developed in this study (Hopfield, 1982). In the same years, the studies about ANN have accelerated with the development of non-instructional learning systems in the studies performed by Kohonen (Kohonen, 1982) and Anderson (Anderson, 1983). In 1986, the back propagation algorithm was developed for multi-layer perceptrons. Since this developed algorithm has a strong structure, it provided resolutions for the previously encountered problems (Rumelhart, Hinton, & Williams, 1986). The back propagation algorithm is

Figure 2. Structure of a neuron
(Şen, 2004)

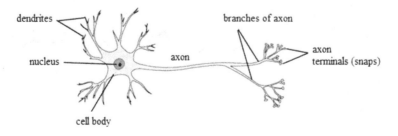

one of the most widely used algorithms. Today, the use of ANN has gained a great intensity in parallel with the developments in computer systems. New learning algorithms are being developed daily, and new studies about the network architectures are being done.

Artificial neural networks can be used in almost every discipline and science. There are many problems solved with ANN, in many areas, such as science, mathematics, medicine, business, and finance. ANN is used in many fields, such as classification, clustering, prediction, pattern recognition, function approach, and optimization. Brief explanations of these fields are stated in the following.

- Classification is the process of determining that one object belongs to which class in more than one class. ANN provides to detect the next data belong to which class by using an existing classification.
- Clustering is different from classification, and it is done notwithstanding a certain limit. There is no need for any information about the data processing. Clustering is performed by including similar data in the same group.
- The most popular application area of artificial neural networks is the prediction. Prediction is producing an idea about the future data with regard to data in the past. Market volatility, sales amount, and weather forecast can be given as examples. ANN gains an important advantage for the estimation of variables, especially in a linear structure (Li, Mehrotra, Mohan, &Ranka, 1990).
- The accurate one of the distorted patterns can be obtained by defining distorted or missing patterns to ANN and comparing the patterns. This method can be used for processes such as face recognition or retina recognition, which are similar. Authentication is possible with those processes.
- There are many calculation methods for which the mathematical expressions are unknown. In other words, although its inputs and outputs are certain, there are situations in which the intermediate function is unknown. The function approach is to define the function that will produce approximately the same output data in response to the same output data. Function and prediction approaches are similar models.
- Optimization is the action of finding the best or most effective performance under the given constraints by maximizing desired factors and minimizing undesired ones for many commercial and scientific events. In engineering, optimization is widely used, especially in the design phase. Optimization can be basically divided into two groups. The first one is the traditional optimization, which is done by creating a function based on the mathematical data. The second one is the optimization technique done by using artificial intelligence. This method is used for the optimization of complex structures that are cannot be expressed as a function. Artificial neural networks also use this method (Civalek & Çatal, 2004).

Structure of ANN

Artificial neural networks consist of many artificial nerve cells. It is desired to combine this structure resembling the biologic nervous system with the process elements on the same direction. The basic architecture consists of three types of neuron layers: input, hidden, and output layers, as shown in Figure 3.

The input layer is the layer where incoming data are received and transferred to the hidden layer. No transaction is performed when transferring data in some network structures (Svozil, Kvasnicka, & Pospichal, 1997). The intermediate layer, also known as the hidden layer, processes the data incoming from the input layer and sends them to the output layer. There can be multiple hidden layers in ANN.

The output layer processes the information incoming from the hidden layer and produces the output that must be generated for the data presented to the input layer.

The information that comes to the cell from the external environment in the structure of ANN is input. It forms an input layer by combining with the cell. The data formed in the input layer go to the processing element with the weights on the connections. The weights indicate the mathematical coefficient of an input acted on a process in the hidden layer. Weights are randomly assigned when the calculation starts. Depending on the procedure, it can continue to change the weights until the error is minimum. The input data combined with the weights are added by the joint function. The joint function can be the maximum or minimum field or multiplication function according to the network structure. The output is defined by processing the input obtained from the joint function with the activation function. There are several types of activation functions. The most appropriate activation function is determined by the trials of the designer. There are several activation functions used in ANN architectures as linear, sigmoid, and tangent hyperbolic activation functions (Ozkan & Sunar, 2003). Since sigmoid activation function is preferred as an activation function for the designed artificial neural network model in order to estimate BoS fatigue lifetimes of 6×36 WS rope, it will be briefly explained with a subhead below.

Sigmoid Activation Function

Sigmoid activation function is differentiable and it is the most used function due to the continuity of both itself and its derivative. The sigmoid function curve looks like a S-shape as depicted in Figure 4. The sigmoid activation function produces only positive values and is defined by the following formula in Equation (1):

$$f(x) = \frac{1}{1 + e^{-x}}$$

(1)

Figure 3. Basic structure of artificial neural network

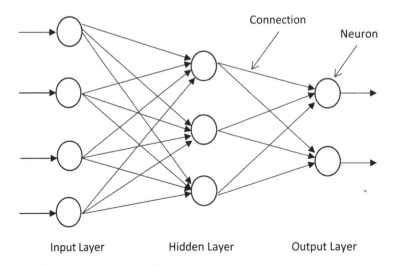

After a brief definition of the sigmoid activation function, Figure 5 is the simple representation of ANN process elements. In Figure 5, *x* is the input, *W* is the weight, *Net* is the adding function, *F(x)* is the activation function, and *Y* is the output value.

Learning Types and Rules in ANN

The most widely used learning rules in artificial neural networks are the Hebb rule, Hopfield rule, Kohonen rule, delta rule, and backpropagation algorithm (Hebb, 1949). Since backpropagation and Levenberg-Marquardt (LM) algorithms are preferred for the designed artificial neural network model in order to estimate BoS fatigue lifetimes of 6×36 WS rope, they will be briefly explained with subheads below.

LM Algorithm

Also known as the damped least squares algorithm, the *LM algorithm* is an approach to Newton's method. The LM algorithm is the method to research the minimum. A parabolic approach is used to

Figure 4. Sigmoid activation function

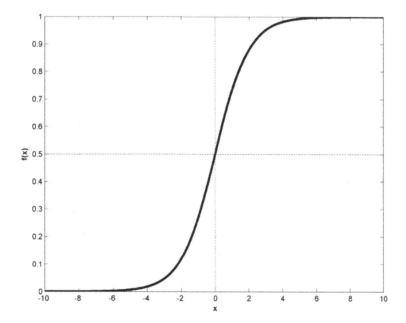

Figure 5. Architecture of ANN

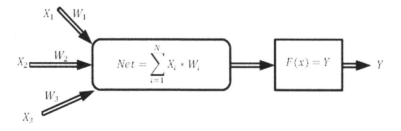

approach the error surface in each iteration and the solution of the parabola for that step is created. The LM learning rule uses the Hessian matrix given by Equation (2).

$$H = J^T J \tag{2}$$

where J represents the Jacobian matrix. The Jacobian matrix as indicated in Equation (3) is the first derivative of the network errors by weights.

$$J(n) = \frac{\delta e(n)}{\delta w(n-1)} \tag{3}$$

In Equation (3), n is the iteration number; δ is the derivative symbol; e is the network errors vector; w is the connection weights. In this case, the gradient is found by using Equation (4),

$$g = J^T e \tag{4}$$

and the connection weight between the neurons is obtained by using Equation (5),

$$w_{k+1} = w_k \left[J^T J + \mu I \right]^{-1} J^T e \tag{5}$$

where w_k, is the weight at iteration k; I is the unit matrix; μ is the Marquardt parameter. The Marquardt parameter is a scalar value. The LM method is named as a Newton algorithm when the Marquardt parameter is zero. The Levenberg-Marquardt learning algorithm produces very fast results when compared with other algorithms (Rojas, 1996).

Back Propagation Algorithm

The back propagation algorithm, or the generalized delta rule, is the most commonly used learning rule. The algorithm produced by Rumelhart, Hinton, and Williams (Rumelhart, Hinton, & Williams, 1986) has given an important acceleration for artificial neural networks. Error is propagated backwards in the back propagation algorithm. The weights are changed by back propagation of the error between the actual outputs and calculated ones. As with all learning algorithms, the goal is to determine the weights to ensure the best fit between the input and output data. The back propagation algorithm is the version of the delta algorithm where momentum term is appended. The momentum term helps to find the point where the error is minimum and to adjust the direction (Caudill, 1988).

ANN by Network Structures

Cells and connections may come together in many different ways. Considering their structures, ANN differ from each other according to the directions of the connections between the neurons and the direction of flow within the network. Artificial neural networks are divided into two groups adhering to their structures, which are feedforward artificial neural networks and feedback artificial neural networks.

The processes are transmitted from input layer to output layer in feedforward artificial neural networks, as shown in Figure 6. The input layer transfers the information received from outside to the cells in the hidden layer without any modification. Data are processed in these layers and transmitted to the output layer. Every output value in the hidden layers is used as an input value for the next layer (Wang, 2007). In Figure 6, X_i and Y are the inputs and output of the network, respectively.

In feedback artificial neural networks, the output of at least one cell is given to itself or to another neuron as input. The information exchange in the feedback network can be not only between cells located at same layer, but also between cells located at an interlayer. The feedback network can be used effectively for the solution of nonlinear problems by exhibiting a dynamic behavior with this structure. Feedback ANN are often used in digital filter designs. The feedback ANN operate more slowly than feedforward ANN (Uçan, Danacı, & Bayrak, 2003). The well-known feedback network types are Hopfield network, cellular artificial neural network, Grossberg network, Adaptive Resonance Theory-1 (ART1), and ART2 networks (Hakimpoor, Arshad, Tat, Khani, & Rahmandoust, 2011).

APPLICATION OF ANN TO DETERMINE ROPE LIFETIME

The Properties of Used Rope

In order to demonstrate the application of ANN to estimate steel wire rope lifetime, steel wire rope presented in Onur and İmrak's work (Onur & İmrak, 2012) is used. The investigated rope construction in (Onur & İmrak, 2012) their study is 6×36 Warrington-Seale (WS) steel wire rope. It has Independent Wire Rope Core (IWRC) and 10 mm diameter (*d*). In addition, it has six strands around a steel core,

Figure 6. Feedforward ANN

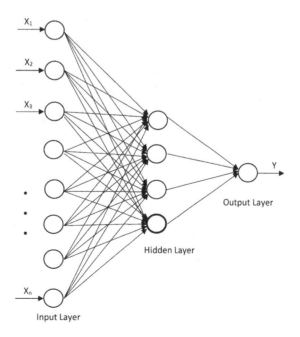

which is a wire rope itself. The 6×36 Warrington-Seale rope offers optimum resistance in fatigue and crushing; thereby, it can be used in mine hoisting, oil industry, cranes, etc. A cross-section of a 6×36 Warrington-Seale rope with IWRC is depicted in Figure 7.

Onur and İmrak (2012) employed eight different tensile loads and two sheaves with 250 mm and 100 mm diameters for BoS fatigue tests to determine the effects of tensile load and sheave diameter to the BoS fatigue lifetimes of 6×36 WS rope with 10 mm diameter in their work. Tensile loads that are 15 kN, 20 kN, 25 kN, and 30 kN have been employed when a sheave with a 250 mm diameter is used, and tensile loads that are 10 kN, 15 kN, 20 kN, and 25 kN have been employed when sheave with a 100 mm diameter is used.

Theoretical Calculation of the Investigated Rope

Theoretical BoS fatigue life estimations have been performed by using the Feyrer equation that is given in Equation (6).

Figure 7. Cross section of 6 x 36 Warrington-Seale rope with IWRC (Onur & İmrak, 2011)

$$\log(N) = b_0 + \left(b_1 + b_3 \log\left(\frac{D}{d}\right) \right) \cdot \left(\log\left(\frac{S}{d^2}\right) - (0.4).\log\left(\frac{R_0}{1770}\right) \right) + b_2 . \log\left(\frac{D}{d}\right) +$$

$$\log\left(\frac{0.52}{-0.48 + \left(\dfrac{d}{16}\right)^{0.3}} \right) + \log\left(\frac{1.49}{2.49 - \left(\dfrac{l/d - 2.5}{57.5}\right)^{-0.14}} \right) + \log(f_c) \qquad (6)$$

where D/d is the diameter ratio; S/d^2 is the specific tensile load; R_o is the wire grade; d is the rope diameter; l is the bending length; f_c is the rope core and number of strands factor. The constants b_i's in Equation (6) are given in Feyrer's book (Feyrer, 2015). Table 1 includes the constants and parameters for 6×36 WS rope.

Feyrer proposes that the numbers of bending cycles calculated by means of constants in Table 1 are valid for up to a few million bending cycles if some conditions are provided. The conditions proposed by Feyrer are: well lubricated with viscous oil or Vaseline for the wire rope samples, having steel grooves for sheaves, groove radius-rope diameter ratio r/d is 0.53, there is no side deflection, and it is in a dry environment. In the case of different conditions in operation, in order to determine the final BoS fatigue life calculation, correction factors must be taken into consideration.

ANN Application

For devising an artificial neural network model in order to estimate the BoS fatigue lifetimes of 6×36 WS rope, experimental test data given in Onur and İmrak's study (Onur & İmrak, 2012) have been used and the back propagation learning algorithm is utilized to predict the data obtained from the data set. In this ANN model, specific tensile load S/d^2 and diameter ratio D/d parameters have been chosen as the input parameters, and the number of bending cycles (N) has been regarded as the output parameter (Onur, İmrak, & Onur, 2016). In order to decide the optimum structure of the neural network, the rate of error convergence was checked by changing the number of hidden neurons. The networks were trained up to cycles where the level of the mean square error (MSE) is satisfactory and further cycles had no significant effect on error reduction. MSE is calculated by using Equation (7),

Table 1. Constants and parameters used for 6x36 Warrington-Seale rope

	Constants and Parameters	Values
Parameters	d (mm)	10
	R_o (N/mm²)	1960
	l (mm)	600
	f_c	0.81
Constants	b_0	1.278
	b_1	0.029
	b_2	6.241
	b_3	-1.613

$$MSE = \frac{1}{m}\sum_{k=1}^{m}\left(de_i - O_i\right)^2 \tag{7}$$

where de_i is the desired or actual value, O_i is the predicted output value, and m is the number of data. For simulating an automated neural network model, MATLAB software has been used. A structure with 18 neurons in the hidden layer is formed and a logistic sigmoid function is preferred as the activation function. During training, the automated neural network is presented by the data thousands of times, which is referred to as cycles. After each cycle, the error between the automated neural network output and desired values are propagated backward to adjust the weight in a manner to mathematically guarantee to converge (Haykin, 1998). MSE is used to stop training automatically. Furthermore, 45% of the data set given in Onur and İmrak's study (Onur & İmrak, 2012) are used for training the network, and 30% are selected randomly to test the performance of the trained network. A final check on the performance of the trained network is made using a validation set (Vadood, Semnani, & Morshed, 2011).

Table 2 includes the results obtained by experimental data given in Onur and İmrak's study (Onur & İmrak, 2012), theoretical results obtained by using Feyrer's equation given in Equation (6), and results acquired by means of simulating designed ANN.

For running ANN simulation, either MATLAB ANN toolbox or commands can be utilized. In order to intervene in the simulation performance and conditions, MATLAB commands can be used in the command window instead of an ANN toolbox. The following command lines are utilized to simulate the designed ANN model in order to predict the BoS fatigue lifetimes of wire rope investigated.

```
%Matlab scripts for calculating the BoS fatigue lifetimes.
%%Input data
p1 = input('Please input input1 data:','s') ; % data of s/d²
p2 = input('Please input input2 data:','s'); %data of D/d
p=[p1; p2]; %input vector includes p1 and p2
%%Output data
t = input('Please input output data:','s'); % Output data N_test
%%Normalization
```

Table 2. Results obtained by Feyrer's theoretical estimation, experimental results (N_{test}) in [47,] and ANN results (N_{ANN}) for 6×36 Warrington-Seale rope

S/d² (N/mm²)	D/d	l/d	N_{ANN}(cycles)	N_{feyrer}(cycles)	N_{test}(cycles)
150	25	60	174693.96	171146	163456
200	25	60	86791.95	90213	86792
250	25	60	70295.48	54898	69619
300	25	60	35504.97	36585	38505
100	10	60	32515.07	25951	32516
150	10	60	28832.41	13653	27774
200	10	60	15469.40	8656	13170
250	10	60	4683.89	6078	4684

```
[pn,minp,maxp,tn,mint,maxt] = premnmx(p,t);
%%Designing, training and simulation of ANN
net=newff(minmax(pn), [N 1], {'logsig' 'purelin'},'trainlm'); % N %is the num-
ber of hidden layers
net.performFcn = 'mse';
net.trainFcn = 'trainlm';
net.layers{1}.transferFcn = 'logsig';
net.layers{1}.initFcn = 'initnw';
net.layers{2}.transferFcn = 'logsig';
[net1, tr] = train(net,pn,tn);
an = sim(net1,pn);
[a] = postmnmx(an,mint,maxt); %Inverse of normalization
y= sim(net1,pn);
e=t-y;
%%Plotting training data and output of ANN
figure(1),plot3(p1(1,:),p2(1,:),t,'o');
hold on,plot3(p(1,:),p2(1,:),a,'r*'),grid on;
legend('Exact value','Output of ANN'),xlabel('p1'),ylabel('p2'),zlabel('t')
view(net)
%%Preparing test data
ptest=[p(1,1:3);p(2,1:3)]
ttest = [t(1,1:3)]
[ptn,minpt,maxpt,ttn,mintt,maxtt] = premnmx(ptest,ttest);
atn = sim(net,ptn); %Simulasyon
[at] = postmnmx(atn,mintt,maxtt);
%%Plotting test data and output of ANN
figure(2),plot3(ptest(1,:),ptest(2,:),ttest,'o');
hold on,plot3(ptest(1,:),ptest(2,:),at,'r*'),grid on;
legend('Exact value','Output of ANN'),xlabel('p1'),ylabel('p2'),zlabel('t'),ti
tle('Test data')
plotregression(t,y,'Regression')
plotfit(net1,p,t);
plottrainstate(tr);
plotperform(tr);
```

In order to provide convenience for comparing the results given in Table 2, a graphic has been plotted and shown in Figure 8.

Results obtained by experimental tests given in Onur and İmrak's study (Onur & İmrak, 2012) and the designed ANN model results are very close to each other, with 0.0036577, 0.0029135, and 0.0001987 minimum square errors for testing, training, and validation, respectively, as shown in Figure 8. Although theoretical estimations of BoS fatigue lifetime can be performed by using Feyrer's equations given in Equation (6), as it can be seen, it requires substantial expertise in the mathematical definitions. Therefore, artificial neural networks can be considered as a useful tool in order to predict the BoS fatigue lifetime of 6×36 Warrington-Seale steel wire ropes.

CONCLUSION

This chapter aimed to evaluate the artificial neural network in predicting Bos fatigue lifetime. Although there are some more theoretical estimations and analysis for BoS fatigue lifetime predictions, they require substantial expertise in the mathematical definitions of equations. As an alternative to the existing references, it is evident from the present study that the ANN model has good prediction capability to predict the BoS fatigue lifetimes of 6×36 Warrington-Seale steel wire ropes with input parameters like specific tensile load S/d^2 and diameter ratio D/d. Based on the study, it is established that the ANN approach seems to be a better option for the appropriate prediction of BoS fatigue lifetimes of 6×36 Warrington-Seale steel wire ropes. The results predicted with this designed ANN model are compared to calculations with Feyrer's equations and with the experimental data presented in Onur and İmrak's study (Onur & İmrak, 2012). Errors obtained are within acceptable limits. It has been concluded that the ANN model can be effectively utilized as a prediction tool for such a purpose. Furthermore, the developed ANN model can be used to estimate the BoS fatigue lifetimes of different types of wire ropes for a wide range of parameters for future research.

Figure 8. BoS fatigue lifetime results given in Table 2

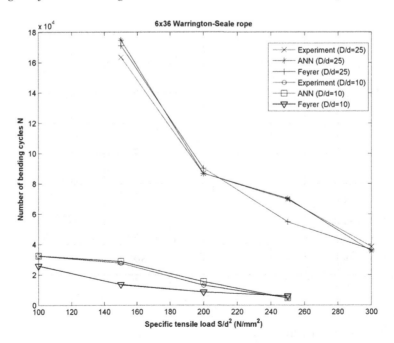

REFERENCES

Adıvar çelik halat. (2010). *Adıvar çelik halat kataloğu*. Retrieved from http://www.adivarcelikhalat.com

Anderson, J. A. (1983). Cognitive and psychological computation with neural models. *IEEE Transactions on Systems, Man, and Cybernetics*, *SMC-13*(5), 799–814. doi:10.1109/TSMC.1983.6313074

Bartels, J. R., McKewan, W. M., & Miscoe, A. J. (1992). *Bending fatigue tests 2 and 3 on 2-Inch 6 x 25 fiber core wire rope*. Retrieved from https://www.cdc.gov/niosh/mining/UserFiles/works/pdfs/ri9429.pdf

Caudill, M. (1988). Neural networks primer, part III. *AI Expert*, *3*(6), 53–59.

Çelik Halat ve Tel Sanayi A.Ş. (1999). *Çelik halat ürün kataloğu*. İzmit, Turkey.

Chaplin, C. R., Ridge, I. M. L., & Zheng, J. (1999, July). *Rope degradation and damage*. Retrieved from http://www.hse.gov.uk/research/otopdf/1999/oto99033.pdf

Civalek, Ö., & Çatal, H. H. (2004). Geriye yayılma yapay sinir ağları kullanılarak elastik kirişlerin statik ve dinamik analizi. *DEÜ Mühendislik Fakültesi Fen ve Mühendislik Dergisi*, *1*, 1–16.

Cookes Limited. (2007). *Wire rope handbook*. Auckland, New Zealand: Cookes Limited.

Cürgül. İ. (1995). Materials handling. İzmit, Turkey, Kocaeli University Publications.

Demirsoy, M. (1991). *Materials Handling* (Vol. I). İstanbul, Turkey: Birsen Publishing House.

Dou, Z., & Wang, M. (2012). *Proceedings of the International Conference on Automatic Control and Artificial Intelligence*, 1614-1616.

Farley, B. G., & Clark, W. A. (1954). Simulation of self-organizing systems by digital computer. *Transactions of the IRE Professional Group on Information Theory*, *4*(4), 76-84. 10.1109/TIT.1954.1057468

Feyrer, K. (2015). *Wire ropes: tension, endurance, reliability*. New York, NY: Springer.

Gibson, P. T. (2001). Operational characteristics of ropes and cables. In J. F. Bash (Ed.), *Handbook of oceanographic winch, wire and cable technology* (pp. 8-3–8-50). USA.

Gibson, P. T., White, F. G., Schalit, L. A., Thomas, R. E., Cote, R. W., & Cress, H. A. (1974, Oct. 31). *A study of parameters that influence wire rope fatigue life*. Retrieved from https://apps.dtic.mil/dtic/tr/fulltext/u2/a001673.pdf

Gorbatov, E. K., Klekovkina, N. A., Saltuk, V. N., Fogel, V., Barsukov, V. K., Barsukov, E. V., ... Kurashov, D. A. (2007). Steel rope with longer service life and improved quality. *Journal of Metallurgist*, *51*(5-6), 279–283. doi:10.100711015-007-0052-y

Hakimpoor, H., Arshad, K. A. B., Tat, H. H., Khani, N., & Rahmandoust, M. (2011). Artificial neural networks applications in management. *World Applied Sciences Journal*, *14*(7), 1008–1019.

Haykin, S. (1998). *Neural networks: a comprehensive foundation*. New York, NY: Prentice Hall.

Hebb, D. (1949). *The organization of behavior: a neuropsychological theory*. New York, NY: Wiley.

Hopfield, J. J. (1982). Neural networks and physical systems with emergent collective computational abilities. *Proceedings of the National Academy of Sciences of the United States of America*, *79*(8), 2554–2558. doi:10.1073/pnas.79.8.2554 PMID:6953413

Janovsky, L. (1999). *Elevator mechanical design*. USA: Elevator World Inc.

Kohonen, T. (1982). Self-organized formation of topologically correct feature maps. *Biological Cybernetics*, *43*(1), 59–69. doi:10.1007/BF00337288

Kohonen, T. (1988). An introduction to neural computing. *Neural Networks*, *1*(1), 3–16. doi:10.1016/0893-6080(88)90020-2

Li, M., Mehrotra, K., Mohan, C. K., & Ranka, S. (1990). Forecasting sunspot numbers using neural networks. *Electrical Engineering and Computer Science Technical Reports*, 67.

McCulloch, W. S., & Pitts, W. (1943). A logical calculus of the ideas immanent in nervous activity. *The Bulletin of Mathematical Biophysics*, *5*(4), 1–19. doi:10.1007/BF02478259

Office of Naval Research Contract. (1960). *Adaptive switching circuits*. Stanford, CA: B. Widrow and M. E. Hoff.

Onur, Y. A. (2010). *Theoretical and experimental investigation of parameters affecting to the rope lifetime* (Unpublished doctoral dissertation). Istanbul Technical University Institute of Science, Turkey.

Onur, Y. A. (2012). Condition monitoring of Koepe winder ropes by electromagnetic non-destructive inspection. *Insight (American Society of Ophthalmic Registered Nurses)*, *54*(3), 144–148.

Onur, Y. A., & İmrak, C. E. (2011). The influence of rotation speed on the bending fatigue lifetime of steel wire ropes. *Proceedings of the Institution of Mechanical Engineers. Part C, Journal of Mechanical Engineering Science*, *225*(3), 520–525. doi:10.1243/09544062JMES2275

Onur, Y. A., & İmrak, C. E. (2012). Experimental and theoretical investigation of bending over sheave fatigue life of stranded steel wire rope. *Indian Journal of Engineering and Materials Sciences*, *19*, 189–195.

Onur, Y. A., İmrak, C. E., & Onur, T. Ö. (2016). Investigation of bending over sheave fatigue life of stranded steel wire rope by artificial neural networks. *International Journal of Mechanical And Production Engineering*, *4*(9), 47–49.

Ozkan, C., & Sunar Erbek, F. (2003). The comparison of activation functions for multispectral landsat TM image classification. *Photogrammetric Engineering and Remote Sensing*, *69*(11), 1225–1234. doi:10.14358/PERS.69.11.1225

Pal, U., Mukhopadhyay, G., Sharma, A., & Bhattacharya, S. (2018). Failure analysis of wire rope of ladle crane in steel making shop. *International Journal of Fatigue*, *116*, 149–155. doi:10.1016/j.ijfatigue.2018.06.019

Ridge, I. M. L., Chaplin, C. R., & Zheng, J. (2001). Effect of degradation and impaired quality on wire rope bending over sheave fatigue endurance. *Journal of Engineering Failure Analysis*, *8*(2), 173–187. doi:10.1016/S1350-6307(99)00051-5

Rojas, R. (1996). *Neural networks: a systematic introduction*. Berlin, Germany: Springer-Verlag. doi:10.1007/978-3-642-61068-4

Rumelhart, D. E., Hinton, G. E., & Williams, R. J. (1986). Learning representations by back propagation error. *Nature*, *32*(6088), 533–536. doi:10.1038/323533a0

Şen, Z. (2004). *Yapay sinir ağları ilkeleri*. Su Vakfı Yayınları.

Suh, J. I., & Chang, S. P. (2000). Experimental study on fatigue behaviour of wire ropes. *International Journal of Fatigue*, *22*(4), 339–347. doi:10.1016/S0142-1123(00)00003-7

Suner, F. (1988). *Crane bridges*. Istanbul, Turkey: Egitim Publications.

Sung-Ho, K., Sung-Hoon, H., & Jae-Do, K. (2014). Bending fatigue characteristics of corroded wire ropes. *Journal of Mechanical Science and Technology*, *28*(7), 2853–2859. doi:10.100712206-014-0639-8

Svozil, D., Kvasnicka, V., & Pospichal, J. (1997). Introduction to multi-layer feed-forward neural networks. *Chemometrics and Intelligent Laboratory Systems*, *39*(1), 43–62. doi:10.1016/S0169-7439(97)00061-0

Torkar, M., & Arzensek, B. (2002). Failure of crane wire rope. *Journal of Engineering Failure Analysis*, *9*(2), 227–233. doi:10.1016/S1350-6307(00)00047-9

Türk Standardları Enstitüsü. (2005). Çelik tel halatlar-güvenlik-bölüm 2: tarifler, kısa gösteriliş ve sınıflandırma. Ankara, Turkey.

Uçan, O. N., Danacı, E., & Bayrak, M. (2003). *İşaret ve görüntü işlemede yeni yaklaşımlar:yapay sinir ağları*. İstanbul Üniversitesi Mühendislik Fakültesi Yayınları.

Vadood, M., Semnani, D., & Morshed, M. (2011). Optimization of acrylic dry spinning production line by using artificial neural network and genetic algorithm. *Journal of Applied Polymer Science*, *120*(2), 735–744. doi:10.1002/app.33252

Verreet, R. (1998). *Calculating the service life of running steel wire ropes*. Retrieved from http://fast-lift.co.za/pdf/CASAR%20%20Calculating%20the%20service%20life%20of%20running%20steel%20wire%20ropes.pdf

Wang, Q. (2007). Artificial neural networks as cost engineering methods in a collaborative manufacturing environment. *International Journal of Production Economics*, *109*(1), 53–64. doi:10.1016/j.ijpe.2006.11.006

Wirerope Works Inc. (2008). *Bethlehem elevator rope technical bulletin 9*. Williamsport, PA.

Zhang, D., Feng, C., Chen, K., Wang, D., & Ni, X. (2017). Effect of broken wire on bending fatigue characteristics of wire ropes. *International Journal of Fatigue*, *103*, 456–465.

KEY TERMS AND DEFINITIONS

Architecture: The structure of a neural network consists of the number and connectivity of neurons. An input layer, one or more hidden layers, and an output layer are the layers that generally form the network.

Back Propagation Learning Algorithm: The algorithm for multi-layer perceptron networks to adjust the connection weights until the optimum network is obtained. In the back propagation algorithm, errors are propagated back through the network and weights are adjusted in the opposite direction to the largest local gradient.

Computer Simulation: It is a simulation that runs on a computer to model the behavior of a system.

Fatigue Testing: It is a type of test performed to determine the behavior of materials under fluctuating loads. In fatigue testing, a specified mean load and an alternating load are applied to a specimen and the number of cycles that are required to produce fatigue life is determined.

Lifetime Prediction: It is the prediction of lifetime by using some methods.

Multi-Layer Perceptron: It is one of the most widely used networks, which consists of multiple layers interconnected in a feedforward way.

Neuron: The basic building block of a neural network. A neuron sums the weighted inputs, processes the weighted inputs by means of an activation function, and produces an output at the last stage.

Rope Lifetime: It is the time for using the rope effectively. In other words, the time until the rope is considered unusable.

Chapter 4
Introduction and Application Aspects of Machine Learning for Model Reference Adaptive Control With Polynomial Neurons

Ivo Bukovsky
Czech Technical University in Prague, Czech Republic

Peter M. Benes
Czech Technical University in Prague, Czech Republic

Martin Vesely
Czech Technical University in Prague, Czech Republic

ABSTRACT

This chapter recalls the nonlinear polynomial neurons and their incremental and batch learning algorithms for both plant identification and neuro-controller adaptation. Authors explain and demonstrate the use of feed-forward as well as recurrent polynomial neurons for system approximation and control via fundamental, though for-practice, efficient machine learning algorithms such as Ridge Regression, Levenberg-Marquardt, and Conjugate Gradients; authors also discuss the use of novel optimizers such as ADAM and BFGS. Incremental gradient descent and RLS algorithms for plant identification and control are explained and demonstrated. Also, novel BIBS stability for recurrent HONUs and for closed control loops with linear plant and nonlinear (HONU) controller is discussed and demonstrated.

DOI: 10.4018/978-1-7998-0301-0.ch004

INTRODUCTION

Machine learning and control algorithms have been amazingly, theoretically developing in recent decades. The data-driven trends of system approximation based on neural networks and this way-based adaptive control is of significant interest to the research community; however, we may observe that these trends are not correspondingly spreading into industrial practice. The main reasons are the relatively high demands on operator qualification, i.e. education in mathematics and dynamical systems and operators' technological competences and ability to understand the concepts and proper applications of machine learning related to neural networks (there are many more parameters and aspects than there are with PID controllers, e.g.). For sure, the stability of nonlinear control loops with neural networks, i.e. the stability of such nonlinear time invariant systems, together with the risk of heavy costs resulting from unstable development within factory lines (or power plant units, for example) is also a crucial consideration, so the control algorithms in the industry remain conservative, preserving comprehensible and analyzable techniques, such as PID control and other linear-based approaches. Thus, we believe that the transition from conservative linear techniques of control toward nonlinear and machine learning–based ones can be successful, if such control principles are well comprehensible both to the academic and industrial community and if the stability analysis is also not acceptable and comprehensible for both sides.

The recent trends of Deep Networks and Deep Learning (Goodfellow, Bengio, & Courville, 2016; LeCun, Bengio, & Hinton, 2015) are not always achievable in industrial practice for control, especially when dynamical systems should be approximated from data, and these neural networks require large training datasets. The networks with Long Short-Term Memory (LSTM) neurons (Hochreiter & Schmidhuber, 1997) are very popular today as well; however, their application to control might be practically limited due to the need for heavier computations for training and due to a relatively complex architecture for analysis. Furthermore, many industrial control tasks would not need to implement a too complex and too nonlinear plant model and controller. When a conventional control (like PID) is doing a less or more sufficient job in practice, then there is a high chance that the control performance can be significantly optimized with reasonably nonlinear neural networks

Thus, the neural computation presented in this chapter relates to polynomial neural networks that can be ranked among shallow neural networks and that, contrary to other shallow networks, feature a linear optimization problem, while the input-output mapping is customable nonlinear. Other shallow neural architectures that shall be mentioned are random vector functional link (RVFL) networks (Zhang & Suganthan, 2016) and recently published their alternatives known as Extreme Learning Machines (ELMs) (Huang, Zhu, & Siew, 2006a, 2006b; Zhang & Suganthan, 2016) and of course also the multilayer perceptrons (MLPs) (Hornik, Stinchcombe, & White, 1989) with a very few hidden layers. Regarding MLPs, it shall be mentioned that novel types of hidden neurons with non-sigmoid somatic operations, i.e. ELU or RELU (Glorot, Bordes, & Bengio, 2011), shall be also considered for their applications due suppressing the issue of vanishing gradients and improving the convergence.

Higher order neural units (Bukovsky, Hou, Bila, & Gupta, 2008; Gupta, Homma, Hou, Solo, & Bukovsky, 2010; Gupta, Bukovsky, Homma, Solo, & Hou, 2013) are standalone architectures stemming from the branch of polynomial neural computation originating from the works, or higher order neural networks (Ivakhnenko, 1971; Kosmatopoulos, Polycarpou, Christodoulou, & Ioannou, 1995; Nikolaev & Iba, 2006; Taylor & Coombes, 1993; Tripathi, 2015).

The adaptive control that is presented in this chapter belongs among the control that utilizes a data-driven model, which is also approximated as a polynomial neural architecture.

Nowadays, the neural network–based control approaches can be classified according to the main principle as follows:

- Model Reference Adaptive Control (MRAC) (Elbuluk, Tong, & Husain, 2002; Narendra & Valavani, 1979; Osburn, 1961; Parks, 1966), where the model of controlled system is required to derive controller learning and the desired behavior is defined by a reference model; the controller learning may be batch offline or it can be incrementally running in real time as well.
- Model Predictive Control (MPC), such (Garcia, Prett, & Morani, 1989; Ławryńczuk, 2009; Morari & Lee, 1999), where the model is necessary for finding such a sequence of control inputs that most suits the desired sequence of system outputs, and it is recalculated at every single sample time.
- Adaptive Dynamic Programming (ADP), also reinforcement learning (Wang, Zhang, & Liu, 2009; Wang, Cheng, & Sun, 2007), where an analytical model is not required, but data for the controller network and for the penalty/award network are obtained from experiments with the system itself.

An advantage of the proposed MRAC-HONU–based control loop design is the rather computationally efficient and fast real-time performance with learning algorithms such as Conjugate Gradients (CG) (Dai & Yuan, 1999; El-Nabarawy, Abdelbar, & Wunsch, 2013; Zhu, Yan, & Peng, 2017) and the recursive least squares (RLS) algorithm, achieving strong error minimization capabilities even for nonlinear process control.

The purpose of our chapter is to review and summarize MRAC as one possible comprehensible approach to nonlinear adaptive control that can be practically considered in cases when more powerful, though also more complex, control schemes are not to be utilized and also when no intensive, real-time computation is required. Also, we newly discuss the BIBS stability approach for these nonlinear adaptive control loops as the method has potential for its comprehensibility and straightforwardness.

Fundamentals on HONUs

This section recalls the background about HONUs for SISO linear and non-linear dynamic systems that we found practical, especially up to the 3rd polynomial order. The very fundamentals of static or dynamic HONUs are their configuration and their Jacobian matrix, which can be defined as follows.

Figure 1. Neural network-based model reference adaptive control (MRAC) loop structure

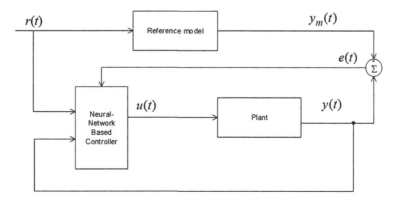

HONU for Plant Approximation

For the MRAC control scheme with HONUs, the plant (Figure 1) is to be approximated from input-output data via static or dynamic HONU. The augmented input vector into dynamic (recurrent) HONU (1) is given as

$$
\mathbf{x}=\begin{bmatrix} x_0=1 \\ x_1 \\ \vdots \\ x_n \end{bmatrix}=\begin{bmatrix} 1 & \tilde{y}(k-1) & \ldots & \tilde{y}(k-n_y) & u(k-1) & \ldots & u(k-n_u) \end{bmatrix}^T=\begin{bmatrix} 1 \\ \tilde{\mathbf{y}}_\mathbf{x} \\ \mathbf{u}_\mathbf{x} \end{bmatrix},
\tag{1}
$$

where $\tilde{y}$ neural output, u is control input, and $x_0=1$ is the augmenting unit allowing HONUs for neural bias and also for lower polynomial orders being subsets of higher order polynomials.

To define a static (feedforward) HONU, the neural output in (1) yields the measured value as

$$
\tilde{y} \leftarrow y \quad \tilde{\mathbf{y}}_\mathbf{x} \leftarrow \mathbf{y}_\mathbf{x},
\tag{2}
$$

so a recurrent HONU was generally defined. Further details of HONUs for plant approximation are given in Table 1.

The Jacobian matrix for a static HONU, i.e. the input vector (1) with only measured values, is as follows

$$
\mathbf{J}_\mathbf{w}(k)=\partial\tilde{y}(k)/\partial\mathbf{w}=\begin{cases} \mathbf{x}^T & for\ r=1 \\ \mathbf{colx}^T & for\ r=2 \\ \mathbf{colx}^T & for\ r=3, \end{cases}
\tag{3}
$$

Table 1. Summary of HONUs for plant approximation of up to 3rd order of nonlinearity

	Polynomial Order of HONU		
	LNU, $r=1$	**QNU, $r=2$**	**CNU, $r=3$**
$\Sigma\Pi$ notation	$\tilde{y}=\sum_{i=0}^{n} w_i \cdot x_i$	$\tilde{y}=\sum_{i=0}^{n}\sum_{j=i}^{n} w_{i,j} \cdot x_i \cdot x_j$	$\sum_{i=0}^{n}\sum_{j=i}^{n}\sum_{\kappa=j}^{n} w_{i,j,\kappa} x_i x_j x_\kappa$
$Col^r()$kernel form	$\tilde{y}=\mathbf{w}\cdot col^{r=1}(\mathbf{x})=\mathbf{w}\cdot\mathbf{x}$	$\tilde{y}=\mathbf{w}\cdot col^{r=2}(\mathbf{x})$	$\tilde{y}=\mathbf{w}\cdot col^{r=3}(\mathbf{x})$
details	$\mathbf{x}=\begin{bmatrix} x_0 \\ x_1 \\ \vdots \\ x_n \end{bmatrix} \mathbf{w}=\begin{bmatrix} w_0\ w_1 \ldots w_n \end{bmatrix}$	$col^2(\mathbf{x})=\left[\{x_i x_j\}\right]$ $\mathbf{w}=\left[\{w_{i,j}\}\right]$ $\forall i,j\ ;\ 0\leq i\leq j\leq n$	$col^3(\mathbf{x})=\left[\{x_i x_j x_\kappa\}\right]$ $\mathbf{w}=\left[\{w_{i,j,\kappa}\}\right]$ $\forall i,j,\kappa\ ;\ 0\leq i\leq j\leq\kappa\leq n$

where colx is a long column vector of polynomial terms as indicated in Tab. 1. Then, equation (3) indicates that Jacobian is constant for all training epochs of static HONUs (i.e. when the input vector is defined as in (1)).

Recurrent HONUs require us to recalculate Jacobian at every sample step and are more difficult to be trained; however, they can be trained for longer output response, i.e., not just for one step ahead prediction as static HONUs do, conditioned by that the real plant has such nonlinearity that can be captured from data by a HONU of given polynomial order for given sampling and configuration.

HONU as a Controller

From previous works as such (Benes & Bukovsky, 2014; Benes, Bukovsky, Cejnek, & Kalivoda, 2014; Benes, Erben, Vesely, Liska, & Bukovsky, 2016; Ivo Bukovsky, Benes, & Slama, 2015) HONUs are initially identified as dynamic plant models, i.e. via recurrent HONUs followed by training of a single HONU feedback controller. In this paper, we extend the classical HONU-MRAC control loop with multiple feedback HONU controllers, in accordance with Figure 2. Usually we assume that the magnitudes of the input and output variables, i.e. of d, y, and y_{ref}, are normalized (z-scored). Further, the input gain r_0 can be also adaptive, and it compensates for the true static gain of the controlled plant. An added advantage of the customizable controller non-linearity of the extended HONU-MRAC control scheme is that the controller computation can be tailored to target different aspects of control for the process dynamics, e.g. one HONU controller can be used to suppress noise and another may be used to minimize steady state error.

A typical input vector into HONU as a controller unit can be state feedback, defined as follows

$$\xi_\iota = \begin{bmatrix} 1 & y_{ref}(k-1) & y_{ref}(k-2) & \dots & y_{ref}(k-m_\iota) \end{bmatrix}^T , \tag{4}$$

where ι is the index of a unit in the controller layer (Figure 2), and m_ι is the number of tapped delay feedbacks forming the controller input. The details of controller units and their aggregation by merely summation are given in Figure 2.

In Figure 2, a single layer network of HONUs serves as feedback controller to calculate control inputs as follows

$$u(k) = r_o \cdot \left(d(k) - \sum^{n_q} q_\iota(k) \right) , \tag{5}$$

where n_q is the number HONUs in the controller layer.

Jacobian matrix $\mathbf{J}_v$ is then via the backpropagation chain-rule (given proper application of the time indexes) as follows

$$\mathbf{J}_{\mathbf{v}_\iota}[k,:] = \frac{\partial \tilde{y}(k)}{\partial \mathbf{v}_\iota} = \frac{\partial \tilde{y}}{\partial u} \frac{\partial u}{\partial q_\iota} \frac{\partial q_\iota}{\partial \mathbf{v}_\iota} \tag{6}$$

Table 2. Summary of HONU as ι^{th} unit within the controller layer (Figure 2) of up to the 3rd order of nonlinearity

	Polynomial Order of HONU		
	LNU, $\gamma=1$	**QNU, $\gamma=2$**	**CNU, $\gamma=3$**
$\Sigma\Pi$ notation	$q=\sum\limits_{i=0}^{m} v_i \cdot \xi_i$	$q=\sum\limits_{i=0}^{m}\sum\limits_{j=i}^{m} v_{i,j} \cdot \xi_i \cdot \xi_j$	$\sum\limits_{i=0}^{m}\sum\limits_{j=i}^{m}\sum\limits_{\kappa=j}^{m} v_{i,j,\kappa} \xi_i \xi_j \xi_\kappa$
$col^\gamma()$kernel form	$q=\mathbf{v}\cdot col^{\gamma=1}(\boldsymbol{\xi})=\mathbf{w}\cdot\boldsymbol{\xi}$	$q=\mathbf{v}\cdot col^{\gamma=2}(\boldsymbol{\xi})$	$q=\mathbf{v}\cdot col^{\gamma=3}(\boldsymbol{\xi})$
details	$\boldsymbol{\xi}=\begin{bmatrix}\xi_0\\\xi_1\\\vdots\\\xi_m\end{bmatrix}$ $\mathbf{w}=[w_0\,w_1\ldots w_n]$	$col^2(\boldsymbol{\xi})=\left[\{\xi_i\xi_j\}\right]$ $\mathbf{v}=\left[\{v_{i,j}\}\right]$ $\forall i,j\,;\,0\leq i\leq j\leq m$	$col^3(\boldsymbol{\xi})=\left[\{\xi_i\xi_j\xi_\kappa\}\right]$ $\mathbf{v}=\left[\{v_{i,j,\kappa}\}\right]$ $\forall i,j,\kappa\,;\,0\leq i\leq j\leq\kappa\leq m$

Figure 2. Discrete time Model Reference Adaptive Control (MRAC) loop with multiple HONU controller: One HONU serves as a plant model and the controller can be a layer of various HONUs of various configurations and even possibly of various learning algorithms

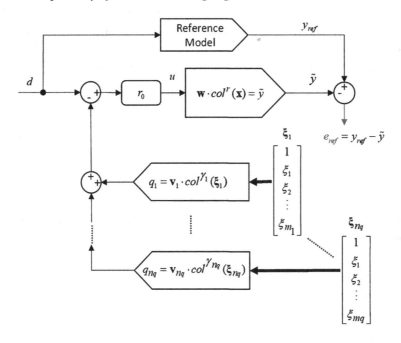

and the last term of Jacobian can be fully recurrently calculated or in many practical cases simplified as follows

$$\frac{\partial q_\iota(k)}{\partial \mathbf{v}_\iota} = col^{\gamma_\iota}(\xi_\iota)^T + \mathbf{v}_\iota \cdot \begin{bmatrix} \mathbf{0} \\ \partial \tilde{y}(k-1)/\partial \mathbf{v}_\iota \\ \partial \tilde{y}(k-2)/\partial \mathbf{v}_\iota \\ \vdots \\ \partial \tilde{y}(k-m_\iota)/\partial \mathbf{v}_\iota \end{bmatrix} \cong col^{\gamma_\iota}(\xi_\iota)^T ,$$ (7)

where m_ι denotes the total number of neural controller weights (i.e. for corresponding length of $\mathbf{v}_\iota$) and

$$\frac{\partial u(k-1)}{\partial \mathbf{v}_\iota} = -r_o \cdot col^{\gamma_\iota}(\xi_\iota)^T .$$ (8)

An essential in controller tuning is to define a reasonable data set of measured values for plant identification and further controller tuning. In this paper, we define several key components of the principle input vector **x** further, **v** in the sense of a HONU feedback controller. The term d denotes the desired value that we can feed into the plant to measure the corresponding plant output data y; further, y_{ref} corresponds to the reference model of the process, i.e. simulating the desired behavior of the process, taking into account the dynamic capabilities and limits of the system properties. The ultimate objective is to modify the control loop so it adopts the reference model behavior once the controller is properly trained. We thus desire that the trained controller provides such output value q so the next sample of the controlled plant y matches with the prescribed reference model output y_{ref}.

Learning Algorithms for HONUs

Both for plant identification and for controller learning, sample-by-sample (incremental) or batch learning can be applied for both static and dynamic HONUs. And of course, the learning algorithms for both plant identification and controller tuning can be:

- manually derived and implemented in your custom code (the must for students and also highly recommended to newcomers for better understanding of the fundamentals of machine learning with dynamical systems and MRAC control), or/and
- existing optimization libraries can be utilized.

These aspects are discussed further with the most fundamental details on the learning algorithms.

Incremental Learning for Plant Approximation and Controller Learning

The most fundamental sample-by-sample adaptation algorithm of both static and dynamic HONUs is the typical backpropagation-family representative algorithm, i.e., the gradient descent rule, e.g. with a normalized learning rate known for linear adaptive filters (Mandic, 2001; Mandic, 2004), that adapts HONU for plant approximation as follows

$$\mathbf{w}(k+1)=\mathbf{w}(k)+\Delta\mathbf{w}=\mathbf{w}(k)+\frac{\mu}{\epsilon+||\mathbf{x}||^2}\cdot\frac{\partial Q(k)}{\partial\mathbf{w}} \tag{9}$$

where ϵ is a small regularization term in denominator that improves adaptation stability (Bukovsky & Homma, 2017), μ is learning rate, and $Q(k)$ is the error criterion usually defined as $Q(k)=e^2(k)$, where

$$e(k)=y(k)-\tilde{y}(k), \tag{10}$$

so $e(k)$ is the neural output error of plant approximation at the last measured time k. Similarly, the adaptation of the ιth controller unit via gradient descent can be derived as follows

$$\mathbf{v}_\iota(k+1)=\mathbf{v}_\iota(k)+\Delta\mathbf{v}_\iota=\mathbf{v}_\iota(k)+\frac{\eta_\iota}{\epsilon+||\xi_\iota||^2}\cdot\frac{\partial Q_{ref}(k)}{\partial\mathbf{v}_\iota}, \tag{11}$$

where η_ι is the learning rate of the ιth controller unit, $Q_{ref}(k)$ is the reference error criterion usually defined as $Q_{ref}=e_{ref}^2(k)$, where the reference error, i.e. the error between the desired output and control loop output is as follows

$$e_{ref}(k)=y(k)-y_{ref}(k), \tag{12}$$

and which is for controller learning implemented via substitution $y\leftarrow\tilde{y}$. The above gradient rules for plant identification and control are easily derived and implemented in a code, and they can be also found well working for many practical control tasks with SISO (stable) plants or for improvement of existing PID control loops (Bukovsky et al., 2010; Bukovsky et al., 2015; Bukovsky, Redlapalli, & Gupta, 2003; Benes & Bukovsky, 2014). Also, interested readers who are new to adaptive control should try to derive and implement this fundamental algorithm to better understand the MRAC control prior to moving forward to more advanced algorithms.

Another efficient learning algorithm advantageous for real-time plant identification is the RLS learning algorithm. Though its application in the field of adaptive filters is quite readily published, its extension to HONUs for plant identification is still a not widely investigated area. The advantages of the RLS algorithm is also its applicability for use in the whole MRAC-HONU closed control loop plant and controller tuning due to its fundamental composition comprising from the covariance matrix of the principle partial derivative for adaptation.

As an initial step, we may recall the classical form of the inverse covariance matrix $\mathbf{R}^{-1}(k)$ to the RLS algorithm as

$$\mathbf{R}^{-1}(k) = \frac{1}{\mu} \cdot \left(\mathbf{R}^{-1}(k-1) - \frac{R_1}{R_2} \right), \tag{13}$$

where $\mathbf{R}$ is the covariance matrix; denoting $\mathbf{I}$ as the identity matrix of corresponding dimension, $\mathbf{R}$ can be initialized at the very first step by

$$\mathbf{R}(0) = \frac{1}{\delta} \cdot \mathbf{I}. \tag{14}$$

for a small initialization constant δ. Then the terms R_1, R_2 follow as

$$R_1 = \mathbf{R}^{-1}(k-1) \cdot \frac{\partial \tilde{y}(k)}{\partial \mathbf{w}} \cdot \frac{\partial \tilde{y}(k)}{\partial \mathbf{w}}^T \cdot \mathbf{R}^{-1}(k-1), \text{ and} \tag{15}$$

$$R_2 = \mu + \frac{\partial \tilde{y}(k)}{\partial \mathbf{w}}^T \cdot \mathbf{R}^{-1}(k-1) \cdot \frac{\partial \tilde{y}(k)}{\partial \mathbf{w}}, \tag{16}$$

simplifying the partial derivative $\dfrac{\partial \tilde{y}(k)}{\partial \mathrm{colW}} = \mathbf{colx}(k)$, we obtain

$$\mathbf{R}^{-1}(k) = \frac{1}{\mu} \cdot (\mathbf{R}^{-1}(k-1) - \frac{\mathbf{R}^{-1}(k-1) \cdot \mathbf{colx}(k) \cdot \mathbf{colx}(k)^T \cdot \mathbf{R}^{-1}(k-1)}{\mu + \mathbf{colx}(k)^T \cdot \mathbf{R}^{-1}(k-1) \cdot \mathbf{colx}(k)}), \tag{17}$$

where the weight update rule yields

$$\Delta \mathbf{w} = e(k) \cdot \mathrm{colx}^T \cdot \mathbf{R}^{-1}(k). \tag{18}$$

The sample-by-sample RLS learning (13)-(18) is also applicable to HONU controller learning based on $\partial \tilde{y}(k)/\partial \mathbf{v}$ instead of $\partial \tilde{y}(k)/\partial \mathbf{w}$. As shown in Figure 2, the RLS usually outperforms the classical gradient descent and L-M algorithm as it rapidly minimizes the sum of square errors (SSE).

In addition to the the straightforward concept and comprehensibility, another advantage of sample-by-sample learning algorithms for MRAC control is that they can provide us with reasonable (and definitely much better than PID) control for many practical systems, as low polynomial-order HONUs (or even linear ones) can temporarily capture the nonlinearity due to real-time adaptation.

Batch Learning for Plant Approximation and Controller Learning

The most straightforward algorithm for plant approximation with static HONUs, i.e. for one sample-ahead predictive model, the Ridge Regression (RR), Levenberg-Marquardt (L-M), and Conjugate Gradients, (CG) can be recommended as efficient and comprehensible algorithms that are easily derived and imple-

Figure 3. MRAC-HONU control loop on Two-Tank liquid system (54)-(55), where one dynamic HONU is as a plant model and second as a nonlinear state feedback controller. The RLS algorithm is superior to classical GD and L-M learning.

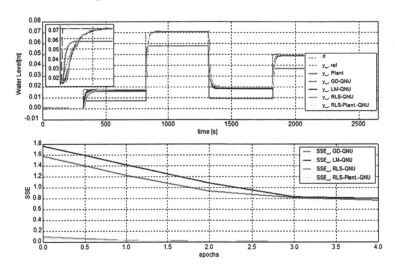

mented in a custom code that can practically work very well, and these are excellent for educational purposes too. The more recently popular optimization algorithms, such as ADAM or BFGS, can be efficiently implemented with existing optimization libraries as well.

For batch learning, we first define the batch matrix of all vectors as follows

$$\mathbf{X}=\begin{bmatrix}\mathbf{x}(k_1)^T \\ \mathbf{x}(k_2)^T \\ \vdots \\ \mathbf{x}(k_N)^T\end{bmatrix},$$

(19)

and by applying the column-wise polynomial kernel for HONUs (Bukovsky & Homma, 2017), we obtain matrix

$$col^r(\mathbf{X})=\begin{bmatrix}col^r\left(\mathbf{x}(k_1)\right)^T \\ col^r\left(\mathbf{x}(k_2)\right)^T \\ \vdots \\ col^r\left(\mathbf{x}(k_N)\right)^T\end{bmatrix},$$

(20)

so all outputs from static HONU are calculated in batch as follows

$$\tilde{\mathbf{y}}=col^r(\mathbf{X})\cdot\mathbf{w}^T,$$

(21)

where $\mathbf{w}^T$ denotes the column-wise vector of HONU weights.

Then, to calculate the weights for plant approximation via static HONU directly, the fundamental variant of the RR algorithm can be derived from error criteria

$$\mathbf{Q} = \|\mathbf{e}\|^2 + \lambda \cdot \|\mathbf{w}\|^2 = (\mathbf{y} - col^r(\mathbf{X}) \cdot \mathbf{w}^T)^T \cdot (\mathbf{y} - col^r(\mathbf{X}) \cdot \mathbf{w}^T) + \lambda \mathbf{w} \cdot \mathbf{w}^T \overset{!}{=} \min, \tag{22}$$

and thus we can obtain weights of the static HONU for plant approximation as follows

$$\frac{\partial \mathbf{Q}}{\partial \mathbf{w}} = \mathbf{0} \Rightarrow \mathbf{w}^T = \left(col^r(\mathbf{X})^T \cdot col^r(\mathbf{X}) + \lambda \cdot \mathbf{I}\right)^{-1} \cdot col^r(\mathbf{X})^T \cdot \mathbf{y}, \tag{23}$$

where λ is the regularization term to suppress the magnitude of neural weights, $\mathbf{I}$ is the identity matrix, and $\mathbf{y}$ is the vector of all real outputs (targets).

In general, $col^r(\mathbf{X}) = \mathbf{J_w}$, i.e., it is the Jacobian of HONUs and as $\mathbf{X}$ is a constant matrix for static HONUs, so the Jacobian is constant for static HONUs too.

The Levenberg-Marquardt algorithm can be recommended also as a first to meet batch algorithm with HONUs. It is applicable for both static and dynamic HONUs where the weight updates $\Delta \mathbf{w}$ can be calculated as follows

$$\Delta \mathbf{w} = \left(\mathbf{J_w}^T \cdot \mathbf{J_w} + \frac{1}{\mu} \cdot \mathbf{I_w}\right)^{-1} \cdot \mathbf{J_w}^T \cdot \mathbf{e}, \tag{24}$$

where J is the Jacobian matrix, I is the identity matrix, and the upper index $^{-1}$ stands for matrix inversion.

Then in the sense of the L-M algorithm, the controller weight can be trained via the following relation

$$\Delta \mathbf{v} = \left(\mathbf{J_v}^T \cdot \mathbf{J_v} + \frac{1}{\mu_{\mathbf{v}}} \cdot \mathbf{I_v}\right)^{-1} \cdot \mathbf{J_v}^T \cdot \mathbf{e}_{ref}, \tag{25}$$

where the subscript $\mathbf{v}$ indicates it is the controller learning rule.

For plant approximation via recurrent HONUs, i.e. to approximate more accurately the long output response knowing initial conditions and only plant input, we should implement the L-M algorithm where $\mathbf{X}$ has to be recalculated for every time sample. To accelerate L-M batch learning, especially for HONUs, the CG algorithm can be suggested (see Figure 4). The principle of CG is to solve a set of equations as follows

$$b - \mathbf{A} \cdot \mathbf{w} = 0, \tag{26}$$

where b is a column vector of constants, $\mathbf{A}$ is a positively semi-definite matrix, and $\mathbf{w}$ is a column vector of unknowns (neural weights). Due to the in-parameter linearity of HONUs, the Jacobian (3) is not directly a function of weights. Thus, the training with respect to a HONU can be restated from (26) as

$$\mathbf{y} - col^r(\mathbf{X}) \cdot \mathbf{w} = \mathbf{0}., \tag{27}$$

which is in fact using the Jacobian of a HONU for all training data (of total length N). Multiplying (6) from the left with the term $col\mathbf{X}^T$ yields that

$$\mathbf{J} \cdot \mathbf{y} - \mathbf{J}^T \cdot \mathbf{J} \cdot \mathbf{w} = \mathbf{0}. \tag{28}$$

This results in a positive definite matrix; therefore, the CG learning form may be directly applied to both static or recurrent HONUs. On the initiation of training, i.e. for the very first epoch, we initiate CG with

$$\mathbf{r_e}(\epsilon=0) = \mathbf{c} - \mathbf{A} \cdot \mathbf{w}(\epsilon=0), \tag{29}$$

where ϵ denotes the index of training epochs from now and further

$$\mathbf{p}(\epsilon=0) = \mathbf{r_e}(\epsilon=0). \tag{30}$$

Then for proceeding training epochs (i.e. $\epsilon > 0$), we calculate the following

$$\alpha(\epsilon) = \frac{\mathbf{r_e}^T(\epsilon) \cdot \mathbf{r_e}(\epsilon)}{\mathbf{p}^T(\epsilon) \cdot \mathbf{A} \cdot \mathbf{p}(\epsilon)}. \tag{31}$$

With the parameter calculation from (31), the following weight update rule yields

$$\mathbf{w}(\epsilon+1) = \mathbf{w}(\epsilon) + \alpha(\epsilon) \cdot \mathbf{p}(\epsilon), \tag{32}$$

where $\Delta \mathbf{w} = \alpha(\epsilon)\mathbf{p}(\epsilon)$ and other CG parameters for the next training epoch are then calculated or updated as follows

$$\mathbf{r_e}(\epsilon+1) = \mathbf{r_e}(\epsilon) - \alpha(\epsilon) \cdot \mathbf{A} \cdot \mathbf{p}(\epsilon), \tag{33}$$

and in a similar sense as the Fletcher-Reeves nonlinear CG method

$$\beta(\epsilon) = \frac{\mathbf{r_e}^T(\epsilon+1) \cdot \mathbf{r_e}(\epsilon+1)}{\mathbf{r_e}^T(\epsilon) \cdot \mathbf{r_e}(\epsilon)}, \tag{34}$$

therefore resulting in

$$\mathbf{p}(\epsilon+1) = \mathbf{r_e}(\epsilon+1) + \beta(\epsilon) \cdot \mathbf{p}(\epsilon). \tag{35}$$

This section derived the extension of the classical Conjugate Gradient learning algorithm for application to HONUs. Due to its structure, is not so suitable for controller weights v training, as the symmetric positive definite matrix is not so achievable, and the CG for a controller then becomes a much more complicated task. As illustrated in Figure 1, the use of pre-training via the L-M algorithm can be enhanced via CG training, following a switch after several epochs. Thus, the proposed CG training is suggested to use in lieu with the L-M training algorithm for rapid acceleration during plant identification, where only L-M can be used as a comprehensible training algorithm for HONUs, including a feedback controller algorithm via the MRAC control scheme Figure 2. As an extension of the work (Bukovsky et al., 2017) the proceeding section extends an efficient incremental training algorithm (RLS) to the presented HONU architectures.

For batch controller tuning, the L-M algorithm can be easily derived and implemented in code with HONUs. More novel algorithms such as ADAM and BFGS and its variants can be efficiently implemented and investigated using existing optimization libraries, where the objective function can be defined as described above. For static HONUs as a plant model, we can efficiently design a controller with these optimizers when we make a substitution for input vector

$$\mathbf{x}=\begin{bmatrix}1\\\mathbf{y_x}\\\mathbf{u_x}\end{bmatrix} \text{ where } y(k)\leftarrow y_{ref}(k) \text{ within } \mathbf{y_x}. \tag{36}$$

Figure 4. Acceleration of training of neural weights of static CNU initiated with L-M learning and changed for CG learning for training epochs $\geq \iota\gamma$

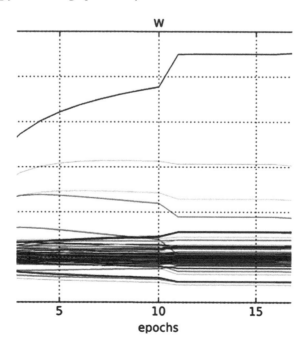

Normalized Mini-Batch Gradient Descent

Batch learning algorithms use the entire training set. That's not usually possible in real-time learning and learning with a small sample time. The training set can grow and then upgrades can be computationally demanding. Incremental learning algorithms use a single training set. Their computational complexity is low, but their learning ability is also lower. Mini-batch learning algorithms use a subset of the training set and seem to be a good compromise for real-time learning. The normalized mini-batch gradient descent is described below.

$$
\mathbf{w}(k+1)=\mathbf{w}(k)-\mu_{norm}(k)\cdot\frac{\partial Q(k)}{\partial \mathbf{w}(k)} \tag{37}
$$

$$
Q_{(k)}=\frac{1}{2m}\sum_{i=1}^{m}e_i^2(k)=\frac{1}{2}\mathbf{e}^T(k)\cdot\mathbf{e}(k)=\frac{1}{2}\sum_{i=1}^{m}\left(y_{refi}(k)-\mathbf{w}(k)\cdot col^r(\mathbf{x}_i)^{\mathrm{T}}(k)\right)^2, \tag{38}
$$

where Q is the cost function, μ_{norm} is the normalized learning rate, and e_i is the reference error between the real output of the system y_{refi} and the output from the model y_i. The mini-batch is chosen as m random training sets from the window in the range $k-h \leq i \leq k-1$. Putting (38) into (37), we get the following formula for weights adaption.

$$
\mathbf{w}(k+1)=\mathbf{w}(k)+\mu_{norm}(k)\cdot\sum_{i=1}^{m}e_i(k)\cdot\frac{\partial y_i(k)}{\partial \mathbf{w}(k)}
$$
$$
=\mathbf{w}(k)+\mu_{norm}(k)\cdot\sum_{i=1}^{m}e_i(k)\cdot col^r(\mathbf{x}_i)^{\mathrm{T}}(k)=\mathbf{w}(k)-\mu_{norm}(k)\cdot\mathbf{e}^T(k)\cdot J(k) \tag{39}
$$

The optimal learning rate μ_{norm} minimizes the cost function after updating weights and can be calculated by the least squares method as follows:

$$
\frac{\partial \tilde{Q}(k)}{\partial \mu}=0, \tag{40}
$$

where $\tilde{Q}$ is the cost function after updating weights to $\mathbf{w}(k+1)$ and can be written as follows

$$\tilde{Q}(k)=\frac{1}{2}\sum_{i=1}^{m}\left(y_{refi}(k)-\mathbf{w}(k+1){\cdot}col^r(\mathbf{x}_i)^{\mathrm{T}}(k)\right)^2$$

$$=\frac{1}{2}\sum_{i=1}^{m}\left(y_{refi}(k)-\left(\mathbf{w}(k)+\mu_{norm}(k){\cdot}\sum_{i=1}^{m}e_i(k){\cdot}col^r(\mathbf{x}_i)^{\mathrm{T}}(k)\right){\cdot}col^r(\mathbf{x}_i)^{\mathrm{T}}(k)\right)^2=$$

$$=\frac{1}{2}{\cdot}\sum_{i=1}^{m}\left(e_i(k)-\mu_{norm}(k){\cdot}\frac{1}{m}\sum_{i=1}^{m}e_i(k){\cdot}col^r(\mathbf{x}_i)^{\mathrm{T}}(k){\cdot}col^r(\mathbf{x}_i)^{\mathrm{T}}(k)\right)^2=$$

$$=\frac{1}{2}{\cdot}\left[\mathbf{e}(k)-\mu_{norm}(k){\cdot}\mathbf{J}{\cdot}\mathbf{J}^T{\cdot}\mathbf{e}(k)\right]^T{\cdot}\left[\mathbf{e}(k)-\mu_{norm}(k){\cdot}\mathbf{J}{\cdot}\mathbf{J}^T{\cdot}\mathbf{e}(k)\right]$$

(41)

Putting (41) into (40), we get the following formula for the optimal learning rate:

$$\mu_{norm}\left(k\right)=\frac{\mathbf{e}^T(k){\cdot}\mathbf{J}(k){\cdot}\mathbf{J}^T(k){\cdot}\mathbf{e}(k)}{\mathbf{e}^T(k){\cdot}\left(\mathbf{J}(k){\cdot}\mathbf{J}^T(k)\right)^2{\cdot}\mathbf{e}(k)}$$

(42)

(42) can be extended of $0{\leq}\mu{\leq}1$, where $\mu{=}1$ is the maximal learning rate. The final formula for μ_{norm} is as follows:

$$\mu_{norm}\left(k\right)=\mu{\cdot}\frac{\mathbf{e}^T(k){\cdot}\mathbf{J}(k){\cdot}\mathbf{J}^T(k){\cdot}\mathbf{e}(k)}{\mathbf{e}^T(k){\cdot}\left(\mathbf{J}(k){\cdot}\mathbf{J}^T(k)\right)^2{\cdot}\mathbf{e}(k)}$$

(43)

The ability to learn depends not only on the learning rate, but also on the batch size *m* and the length of the training sets window *h*. Testing data and trained LNU for *m=256* are shown in Figure 5.

The MSE function for different sizes of the batch *m* is shown in Figure 6. It is shown that the higher learning ability is better for a bigger batch and with smaller fluctuations. Fluctuations during learning were given by noise in reference signal *y*.

Stability Analysis

Decomposition Approach to HONU-MRAC Control Loop Stability

This section analyzes the dynamical stability of the proposed MRAC–based closed loop via the decomposition method presented in the 2016 work of Benes and Bukovsky (Benes & Bukovsky, 2016) and, further, in their 2018 study (Benes & Bukovsky, 2018), where its extension as a MRAC control loop was presented. For better comprehensibility, let us assume a single static LNU as a plant, with extension of the control law (5). With this statement, the HONU, as per Table 1, can be redefined as

$$\tilde{y}(k)=\sum_{i=1}^{n_y+n_u}\hat{x}_i(k-i){\cdot}\hat{\alpha}_i+\sum_{i=1}^{n_u+n_u}\hat{u}_i(k-i)\hat{\beta}_i+C_i(\mathbf{w}_0),$$

(44)

Figure 5. Mini-batch identification LNU from data with a noise

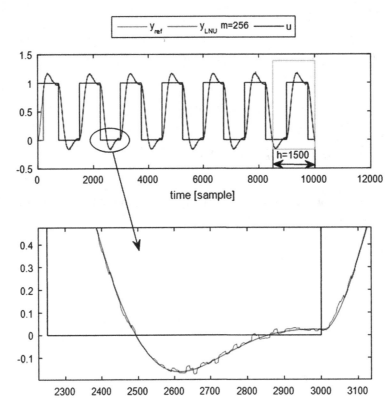

Figure 6. MSE for different length of the batch m

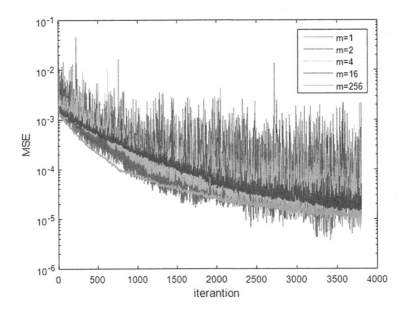

where we introduce a new vector of state variable terms $\hat{\mathbf{x}}(k\text{-}1)$, corresponding to previous step-delayed output terms of the HONU-MRAC based control loop. Similarly, $\hat{\mathbf{u}}(k\text{-}1)$ corresponds to the vector of step-delayed input terms; both can be explicitly defined as

$$\hat{\mathbf{x}}(k)=[\tilde{y}(k-(n_y+n_u)+1) \,\, .. \,\, \tilde{y}(k\text{-}1) \,\, \tilde{y}(k)]^T, \text{ and} \tag{45}$$

$$\hat{\mathbf{u}}(k)=[d(k-(n_u+n_u)+1) \,\, .. \,\, d(k\text{-}1) \,\, d(k)]^T. \tag{46}$$

Then, for simplification, the operator $C_i(.)$ is now introduced to denote the sum of constant neural bias weight terms. Then the local characteristic coefficients $\hat{\alpha}_i$ and may be computed via the following sub-polynomial expressions:

$$\hat{\alpha}_i=-r_o\cdot\sum_{j=1}^{n_u}w_{n_y+j}\cdot\sum_{l=1}^{n_q}C_i(q_l(k-j)) \,\,\, for \,\,\, i=1,2,..,n_y+n_u, \tag{47}$$

where, in the sense of a dynamic LNU plant, the coefficient $\hat{\alpha}_i$ for $i=1,2,3,\ldots,n_y$ is given as

$$\hat{\alpha}_i=w_i-r_o\cdot\sum_{j=1}^{n_u}w_{n_y+j}\cdot\sum_{l=1}^{n_q}C_i(q_l(k-j)) \,\,\, for \,\,\, i=1,2,3,..,n_y. \tag{48}$$

The corresponding coefficient terms $\hat{\beta}_i$ for the input vector $\hat{\mathbf{u}}(k\text{-}1)$ may then be computed as

$$\hat{\beta}_i=\begin{cases} r_o\cdot\left[w_{n_y+i}-\sum_{j=1}^{n_u}w_{n_y+j}\cdot\sum_{l=1}^{n_q}C_i(q_l(k-j)) \,\,\, for \,\,\, i=1,..,n_u\right] \\ -r_o\cdot\sum_{j=1}^{n_u}w_{n_y+j}\cdot\sum_{l=1}^{n_q}C_i(q_l(k-j)) \,\,\, for \,\,\, i=n_u+1,..,n_u+n_u, \end{cases} \tag{49}$$

where the resulting expressions (47)-(48) as well as (49) may be applied to the more classical HONU-MRAC configuration as in the work of Benes and Bukovsky (Benes & Bukovsky, 2018) and Bukovsky et al. (Bukovsky et al., 2015). Then, we may express the resulting canonical state space form as

$$\hat{\mathbf{x}}(k)=\hat{\mathbf{M}}(k\text{-}1)\cdot\hat{\mathbf{x}}(k-1)+\hat{\mathbf{N}}_\mathbf{a}\cdot\hat{\mathbf{u}}_\mathbf{a}(k-1) \,\, ; \,\, \tilde{y}(k)=\hat{\mathbf{C}}\cdot\hat{\mathbf{x}}(k), \tag{50}$$

$$\hat{\mathbf{M}}=\begin{bmatrix} 0 & 1 & 0 & \cdots & 0 \\ 0 & 0 & 1 & \cdots & \vdots \\ 0 & 0 & \ddots & \ddots & 0 \\ 0 & 0 & \ddots & 0 & 1 \\ \hat{\alpha}_{n_y+n_u} & \hat{\alpha}_{n_y+n_u-1} & \cdots & \hat{\alpha}_2 & \hat{\alpha}_1 \end{bmatrix}, \hat{\mathbf{N}}=\begin{bmatrix} 0 & 0 & \cdots & 0 \\ \vdots & \vdots & \ddots & \vdots \\ 0 & 0 & \cdots & 0 \\ \hat{\beta}_{n_u+n_u} & \hat{\beta}_{(n_u+n_u)-1} & \cdots & \hat{\beta}_1 \end{bmatrix}. \tag{51}$$

For further reference, we term $\hat{M}$ as the local matrix of dynamics (LMD). Further, the augmented input matrix and input vector may be defined as

$$\hat{\mathbf{N}}_{\mathbf{a}}=\begin{bmatrix} 0 \\ \vdots \\ \hat{\mathbf{N}} & 0 \\ 1 \end{bmatrix}, \hat{\mathbf{u}}_{\mathbf{a}}(k\text{-}1)=\begin{bmatrix} \hat{\mathbf{u}}(k\text{-}1) & C_i(\mathbf{w}_0) \end{bmatrix}^T. \tag{52}$$

According to the definitions of BIBO and ISS stability (Wang & Liu, 2014), the forms (50)-(51) for the HONU-MRAC closed control loop may be justified for ISS (further BIBS) stability (Benes & Bukovsky, 2018), if the following holds from an initial state sample k_0 until..

$$S=\left\| \hat{\mathbf{x}}(k) \right\| - \left[\left\| \prod_{\kappa=k_0}^{k-1} \hat{\mathbf{M}}(\kappa) \right\| \cdot \left\| \hat{\mathbf{x}}(k_0) \right\| + \sum_{\kappa=k_0}^{k-1} \left\| \prod_{i=\kappa}^{k-1} \hat{\mathbf{M}}(i) \cdot \hat{\mathbf{N}}_{\mathbf{a}}(\kappa) \right\| \cdot \left\| \hat{\mathbf{u}}_{\mathbf{a}}(\kappa) \right\| \right] \leq 0, \tag{53}$$

where a sufficient condition for maintaining BIBS yields if $\Delta S(k)=S(k)-S(k\text{-}1)\leq 0$, given the relation (53) in sample k is not violated.

Two-Tank Liquid Level System

To investigate the application of the decomposed stability approach described in section 5.1, let us consider a weakly non-linear two-tank liquid level system described via the following balancing equations:

$$A\cdot\frac{dh_1}{dt}=Q_t-C_{db}\cdot s_1\cdot\sqrt{2\cdot g\cdot(h_1-h_2)}, \text{ and} \tag{54}$$

$$A\cdot\frac{dh_2}{dt}=C_{db}\cdot s_1\cdot\sqrt{2\cdot g\cdot(h_1-h_2)}-C_{dc}\cdot s_2\cdot\sqrt{2\cdot g\cdot h_2}, \tag{55}$$

where $Q_t[m^3 s^{-1}]$ denotes the inlet flow rate of the system. The tank cross-sectional area is $A=0.002[m^2]$, the orifice cross-sectional areas are $s_1=s_2=0.000785[m^2]$, the orifice discharge coefficients are $C_{db}=C_{dc}=0.60$, the density of water is $\rho=1000[kg/m^3]$, and the gravitational constant of acceleration $g=9.81[ms^{-2}]$. To add discussion on the application of dynamic HONUs, the approach (48)-(49) is investigated with two

HONUs, i.e. one dynamic HONU as a plant and the second as a feedback controller as in the work of Benes and Bukovsky (Benes & Bukovsky, 2018).

For the offline tuned HONU-MRAC control loop, a single dynamic HONU is identified via RLS training with five previous model output values. Four previous process inputs are further incorporated into the input vector. The HONU feedback controller consists of a single HONU feedback controller with the same input vector length, i.e. n_y=5 and n_u=4 and the feedback gain is r_0=0.01. The offline tuned HONU-MRAC control loop tuned after 200 epochs is in its last training epoch, introduced with a large increase in its learning rate to μ=0.9998 at time t>488[s]. It is evident that from t>490[s] the condition (53) switches from a monotonic decrease to $\Delta S(\Delta S>0)$ and hence signifies the onset of instability where the BIBS condition is violated for t>492[s]. This is reflected in Figure 8 b) and c) via the violation of BIBO stability corresponding to spectral radii ρ(.)>1; however, due to the relation (53) accounting for the previous samples of HONU-MRAC state transitions, $\Delta S(\Delta S>0)$ yields a stronger condition that clearly pronounces that onset of instability. Due to the relation (53) accounting for the previous samples of HONU-MRAC state transitions, $\Delta S(\Delta S>0)$ yields a stronger condition than a standalone assessment of the spectral radii that clearly pronounces the onset of instability. It further justifies the trajectory in state space for the given control input is becoming unstable, as opposed to the local dynamics in the vicinity of the discrete state point. In such a case, though the LMD eigenvalues may locally recover, i.e.

Figure 7. Comparison of offline tuned and real-time application of HONU-MRAC as constant parameter control loop on real two-tank liquid level system: One as a plant model and second as a nonlinear state feedback controller

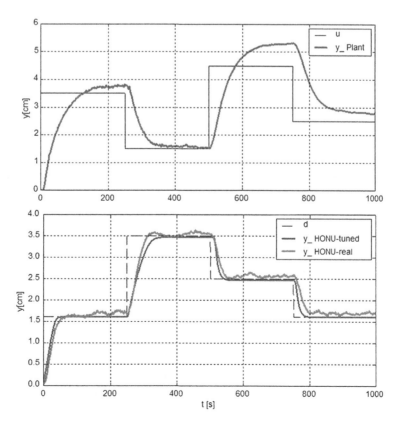

Figure 8. a) Adaptive (RLS) LNU-QNU control loop becomes unstable soon after learning rate μ t>488)=0.9998. b) Spectral radii through time of LNU-QNU closed loop LMD. c) BIBS condition (53) through time. d) Showing the positive difference of (53), i.e. ΔS(ΔS>0) reveals instability onset soon after (53) learning rate becomes changed for t>488.

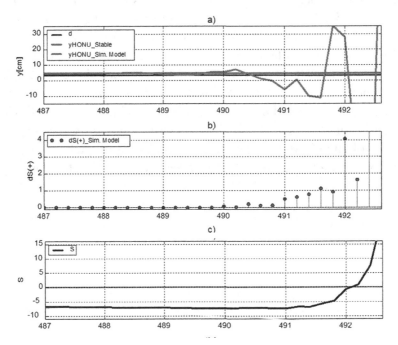

$\rho(.)\leq 1$ in the sense of an adaptive control loop, in a global sense, the HONU-MRAC response may be dynamically unstable with respect to transition from neighboring states.

Experimental Analysis

Torsional Pendulum

To provide further analysis on the incremental control algorithms, namely normalized gradient descent (NGD) and RLS, and further, to emphasis the limits for applicability of the discussed approach, let us consider a torsional pendulum (56) adopted from the work of Liu and Wei (Liu & Wei, 2014). As a modification, we introduce an increased friction coefficient for stabilization of the system to obtain training data. It is thus a key point to note that our MRAC-HONU control loop scheme requires sufficiently stable training data in order to properly identify the corresponding process dynamics and extend controller tuning. With this note, the modified (stabilized) discrete time inverted pendulum model is then as follows:

$$\chi_1(k+1)=\chi_1(k)+0.1\cdot\chi_2(k)$$
$$\chi_2(k+1)=-0.49\cdot sin(\chi_1(k))+(1-0.1f_d)\cdot\chi_2(k)+0.1\cdot u(k),$$

(56)

where the measured output is simulated as $y=\chi_1$, $f_d=0.6$ is the friction and u is control input. Given the successful setup in Figure 8, in a similar manner, a static CNU featuring $n_y=n_u=2$ is trained via L_M over 20 epochs, followed by an additional 20 epochs of CG training to accelerate the plant identification.

Following a successful plant identification, a single static QNU (i.e. HONU, $r=2$) is applied as a feedback controller for controller input vector lengths $n_{qy}=n_{qe}=3$, respectfully. Two setups are tuned on the previously identified HONU model (as a dynamically trained HONU) on the first epoch via NGD and RLS tuning, respectively, for comparison. Figure 10 illustrates the performance of both approaches on the same HONU-MRAC control loop configuration. Although the NGD algorithm towards the final samples of training data starts to fit closer to the desired control loop response, as can be seen in the HONU feedback controller weights, the tuned values are more erratic during training and not so marginally retuned as compared to the neural weights shown for RLS training of the same control loop. Figure 9 and Figure 10 illustrate the performance of both approaches on the same HONU-MRAC control loop configuration. Figure 9 shows that, although the NGD algorithm towards the final samples of training data starts to fit closer to the desired control loop response, as can be seen in the HONU feedback controller weights, the tuned values are more erratic during training and not so marginally retuned as compared to the neural weights shown in Figure 10 for RLS training of the same control loop. In the final samples of training data, the RLS algorithm, even after one epoch, is able to adequately minimize the steady state error to fit the desired control loop response, which is reflected in the substantially retuned neural weights across the whole set of training data.

However, for higher amplitudes at the applied plant input, a strong sinusoidal non-linearity can be seen, as depicted in Figure 11. On application of higher amplitudes of the plant input, in this case, the desired value of the control loop, it is evident that for increased amplitudes, the HONU-MRAC closed

Figure 9. A single static HONU trained via L-M with CG algorithm after 20 epochs for plant identification of (56). Used for investigation of performance in control of plant between NGD and RLS feedback controller training (see Figure 10)

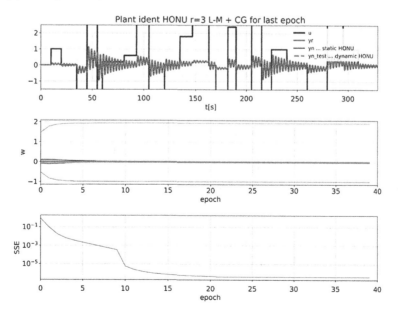

Figure 10. Superior controller tuning performance via RLS compared to NGD after one training epoch for control of plant. A single static HONU controller is tuned on identified HONU model via RLS and applied as an extension to a P-control loop.

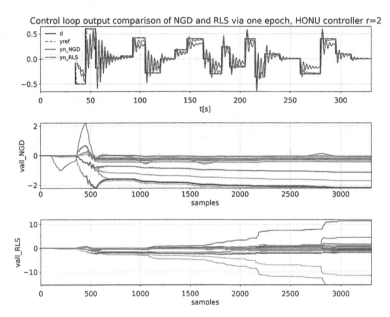

control loop struggles to maintain accurate control of such strong nonlinearity within the system, hence highlighting the boundaries for the application of such a control approach to nonlinear systems.

CONCLUSION

This paper presented a HONU-MRAC control strategy, with a focus on static HONUs to avoid recurrent computations and further improve convergence of the controller training and multiple HONU feedback controller configurations. The CG algorithm was presented as an efficient technique for accelerating plant identification in lieu with the L-M training algorithm. This adaptive control technique was shown to be easily applied to weakly nonlinear system,s which can be well approximated with HONUs of appropriate polynomial order (here < 3). In addition to the work of Bukovsky et al. (Bukovsky et al., 2017), a deeper investigation behind incremental training algorithms, namely GD, NGD, and RLS, were presented, and the capabilities of RLS training with application to the presented HONU-MRAC control loop approach were highlighted. Furthermore, a study into stability analysis of the proposed control loop configurations was presented, whereas in any adaptive control loop, design plays a paramount role in ensuring the adaptive control loop maintains dynamical stability along its trajectory in state-space, especially in the sense of real-time application. Connotations to practical industrial processes were highlighted via this straightforward control approach; however, as seen in the sense of the torsional pendulum example for dynamical systems with such strong nonlinearity, maintaining accurate control across all operating points remains a challenge of the presented HONU-MRAC based design.

Figure 11. Control of plant with the increasing desired value (dashed); the accurate control is more difficult to achieve as the sinusoidal nonlinearity becomes stronger with increasing magnitude of desired value, so it demonstrates current limits and challenges of the identification and control of strongly nonlinear systems with HONUs

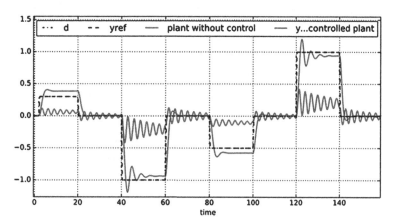

ACKNOWLEDGMENT

Authors acknowledge support from the EU Operational Programme Research, Development and Education, and from the Center of Advanced Aerospace Technology (CZ.02.1.01/0.0/0.0/16_019/0000826), Faculty of Mechanical Engineering, Czech Technical University in Prague.

REFERENCES

Benes, P. & Bukovsky, I. (2016). On the intrinsic relation between linear dynamical systems and higher order neural units. In R. Silhavy, R. Senkerik, Z. K. Oplatkova, Z. Prokopova, & P. Silhavy (Eds.), *Intelligent Systems in Cybernetics and Automation Theory*. doi:10.1007/978-3-319-18503-3_27

Benes, P., & Bukovsky, I. (2018). An input to state stability approach for evaluation of nonlinear control loops with linear plant model. In R. Silhavy, R. Senkerik, Z. K. Oplatkova, Z. Prokopova, & P. Silhavy (Eds.), *Cybernetics and Algorithms in Intelligent Systems* (pp. 144–154). Springer International Publishing.

Benes, P. M. & Bukovsky, I. (2014). Neural network approach to hoist deceleration control. In *Proceedings 2014 International Joint Conference on Neural Networks (IJCNN)*, 1864–1869. IEEE. Retrieved from http://ieeexplore.ieee.org/xpls/abs_all.jsp?arnumber=6889831

Benes, P. M., Bukovsky, I., Cejnek, M., & Kalivoda, J. (2014). Neural network approach to railway stand lateral skew control. Computer Science & Information Technology (CS& IT), 4, 327–339. Sydney, Australia: AIRCC.

Benes, P. M., Erben, M., Vesely, M., Liska, O., & Bukovsky, I. (2016). HONU and supervised learning algorithms in adaptive feedback control. In Applied Artificial Higher Order Neural Networks for Control and Recognition (pp. 35–60). Hershey, PA: IGI Global.

Bukovsky, I., Benes, P., & Slama, M. (2015). Laboratory systems control with adaptively tuned higher order neural units. In R. Silhavy, R. Senkerik, Z. K. Oplatkova, Z. Prokopova, & P. Silhavy (Eds.), *Intelligent Systems in Cybernetics and Automation Theory* (pp. 275–284). doi:10.1007/978-3-319-18503-3_27

Bukovsky, I., & Homma, N. (2017). An approach to stable gradient-descent adaptation of higher order neural units. *IEEE Transactions on Neural Networks and Learning Systems*, 28(9), 2022–2034. doi:10.1109/TNNLS.2016.2572310 PMID:27295693

Bukovsky, I., Homma, N., Smetana, L., Rodriguez, R., Mironovova, M., & Vrana, S. (2010, July). Quadratic neural unit is a good compromise between linear models and neural networks for industrial applications. 556–560. doi:10.1109/COGINF.2010.5599677

Bukovsky, I., Hou, Z.-G., Bila, J., & Gupta, M. M. (2008). Foundations of nonconventional neural units and their classification. *International Journal of Cognitive Informatics and Natural Intelligence*, 2(4), 29–43. doi:10.4018/jcini.2008100103

Bukovsky, I., Redlapalli, S., & Gupta, M. M. (2003). Quadratic and cubic neural units for identification and fast state feedback control of unknown nonlinear dynamic systems. In *Proceedings Fourth International Symposium on Uncertainty Modeling and Analysis, 2003. ISUMA 2003*, 330–334. 10.1109/ISUMA.2003.1236182

Bukovsky, I., *Voracek, J., Ichiji, K., & Noriyasu, H.* (2017). Higher order neural units for efficient adaptive control of weakly nonlinear systems. (pp. 149–157). doi:10.5220/0006557301490157

Dai, Y. H. & Yuan, Y. (1999). *A nonlinear conjugate gradient method with a strong global convergence property. 10(1)*, 177–182. doi:10.1137/S1052623497318992

El-Nabarawy, I., Abdelbar, A. M., & Wunsch, D. C. (2013, Aug. 4). *Levenberg-Marquardt and Conjugate Gradient methods applied to a high-order neural network. 1*–7. doi:10.1109/IJCNN.2013.6707004

Elbuluk, M. E., Tong, L., & Husain, I. (2002). Neural-network-based model reference adaptive systems for high-performance motor drives and motion controls. *IEEE Transactions on Industry Applications*, 38(3), 879–886. doi:10.1109/TIA.2002.1003444

Garcia, C. E., Prett, D. M., & Morani, M. (1989). Model predictive control: Theory and practice—. *Survey (London, England)*, 25(3), 335–348. doi:10.1016/0005-1098(89)90002-2

Glorot, X., Bordes, A., & Bengio, Y. (2011). Deep sparse rectifier neural networks. In G. Gordon, D. Dunson, & M. Dudík (Eds.), *Proceedings of the Fourteenth International Conference on Artificial Intelligence and Statistics* (pp. 315–323). Retrieved from http://proceedings.mlr.press/v15/glorot11a.html

Goodfellow, I., Bengio, Y., & Courville, A. (2016). *Deep learning*. Retrieved from http://www.deeplearningbook.org/

Gupta, M. M., Bukovsky, I., Homma, N., Solo, A. M. G., & Hou, Z.-G. (2013). Fundamentals of higher order neural networks for modeling and simulation. In M. Zhang (Ed.), *Artificial Higher Order Neural Networks for Modeling and Simulation* (pp. 103–133). doi:10.4018/978-1-4666-2175-6.ch006

Gupta, M. M., Homma, N., Hou, Z.-G., Solo, M., & Bukovsky, I. (2010). Higher order neural networks: Fundamental theory and applications. *Artificial Higher Order Neural Networks for Computer Science and Engineering: Trends for Emerging Applications*, 397–422.

Hochreiter, S., & Schmidhuber, J. (1997). Long short-term memory. *Neural Computation*, *9*(8), 1735–1780. doi:10.1162/neco.1997.9.8.1735 PMID:9377276

Hornik, K., Stinchcombe, M., & White, H. (1989). Multilayer feedforward networks are universal approximators. *Neural Networks*, *2*(5), 359–366. doi:10.1016/0893-6080(89)90020-8

Huang, G.-B., Zhu, Q.-Y., & Siew, C.-K. (2006a). Extreme learning machine: Theory and applications. *Neurocomputing*, *70*(1–3), 489–501. doi:10.1016/j.neucom.2005.12.126

Huang, G.-B., Zhu, Q.-Y., & Siew, C.-K. (2006b). Extreme learning machine: Theory and applications. *Neurocomputing*, *70*(1–3), 489–501. doi:10.1016/j.neucom.2005.12.126

Ivakhnenko, A. G. (1971). Polynomial theory of complex systems. *IEEE Transactions on Systems, Man, and Cybernetics*, *SMC-1*(4), 364–378. doi:10.1109/TSMC.1971.4308320

Kosmatopoulos, E. B., Polycarpou, M. M., Christodoulou, M. A., & Ioannou, P. A. (1995). High-order neural network structures for identification of dynamical systems. *IEEE Transactions on Neural Networks*, *6*(2), 422–431. doi:10.1109/72.363477 PMID:18263324

Ławryńczuk, M. (2009). *Neural networks in model predictive control*. 31–63. doi:10.1007/978-3-642-04170-9_2

LeCun, Y., Bengio, Y., & Hinton, G. (2015). Deep learning. *Nature*, *521*(7553), 436–444. doi:10.1038/nature14539 PMID:26017442

Liu, D., & Wei, Q. (2014). Policy iteration adaptive dynamic programming algorithm for discrete-time non-linear systems. *IEEE Transactions on Neural Networks and Learning Systems*, *25*(3), 621–634. doi:10.1109/TNNLS.2013.2281663 PMID:24807455

Mandic, D. P. (2001). *Recurrent neural networks for prediction: Learning algorithms, architectures, and stability*. Chichester, NY: John Wiley. doi:10.1002/047084535X

Mandic, D. P. (2004). A generalized normalized gradient descent algorithm. *IEEE Signal Processing Letters*, *11*(2), 115–118. doi:10.1109/LSP.2003.821649

Morani, M. & Lee, J. H. (1999). *Model predictive control: Past, present and future*. 23(4–5), 667–682. doi:10.1016/S0098-1354(98)00301-9

Narendra, K. S., & Valavani, L. S. (1979). Direct and indirect model reference adaptive control. *Automatica*, *15*(6), 653–664. doi:10.1016/0005-1098(79)90033-5

Nikolaev, N. Y. & Iba, H. (2006). *Adaptive learning of polynomial networks genetic programming, backpropagation and Bayesian methods*. Retrieved from http://public.eblib.com/choice/publicfullrecord.aspx?p=303002

Osburn, P. V. (1961). *New developments in the design of model reference adaptive control systems*. Institute of the Aerospace Sciences.

Parks, P. C. (1966). Liapunov redesign of model reference adaptive control systems. *IEEE Transactions on Automatic Control, 11*(3), 362–367.

Taylor, J. G., & Coombes, S. (1993). Learning higher order correlations. *Neural Networks, 6*(3), 423–427. doi:10.1016/0893-6080(93)90009-L

Tripathi, B. K. (2015). Higher-order computational model for novel neurons. In B. K. Tripathi (Ed.), *High Dimensional Neurocomputing* (Vol. 571, pp. 79–103). doi:10.1007/978-81-322-2074-9_4

Wang, F. Y., Zhang, H., & Liu, D. (2009, May). Adaptive dynamic programming: an introduction. *IEEE Computational Intelligence Magazine, 4*(2), 39–47. doi:10.1109/MCI.2009.932261

Wang, X., Cheng, Y., & Sun, W. (2007). *A proposal of adaptive PID controller based on reinforcement learning, 17*(1), 40–44. doi:10.1016/S1006-1266(07)60009-1

Wang, Z., & Liu, D. (2014). Stability analysis for a class of systems: from model-based methods to data-driven methods. *IEEE Transactions on Industrial Electronics, 61*(11), 6463–6471. doi:10.1109/TIE.2014.2308146

Zhang, L., & Suganthan, P. N. (2016). A comprehensive evaluation of random vector functional link networks. *Information Sciences, 367–368*, 1094–1105. doi:10.1016/j.ins.2015.09.025

Zhu, T., Yan, Z., & Peng, X. (2017). *A modified nonlinear conjugate gradient method for engineering computation*. 1–11. doi:10.1155/2017/1425857

Chapter 5
Optimum Design of Carbon Fiber–Reinforced Polymer (CFRP) Beams for Shear Capacity via Machine Learning Methods:
Optimum Prediction Methods on Advance Ensemble Algorithms – Bagging Combinations

Melda Yucel
Istanbul University-Cerrahpaşa, Turkey

Aylin Ece Kayabekir
Istanbul University-Cerrahpaşa, Turkey

Sinan Melih Nigdeli
Istanbul University-Cerrahpaşa, Turkey

Gebrail Bekdaş
Istanbul University-Cerrahpaşa, Turkey

ABSTRACT

In this chapter, an application for demonstrating the predictive success and error performance of ensemble methods combined via various machine learning and artificial intelligence algorithms and techniques was performed. For this reason, two single methods were selected, and combination models with a Bagging ensemble were constructed and operated with the goal of optimally designing concrete beams covering with carbon-fiber-reinforced polymers (CFRP) by ensuring the determination of the design variables. The first part was an optimization problem and method composing an advanced bio-inspired

DOI: 10.4018/978-1-7998-0301-0.ch005

metaheuristic called the Jaya algorithm. Machine learning prediction methods and their operation logics were detailed. Performance evaluations and error indicators were represented for the prediction models. In the last part, performed prediction applications and created models were introduced. Also, the obtained predictive success of the main model, as generated with optimization results, was utilized to determine the optimal predictions of the test models.

INTRODUCTION

Artificial Intelligence (AI) methods are effective in solving multidisciplinary engineering problems. Also, AI methods can be trained with optimization methodologies to provide the prediction of optimization results. In this chapter, the authors present a study showing the application of the predictive success and error performance of ensemble methods employing various machine learning and artificial intelligence algorithms. Two single methods were selected, and combination models with a Bagging ensemble were constructed. The optimal design is that of using concrete beams with a covering of carbon-fiber-reinforced polymers (CFRP) by ensuring the determination of design variables for the minimization of CFRP material in order to increase the shear capacity of the beam. For an RC beam using CFRP, the width, spacing, and application angle of the CFRP strip are the design variables. Their optimization has previously been done (Kayabekir, Sayin, Bekdas, & Nigdeli, 2017; Kayabekir, Sayin, Nigdeli, & Bekdas, 2017; Kayabekir, Sayin, Bekdas, & Nigdeli, 2018; Kayabekir, Bekdaş, Nigdeli, & Temür, 2018) by using several metaheuristic algorithms—namely, Flower Pollination Algorithm (FPA) (Yang, 2012), Teaching-Learning-Based Optimization (TLBO) (Rao, Savsani, & Vakharia, 2011), and Jaya Algorithm (JA) (Rao, 2016).

CARBON-FIBER-REINFORCED POLYMER (CFRP) BEAM MODEL

The Optimization Problem

The capacity of reinforced concrete elements may be insufficient due to reasons such as a change in the purpose of use of the structure (for examples, adding a new floor to the existing structure or retrofitting it for a capacity increase due to earthquake force mitigation; etc.). In such cases, various retrofit methods are utilized to increase the shear force, flexural moment, or axial force capacities. These methods generally necessitate the partial destruction of existing members; and the use of such structures may not always be possible in such case. Furthermore, since the total weight and rigidity of the structure are changed, a structural re-analysis is required. Another option is to use carbon-fiber-reinforced polymer (CFRP), having a linear deformation behavior with a large strain capacity, without changing the existing behavior of the structure. This method can be easily applied and provides for the use of the structure during its application.

In this chapter, optimal carbon-fiber-reinforced polymer design is presented with the goal of increasing the shear capacity of T-shaped RC beam members. This is done by considering the rules of regulation ACI 318 (Building Code Requirements for Structural Concrete); and by following various advanced machine learning applications that were carried out. It was likewise done by determining these designs rapidly and effectively in a short time—with the purpose of preventing the loss of effort, time, and cost while generating structural designs. In the optimization process, the Jaya Algorithm (JA)—a metaheuristic method—was utilized. Objective function was defined as being the minimization of the required CFRP area per meter. In other words, optimal design variables providing the minimum required CFRP area per meter and providing for structural safety are searched. The mentioned design variables are the spacing (s_f), width (w_f), and angle of CFRP strips (β). Structural model and design variables of the problem are shown in Figure 1.

In the figure, d_f represents the depth of beam, where the wrapping of CFRP is applied. It can be calculated as

$$d_f = d - h_f \tag{1}$$

in which h_f represents the thickness of the slab, and d represents the effective depth of the beam.

In addition, objective function of the problem can be written as in Equation (2). Here, the total area covered with CFRP within the beam unit length (m), which is defined by *A*, is minimized.

$$A = \frac{w_f\left(\dfrac{2d_f}{\sin^2} + b_w\right)}{s_f} x1000 \tag{2}$$

Here, b_w represents the breadth of the beam.

Design constraints are determined according to the regulation of ACI 318. These constraints can be given as given in Eqs. (3-5);

$$g_1(x): s_f \leq \frac{d}{4} \tag{3}$$

$$g_2(x): 0.7\text{R} \frac{\left(2t_f w_f f_{fe}\right)\left(\sin^2 + \cos^2\right)d_f}{s_f + w_f} > V_{add} \tag{4}$$

Figure 1. Structural model and design variables of the problem
(Khalifa & Nanni, 2000)

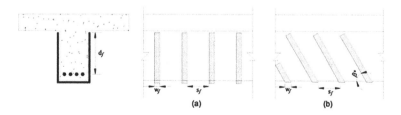

$$g_3(x) : \frac{\left(2t_f w_f f_{fe}\right)\left(\sin^2 + \cos^2\right)}{s_f + w_f} \leq \frac{2\sqrt{f'_c} b_w d}{3} - V_s \tag{5}$$

In these equations, V_{add}, R, t_f, f_{fe}, f_c, and V_s indicate the required additional shear capacity, the reduction factor, the thickness of the CFRP, the effective tensile strength of the CFRP, and the compression strength of the concrete total shear force capacity of the rebar, respectively. In addition, the design constants and ranges of the design variables are summarized in Table 1.

Metaheuristic Method for Optimization: Jaya Algorithm

Jaya was originally a Sanskrit word, and its meaning (in English) is "victory." After testing the algorithm, Rao (2016) probably was the one who gave it this name because of its having achieved such successful results. The most important feature of this algorithm is that it does not require the user to enter any user-defined parameters. For that reason, it is easy to apply it to optimization problems. Similarly, as with other metaheuristics, the optimization process of Jaya can be explained in five steps. These steps are summarized in following section.

In the first step—that of optimization—the design constants, the ranges of the design variables, and population and termination criterion are determined. In this optimization problem, termination criterion is defined as being the maximum iteration number.

In the second step, the initial solution matrix—including candidate solution vectors (in other words, a set of candidate solutions)—is generated. These candidate solution vectors include randomly generated design variables. The initial solution matrix can be indicated symbolically as follows, in Equation (6).

Table 1. Design constants and variables

Components	Definition	Symbol	Unit	Value
Design constants	Breadth of the beam section	b_w	mm	200–500
	Height of the beam section	h	mm	300–800
	Effective depth of the beam	d	mm	0.9h
	Thickness of the CFRP	t_f	mm	0.165
	Reduction factor	R	mm	0.5
	Thickness of the slab	h_f	-	80–120
	Compression strength of the concrete	f'_c	MPa	20
	Effective tensile strength of the CFRP	f_{fe}	MPa	3790
	Additional shear force	V_{add}	kN	50–200
	Shear force capacity of the rebar	V_s	kN	50
Design variables	Width of the CFRP	w_f	mm	10–1000
	Spacing between CFRPs	s_f	mm	0–d/4
	Covering angle of the CFRP	β	°	0–90

$$CL = \begin{bmatrix} X_{1,1} & X_{1,2} & \cdot & \cdot & \cdot & X_{1,vn} \\ X_{2,1} & X_{2,2} & \cdot & \cdot & \cdot & X_{2,vn} \\ \cdot & \cdot & \cdot & \cdot & & \cdot \\ \cdot & \cdot & \cdot & \cdot & & \cdot \\ X_{pn-1,1} & X_{pn-1,2} & \cdot & \cdot & \cdot & X_{pn-1,vn} \\ X_{pn,1} & X_{pn,2} & \cdot & \cdot & \cdot & X_{pn,vn} \end{bmatrix} \tag{6}$$

Each column of the *CL* matrix represents a candidate solution set. The total number of these candidate solutions is a user-defined value, and it is called a vector number (*vn*). In the matrix, $X_{i,j}$ characterizes the value of the ith design variable in the jth solution vector. The values of the design variables are generated via Equation (7).

$$X_I = X_{I(\min)} + rand(X_{I(\max)} - X_{I(\min)}) \tag{7}$$

$X_{i(min)}$ and $X_{i(max)}$ indicate the ultimate limits of the design variables. Rand is a function producing a random number between 0 and 1.

In the third step, the objective function is calculated (Equation (2)) for each of candidate solution vectors. The calculated objective function values are stored in a vector for later comparison. Also, candidate solution vectors must provide design constraints in order to ensure structural safety. Therefore, for candidate solutions that do not provide design constraints, the objective function is punished with very large values.

These first three steps are similar for optimization processes of all metaheuristic algorithms. The differences start with the fourth step. In this step, a new solution matrix is generated, considering the rules of the algorithm. According to the rules of the Jaya algorithm, new values of design variables are generated by utilizing the best and worst solution vectors in the existing solution matrix.

The best and worst solution vectors are determined according to values of the objective function. In this optimization problem, since the design of the CFRP having a minimum area for 1 meter is searched for, the best solution represents a candidate solution vector providing a minimum value of the objective function. The worst solution represents a candidate solution vector giving a maximum value of the objective function. This situation can be expressed numerically via Equation (8).

$$X_{i,j}^{t+1} = X_{i,j}^{t} + rand\left(g^{*} - \left|X_{i,j}^{t}\right|\right) - rand\left(g^{w} - \left|X_{i,j}^{t}\right|\right) \tag{8}$$

Here, $X_{i,j}^{t+1}$, $X_{i,j}^{t}$ and t represent a new (or a next) candidate design variable, an existing (or an old) design variable, and an iteration number, respectively. On the other hand, g* indicates the values of ith design variables in the best candidate solution vector; and g^{w} indicates the values of ith design variables in the worst candidate solution vector.

New candidate solution vectors (including new design variables) which are generated are stored in a matrix. Candidate solutions that do not meet design constraints are penalized with large objective function values, as in Step Three. Then, comparisons are done between the existing solution matrix and the new solution matrix. This part of the comparison constitutes the last step of the optimization process. In this step, if the new candidate solution or solutions provides a better solution, the existing solution matrix is modified with a new candidate solution. Otherwise, no changes are made in the existing solution matrix.

The last two steps are repeated until the provision of termination criterion and the optimization process of JA can be summarized as follows, in Figure 2.

MACHINE LEARNING METHODS

Multilayer Perceptrons (MLP)

Artificial Neural Networks (ANNs) can be seen as being a machine learning method that ensures the mimesis of various behaviors unique to the human brain, in regard to different systems (Yucel, Bekdaş,

Figure 2. The optimization process of JA

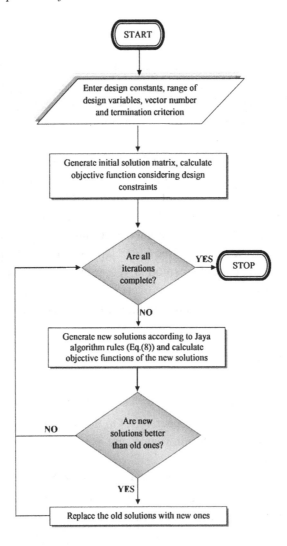

Nigdeli & Sevgen, 2019). There are so many models and types of ANNs in the literature. One such ANN is that of multilayer perceptrons (MLPs), which have input, output, and one or more hidden layers.

These MLP neural networks can learn from samples. The input layer, hidden layers, and output layer includes input nodes, calculation nodes, and output node, respectively. The input and output nodes store the data samples and result predictions, respectively (Chou, Tsai, Pham & Lu, 2014). In these respects, MLPs have an ability to determine the effects of input parameters onto output parameters. This can result in the occurrence of a map of relation for input-output parameters (Maimon & Rokach, 2010). A multilayer perceptron structure is presented in Figure 3 for the prediction of two different FRP beam design outputs in this study.

While realizing this process, a set of training stages is carried out with different algorithms with respect to the aim of determining the output result. The reason for this is the input data that can be nonlinear or unsuitable to operate by using a specific curve or an equation for predicting the output values directly. With this aim, MLP are trained by means of numerous input samples.

In this issue, an algorithm called back-propagation, which is one of the most frequently preferred algorithms nowadays, was proposed by G. E. Hinton, Rumelhart and R. O. Williams in 1986 in order to train neural networks (Raza & Khosravi, 2015).

This training process is realized after the spreading of input data, as a feed-forward mechanism. At this stage, firstly, there is an independent multiplication of each input with the connection weights belonging to nodes within the next layer; and a summation of these multiplications is obtained.

Moreover, this summation is carried out with the sum function (Σ), which benefits to determine the net input for any single node. Net input (net$_j$) is obtained as a combination of the weighted summation and the bias value (b) for any node j as follows, in Equation (9). Also, activation function (*f*) is beneficial for the calculation of output prediction by means of processing the net input (Equation (10)). Obtained prediction output and real output differences are named as being an error, and this value is used to update

Figure 3. Multilayer perceptron model generated for optimal design variables

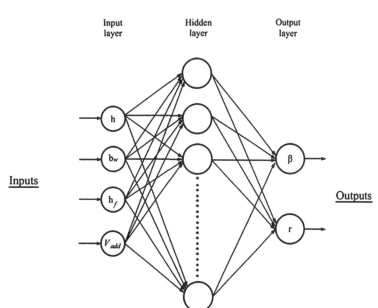

the connection weights via transference to the general network. These components and operations can be seen in Figure 3.

In addition to this, expressed in these equations and in Figure 4, i is input node, in other words, outputs existing in previous layer; j is current node that targets the determination of its net input; W_{ij} represents the connection weights between nodes i and j; X_i is the output value from the previous layer, and α is a constant value which ensures the function gradient (Topçu & Saridemir, 2008).

$$net_j = \sum_{i=1}^{n} W_{ij} X_i + b \qquad (9)$$

$$f\left(net_j\right) = \frac{1}{1 + e^{-\alpha net_j}} \qquad (10)$$

On the other hand, this activation function expressed in the above equation is a sigmoid/logistic function and is known as being a transfer function, too. Additionally, there are various types of functions such as a hyperbolic tangent, sine, cosine, linear, or identity function. However, the most-used one from among these is the sigmoid function, in virtue of advantages related to nonlinearity and good simulation with respect to the operation of human brain neurons intended for prediction applications (Maimon & Rokach, 2010).

Random Tree

The random tree method is a decision tree algorithm. In this respect, it is an algorithm applied to obtain an ultimate prediction result by splitting the tree according to defined labels of parameters via decisions made in the direction of the values of these labels. On the other hand, this algorithm is generally applied for generating random forest ensembles. Therefore, each of the random trees can be considered as being a member existing in forest. This decision tree algorithm is called Random Tree. It can be preferred/ intended for the application of solving classification and regression problems. It was proposed by Leo Breiman and Adele Cutler (Kalmegh, 2015). During the constructing of the random decision tree model, root and internal labels must first be determined in order to branch the tree structure. However, there

Figure 4. Calculation of the output for any node via activation of the sigmoid function
(Topçu & Saridemir, 2008)

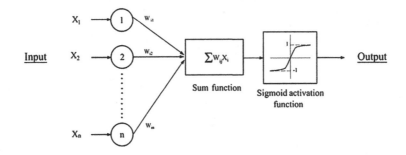

is no obligation to do that in an analysis of the problem data for the identification of attributes, which will be taking part in the interior nodes of the decision tree, because of the renewable and exchangeable structure of the tree (Jagannathan, Pillaipakkamnatt & Wright, 2009).

Bagging Learners

Bagging method is one of the independent ensemble learning techniques developed by Breiman, and its name occurred with respect to a combination of bootstrap and aggregation approaches (Cichosz, 2015; Skurichina & Duin, 2002). Ensemble models have an ability related to generalizing learning due to making mistakes less than that which base learners perform independently (Jagannathan et al., 2009).

This method applies an approach which is related to the fact that one basis prediction value is obtained by integrating the output results produced by different learner methods or classifiers, with the purpose of enhancing prediction accuracy (Rokach & Maimon, 2015). In this respect, this approach is carried out so that the major dataset is split into different subgroups by those data samples which are selected randomly. Hence, these can include the same data more than once. Also, these sub data samples can be trained via any method, which is named as the basic/main learning algorithm or classifier.

This random selection, which is realized for its determination of data belonging to K different subgroups, is performed by means of sampling with replacement (putting sampled data back into the dataset) of each data sample from the training set composed of n samples. As has been said before, some data can be selected many times. Therefore, some of these may not take part in the training set. On the other hand, each basic learner method in ensemble is trained with training sets/subgroups including n samples different from each other, and the results (predictions) are combined with voting of majority (Kılınç, Borandağ, Yücalar, Özçift & Bozyiğit, 2015). However, these prediction results of sub data groups can be combined by using the other rules, too (Skurichina & Duin, 2002). A chart about prediction process via the Bagging method is given as Figure 5.

PREDICTION AND ERROR EVALUATION INDICATORS

After performing output prediction, the evaluation of obtained values in comparison with real data is required. The aim of this operation is that validation of accuracy, reliability, and precision belonging to the generated prediction model is provided. By means of this, the possibility and measure of making mistakes by the prediction model can be observed, and the success and performance of the model can be sensed more clearly.

All of the presented evaluation criteria and error performance indicators, their formulations, and their components acting as these indicators are shown in Table 2.

Correlation Coefficient (R)

The *Correlation Coefficient (R)* metric is a measure of compatibility between real/actual and predicted values belonging to data samples. It shows how much these values are close or similar to each other. This coefficient value can take the form of values between -1 and 1. It can be said that if R is 0, then there is no similarity of values. In addition, if its value is 1 or close to 1, there is strong fitness between the actual and predicted results of the data.

Figure 5. Prediction process via the bagging method

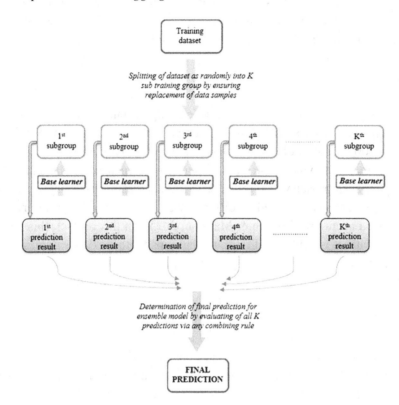

Mean Absolute Error (MAE)

Absolute error means the absolute value of the error occurring between the actual value and the predicted result of the data sample. For this reason, the *Mean Absolute Error (MAE)* metric is the mean value of these errors observed for all samples. In addition, the absolute function is beneficial when used for preventing the incorrect processing of negative error, which is generated in the case that the actual value is smaller than predicted.

It is possible that the MAE metric, which is an important indicator with respect to determining the best and most promotive solution in any case such as a regression problem, can be called an absolute loss (Cichosz, 2015).

Mean Square Error (MSE)

Square error expresses the squared error of difference. In other words, it shows error which is observed when actual values are compared with predicted values. Therefore, *Mean Square Error (MSE)* is an average value of the error squares of all samples. On the other hand, MSE error indicator benefits from enabling noticing and evaluating of errors more carefully, owing to the fact that the square function can perform an amplification of the observed error rate.

Also, increasing the success of prediction can be carried out by means of reducing the value of the MSE metric (Hill, Malone & Trocine, 2004).

Table 2. Error performance indicators and various details

Performance Indicator	Abbreviation	Formulation	Components
Correlation Coefficient	R	$$\dfrac{n\sum_{i=1}^{n} a_i p_i - \sum - \left(\sum_{i=1}^{n} a_i\right)\left(\sum_{i=1}^{n} p_i\right)}{\sqrt{n\left[\sum_{i=1}^{n} a_i^2\right] - \left(\sum_{i=1}^{n} a_i\right)^2} - \sqrt{n\left[\sum_{i=1}^{n} p_i^2\right] - \left(\sum_{i=1}^{n} p_i\right)^2}}$$	
Mean Absolute Error	MAE	$$\dfrac{1}{n}\sum_{i=1}^{n}\left| a_i - p_i \right|$$	
Mean Square Error	MSE	$$\dfrac{1}{n}\sum_{i=1}^{n}\left(p_i - a_i\right)^2$$	
Root Mean Square Error	RMSE	$$\sqrt{\dfrac{1}{n}\sum_{i=1}^{n}\left(p_i - a_i\right)^2}$$	a_i: ith actual value p_i: ith predicted value m_a: mean of the actual data n: data number
Relative Absolute Error	RAE	$$\dfrac{\sum_{i=1}^{n}\left| p_i - a_i \right|}{\sum_{i=1}^{n}\left| m_a - a_i \right|}$$	
Root Relative Square Error	RRSE	$$\sqrt{\dfrac{\sum_{i=1}^{n}\left(p_i - a_i\right)^2}{\sum_{i=1}^{n}\left(m_a - a_i\right)^2}}$$	

Root Mean Square Error (RMSE)

The *Root Mean Square Error (RMSE)* metric is the root value of the MSE indicator. In other words, it shows the root of the mean of all occurred errors' squares, according to a specifically adjusted line. For this reason, this root value expresses ultimate error, and actually expresses the mean distance between actual and predicted results (Chou & Tsai, 2012).

Relative Absolute Error (RAE)

The *Relative Absolute Error (RAE)* metric, as seen in its formulation, is related to deviation from the mean value of actual data. Therefore, this metric is beneficial in that it helps users in noticing the mean deviation of the predicted results. Moreover, a case is suitable when relative absolute error value is smaller than 1, or when—in a good model—it even equals 0 (Cichosz, 2015). In addition, RAE metric generates the total absolute error value (Witten & Frank, 2005).

Root Relative Square Error (RRSE)

The *Relative Square Error (RSE)* metric, as in relative absolute error, is performed with normalization of errors. Therefore, RSE metric square errors of all samples are summed up and determined by dividing the squares of deviation from the mean of actual values belonging to that data. Also, the *Root Relative Square Error (RRSE)* metric is a root value of the RSE metric (Witten & Frank, 2005).

All of the evaluation criteria and error performance indicators, their formulations, and their components acting as these indicators are shown in Table 2.

PREDICTION APPLICATIONS VIA MACHINE LEARNING METHODS FOR OPTIMAL FRP BEAM DESIGNS

The Main Model

The optimization process was carried out with the aim of determining the design variables belonging to beams covered with carbon-fiber-reinforced polymers (CFRP). Values obtained for variables and design properties, which will be dealt with in prediction applications by machine learning algorithms, were used to generate a dataset.

As expressed in the optimization process, design constants composed of various numerical expressions and beam properties. These include h (height), b_w (breadth), h_f (the slab thickness of the beam), and V_{add} (additional shear force values). These were considered as being input parameters for the prediction model. In addition, design variables that engineers aim to optimize are β (CFRP covering angle), w_f (FRP width), and s_f (the amount of space between two consecutive CFRPs) values. However, a new parameter was generated by using w_f and s_f variables besides β, with the aim of its being used in a training dataset as one of the output parameters. This output is CFRP rate (r) and expresses CFRP width existing in length until the starting point of the next CFRP. The CFRP rate equation can be formulated as follows:

$$r = \frac{w_f}{w_f + s_f} \qquad (11)$$

Also, machine learning algorithms and combinations of them were determined, together with the creating of dataset intended for prediction operations, owing to the performance of model training.

During predictions of optimal variables for CFRP beam designs, four different machine learning approaches as two single and two hybrid prediction models, generated from (and comprising) multi-layer perceptrons (MLPs), Random tree, and Bagging methods combined with MLP and Random Tree methods as separate base learner methods, were proposed. In this respect, the dataset—which includes four inputs (h, b_w, h_f and V_{add}) and two outputs (β and r)—was trained in the direction of these prediction models using Weka machine learning software (Witten et al., 2016).

For the main training model, the results of the optimal variables' predictions and indicators for error evaluations are shown in Tables 3 and 4 for β and CFRP rate, respectively.

Test Models

After the training process, test models were handled for evaluation for the main model. These models are composed of 10 different design combination data, and optimal design parameters and variables for these combinations are expressed in Table 5.

The obtained results for optimal variable predictions, made by means of each machine learning training models, are shown in Tables 6–9 for β, together with the CFRP rate. Furthermore, in these tables, various performance error indicators—mean absolute (MAE), absolute percentage (MAPE), and square error (MSE)—are shown, with the purpose of comparing them according to JA used in optimization for analyzed prediction models.

CONCLUSION

For the main model, according to the results, it is seen that the *correlation coefficient (R)*—which expresses how well actual values are explained by predicted values—is observed as being extremely high in all models for both CFRP rate and β outputs. Moreover, the success of the ensemble models is (to a large extent) remarkable in terms of increasing predictive success for β via each hybridization of the bagging method. However, for the CFRP rate only, the ensemble model that operated by combining with

Table 3. Error indicators and performance of prediction for β output

β (°)	Prediction Models			
	MLP	*Bagging+MLP*	*Random Tree*	*Bagging+Random Tree*
Correlation coefficient (R)	99.69%	99.77%	98.89%	99.38%
Mean absolute error (MAE)	0.0908	0.0776	0.1305	0.1069
Root mean squared error (RMSE)	0.1155	0.0998	0.2176	0.1684
Relative absolute error (RAE)	8.0829%	6.9034%	11.6068%	9.5150%
Root relative squared error (RRSE)	7.9472%	6.8622%	14.9707%	11.5845%
Total Number of Instances	231	231	231	231

Table 4. Error indicators and performance of prediction for CFRP rate (r) output

FRP Rate (r)	Prediction Models			
	MLP	*Bagging+MLP*	*Random Tree*	*Bagging+Random Tree*
Correlation coefficient (R)	99.70%	99.49%	99.36%	99.59%
Mean absolute error (MAE)	0.0121	0.0176	0.0123	0.0139
Root mean squared error (RMSE)	0.0175	0.0249	0.0258	0.0215
Relative absolute error (RAE)	6.4224%	9.3321%	6.5124%	7.3985%
Root relative squared error (RRSE)	7.6681%	10.9200%	11.3089%	9.4058%
Total Number of Instances	231	231	231	231

Table 5. Optimization results for designs in test model determined with the JA metaheuristic algorithm

h (mm)	b_w (mm)	h_f (mm)	V_{add} (kN)	β (°)	w_f (mm)	s_f (mm)	CFRP Rate (r)
350	200	120	70000	65.016	35.170	61.271	0.365
530	420	85	55000	63.324	18.639	54.053	0.256
750	370	100	120000	63.593	83.198	59.184	0.584
680	500	95	110000	62.608	58.911	53.272	0.525
420	450	120	70000	62.753	61.816	110.000	0.360
300	300	100	100000	64.259	99.541	104.068	0.489
770	220	90	80000	64.974	58.013	93.878	0.382
400	510	85	90000	62.683	76.496	106.484	0.418
550	200	80	50000	64.839	31.058	102.775	0.232
800	300	100	120000	64.217	97.372	68.646	0.587

Random Tree became successful in terms of moving in the right direction towards the achievement of this purpose.

According to the error evaluations for these four different prediction models, *mean absolute error (MAE)* values are extremely low for both outputs. However, MAE values decreased for β, and increased for the CFRP rate, from the point of view of the base learner methods (the MLP and Random Tree algorithms alone) in comparison to their combination with the Bagging algorithm. In addition, the *root mean square error (RMSE)* metric showed a similar behavior as MAE with respect to both design

Table 6. β° and CFRP rate (r) predictions for the test model via the MLP algorithm

MLP Model						
β (°)	Error Values for JA		CFRP Rate (r)	Error Values for JA		
	Absolute Error	Square Error		Absolute Error	Square Error	
63.657	1.359	1.847	0.604	0.239	0.057	
63.644	0.320	0.102	0.240	0.016	0.000	
64.902	1.309	1.714	0.377	0.207	0.043	
63.930	1.322	1.747	0.372	0.153	0.023	
61.858	0.895	0.801	0.456	0.096	0.009	
61.944	2.315	5.358	0.910	0.421	0.177	
65.969	0.995	0.991	0.220	0.162	0.026	
61.508	1.175	1.382	0.545	0.127	0.016	
65.498	0.659	0.435	0.212	0.020	0.000	
65.531	1.314	1.726	0.353	0.234	0.055	
	MAE	MSE		MAE	MSE	
Mean	1.166	1.610		0.168	0.041	
RMSE		1.269			0.202	

Table 7. β° and CFRP rate (r) predictions for the test model via the Bagging algorithm in conjunction with the MLP algorithm

Bagging+MLP Model					
β (°)	Error Values for JA		CFRP Rate (r)	Error Values for JA	
	Absolute Error	Square Error		Absolute Error	Square Error
63.556	1.460	2.132	0.621	0.256	0.066
63.634	0.310	0.096	0.236	0.020	0.000
64.917	1.324	1.754	0.370	0.214	0.046
63.880	1.272	1.618	0.367	0.158	0.025
61.780	0.973	0.947	0.460	0.100	0.010
61.846	2.413	5.822	0.945	0.456	0.208
65.977	1.003	1.007	0.210	0.172	0.030
61.506	1.177	1.386	0.551	0.133	0.018
65.487	0.648	0.420	0.208	0.024	0.001
65.558	1.341	1.798	0.343	0.244	0.059
	MAE	MSE		MAE	MSE
Mean	1.192	1.698		0.178	0.046
RMSE		1.303			0.214

Table 8. β° and CFRP rate (r) predictions for the test model via the Random Tree algorithm

Random Tree Model					
β (°)	Error Values for JA		CFRP Rate (r)	Error Values for JA	
	Absolute Error	Square Error		Absolute Error	Square Error
64.239	0.777	0.604	0.353	0.012	0.000
63.597	0.273	0.075	0.237	0.019	0.000
64.944	1.351	1.826	0.277	0.307	0.094
63.940	1.332	1.774	0.323	0.202	0.041
61.143	1.610	2.592	0.353	0.007	0.000
61.733	2.526	6.380	0.987	0.498	0.248
66.158	1.184	1.402	0.279	0.103	0.011
61.753	0.930	0.866	0.607	0.189	0.036
65.762	0.923	0.852	0.191	0.041	0.002
65.458	1.241	1.539	0.279	0.308	0.095
	MAE	MSE		MAE	MSE
Mean	1.215	1.791		0.169	0.053
RMSE		1.338			0.230

Table 9. $\beta°$ and CFRP rate (r) predictions for the test model via the Bagging algorithm in conjunction with the Random Tree algorithm

	Bagging+Random Tree Model				
	Error Values for JA			Error Values for JA	
β (°)	Absolute Error	Square Error	CFRP Rate (r)	Absolute Error	Square Error
64.335	0.681	0.464	0.346	0.019	0.000
63.595	0.271	0.073	0.235	0.021	0.000
64.892	1.299	1.688	0.282	0.302	0.091
63.942	1.334	1.779	0.326	0.199	0.040
61.227	1.526	2.328	0.372	0.012	0.000
61.707	2.552	6.512	0.955	0.466	0.217
66.115	1.141	1.302	0.290	0.092	0.008
61.818	0.865	0.749	0.605	0.187	0.035
65.708	0.869	0.756	0.191	0.041	0.002
65.468	1.251	1.564	0.285	0.302	0.091
	MAE	MSE		MAE	MSE
Mean	1.179	1.722		0.164	0.049
RMSE		1.312			0.221

variables. Nevertheless, one difference was realized in CFRP rate prediction performance via the Bagging algorithm combined with Random Tree algorithm. This is a decreasing of the RMSE value for prediction in the CFRP rate model.

Following this, for the test model comprised of ten design samples, the prediction of optimal values for β and CFRP rate outputs were carried out by means of predictive success and low error of proposed machine learning methods, and their combinations created by using the bagging ensemble algorithm with respect to the main model. In this direction, two different error metrics comprised of MAE and mean square error (MSE) were calculated from the obtained results via Weka software, comparing this to the Jaya metaheuristic optimization algorithm.

For β output, MAE error is 1.166 and 1.192, respectively, in MLP and Bagging combined with MLP. Also, for CFRP rate, these values were 0.168 and 0.178, respectively in both models. In addition to this, MSE values showed increases from 1.610 to 1.698 for β, and 0.041 to 0.046 for CFRP rate, despite the main model not being successful only for the CFRP rate in terms of the MLP algorithm when comparing it to the Bagging algorithm according to correlation.

Both error metric values increased in comparison to real optimal design values for variables, too. Actually, these error values were pretty low, but if combinations for Bagging models are evaluated, the model/method that combined the Bagging and MLP algorithms can be considered as having failed in terms of its prediction of test model designs. Nevertheless, it can be seen that the increasing error rate for the β model is higher than that of the CFRP rate for the model that combined the Bagging and MLP algorithms. In addition, RMSE, which expresses ultimate (or what can be considered as being average) error for all models, is 1.269–1.303 for β, and 0.202–0.214 for the CFRP rate in MLP and Bagging with

MLP hybridization, respectively. The obtained increase in this metric was lower in the CFRP rate, as was likewise the case for the other metrics.

On the other hand, another machine learning technique—and one used in combination with the Bagging algorithm—is the Random Tree algorithm.

By combining the Random Tree and Bagging algorithms relative to this method, this ensured that MAE values decreased for both β (1.215 to 1.179) and CFRP rate (0.169 to 0.164) outputs. Moreover, MSE errors were decreased similarly to MAE. MSE values changed from 1.791 to 1.722 for β, and from 0.053 to 0.049 for the CFRP rate. Thus, due to this model's low error rate, it can be said that the Bagging with Random Tree hybridization model is successful for the prediction of values. Also, errors could be reduced in a major part of design samples for the prediction of optimal results via this model. RMSE values were decreased, too.

As a result, it can be concluded that all four models are successful for the prediction of optimal values belonging to design variables by means of their good predictive performance and low error performance. As was especially the case for the CFRP rate, the error values obtained were very close to 0.

Differences were discovered between the two models and their combinations. The first combination, created via the Bagging algorithm in comparison with its base learner (MLP), was not successful with regard to outputs. The second one, though, performed better than the first one did, in terms of its achievement of its purpose of predicting β and the CFRP rate. However, if general evolution were to improve these four models even further, then MLP as a single algorithm, and Bagging with Random Tree as a combination of algorithms, would become considered as being even more suitable and efficient means of making output predictions.

REFERENCES

American Concrete Institute. (2005). *Building code requirements for structural concrete and commentary* (ACI 318M-05).

Chou, J. S., & Tsai, C. F. (2012). Concrete compressive strength analysis using a combined classification and regression technique. *Automation in Construction, 24*, 52–60. doi:10.1016/j.autcon.2012.02.001

Chou, J. S., Tsai, C. F., Pham, A. D., & Lu, Y. H. (2014). Machine learning in concrete strength simulations: Multi-nation data analytics. *Construction & Building Materials, 73*, 771–780. doi:10.1016/j.conbuildmat.2014.09.054

Cichosz, P. (2015). *Data mining algorithms: explained using R*. Chichester, UK: John Wiley & Sons. doi:10.1002/9781118950951

Hill, C. M., Malone, L. C., & Trocine, L. (2004). Data mining and traditional regression. In H. Bozdogan (Ed.), *Statistical data mining and knowledge discovery* (p. 242). Boca Raton, FL: CRC Press.

Jagannathan, G., Pillaipakkamnatt, K., & Wright, R. N. (2009). A practical differentially private random decision tree classifier. In *Proceedings of 2009 IEEE International Conference on Data Mining Workshops*. Miami, FL: IEEE. 10.1109/ICDMW.2009.93

Kalmegh, S. (2015). Analysis of weka data mining algorithm reptree, simple cart and randomtree for classification of Indian news. *International Journal of Innovative Science, Engineering & Technology, 2*(2), 438–446.

Kayabekir, A. E., Bekdaş, G., Nigdeli, S. M., & Temür, R. (2018). Investigation of cross-sectional dimension on optimum carbon-fiber-reinforced polymer design for shear capacity increase of reinforced concrete beams. In *Proceedings of 7th International Conference on Applied and Computational Mathematics (ICACM '18)*. Rome, Italy: International Journal of Theoretical and Applied Mechanics.

Kayabekir, A. E., Sayin, B., Bekdas, G., & Nigdeli, S. M. (2017). Optimum carbon-fiber-reinforced polymer design for increasing shear capacity of RC beams. In *Proceedings of 3rd International Conference on Engineering and Natural Sciences (ICENS 2017)*. Budapest, Hungary.

Kayabekir, A. E., Sayin, B., Bekdas, G., & Nigdeli, S. M. (2018) The factor of optimum angle of carbon-fiber-reinforced polymers. In *Proceedings of 4th International Conference on Engineering and Natural Sciences (ICENS 2018)*. Kiev, Ukraine.

Kayabekir, A. E., Sayin, B., Nigdeli, S. M., & Bekdaş, G. (2017). Jaya algorithm based optimum carbon-fiber-reinforced polymer design for reinforced concrete beams. In *Proceedings of 15th International Conference of Numerical Analysis and Applied Mathematics*. Thessaloniki, Greece: AIP Conference Proceedings 1978.

Khalifa, A., & Nanni, A. (2000). Improving shear capacity of existing RC T-section beams using CFRP composites. *Cement and Concrete Composites, 22*(3), 165–174. doi:10.1016/S0958-9465(99)00051-7

Kılınç, D., Borandağ, E., Yücalar, F., Özçift, A., & Bozyiğit, F. (2015). Yazılım hata kestiriminde kolektif sınıflandırma modellerinin etkisi. In *Proceedings of IX. Ulusal Yazılım Mühendisliği Sempozyumu*. İzmir, Turkey: Yaşar Üniversitesi.

Maimon, O., & Rokach, L. (Eds.). (2010). *Data mining and knowledge discovery handbook* (Vol. 14). Springer Science & Business Media. doi:10.1007/978-0-387-09823-4

Rao, R. (2016). Jaya: A simple and new optimization algorithm for solving constrained and unconstrained optimization problems. *International Journal of Industrial Engineering Computations, 7*(1), 19–34.

Rao, R. V., Savsani, V. J., & Vakharia, D. P. (2011). Teaching–learning-based optimization: A novel method for constrained mechanical design optimization problems. *Computer Aided Design, 43*(3), 303–315. doi:10.1016/j.cad.2010.12.015

Raza, M. Q., & Khosravi, A. (2015). A review on artificial intelligence-based load demand forecasting techniques for smart grid and buildings. *Renewable & Sustainable Energy Reviews, 50*, 1352–1372. doi:10.1016/j.rser.2015.04.065

Rokach, L., & Maimon, O. (2015). *Data mining with decision trees: theory and applications* (2nd ed., Vol. 81). Singapore: World Scientific Publishing.

Skurichina, M., & Duin, R. P. (2002). Bagging, boosting and the random subspace method for linear classifiers. *Pattern Analysis & Applications, 5*(2), 121–135. doi:10.1007100440200011

Topçu, İ. B., & Sarıdemir, M. (2008). Prediction of mechanical properties of recycled aggregate concretes containing silica fume using artificial neural networks and fuzzy logic. *Computational Materials Science, 42*(1), 74–82. doi:10.1016/j.commatsci.2007.06.011

Witten, I. H., & Frank, E. (2005). *Data mining: Practical machine learning tools and techniques* (2nd ed.). San Francisco, CA: Morgan Kaufmann.

Witten, I. H., Frank, E., Hall, M. A., & Pal, C. J. (2016). *Data mining: Practical machine learning tools and techniques* (4th ed.). Cambridge, MA: Morgan Kaufmann.

Yang, X. S. (2012). Flower pollination algorithm for global optimization. In Jérôme Durand- Lose & Nataša Jonoska (Eds.), *International conference on unconventional computation and natural computation* (pp. 240-249). Berlin, Germany: Springer.

Yucel, M., Bekdaş, G., Nigdeli, S. M., & Sevgen, S. (2019). Estimation of optimum tuned mass damper parameters via machine learning. *Journal of Building Engineering*, 100847.

Chapter 6
A Scientometric Analysis and a Review on Current Literature of Computer Vision Applications

Osman Hürol Türkakın

https://orcid.org/0000-0003-2241-3394

Istanbul University-Cerrahpaşa, Turkey

ABSTRACT

Computer vision methods are widespread techniques mostly used for detecting cracks on structural components, extracting information from traffic flows, and analyzing safety in construction processes. In recent years, with the increasing usage of machine learning techniques, computer vision applications have been supported by machine learning approaches. So, several studies have been conducted which apply machine learning techniques to image processing. As a result, this chapter offers a scientometric analysis for investigating the current literature of image processing studies for the civil engineering field in order to track the scientometric relationship between machine learning and image processing techniques.

INTRODUCTION

Construction works need frequent monitoring and investigation. Computer vision applications are becoming essential techniques for monitoring components autonomously. Besides investigating construction components, image processing methods are suitable for monitoring traffic-flows and for making safety-level valuations. As far as developing various measurement techniques, image processing tools have become a versatile technique for various applications. In the field of civil engineering, image processing techniques are used for several applications. These include object detection, object tracking, action recognition (Seo, Han, Lee, & Kim, 2015), and crack detection (Yamaguchi & Hashimoto, 2010). Object detection is a comparatively primitive method among these methodologies. Object detection techniques are convenient algorithms for detecting the degree to which a construction site is unsafe, by detecting data regarding construction workers and their equipment. The basic principle of object detection methodology is dividing up the image into several mini-sized frames, and then detecting the searched-for object by

DOI: 10.4018/978-1-7998-0301-0.ch006

scanning these mini frames. (Murphy, Torralba, Eaton, & Freeman, 2006). One of the other applications for object detection is vehicle detection transportation video records (Bas, Tekalp, & Salman, 2007). A second methodology, object tracking, is a more sophisticated methodology than the object detection method. To implement object tracking methodology, video records are needed. In the first frame of the video record, the object is detected; and in the subsequent frames, the same object is tracked to obtain the object's trajectory. Three type of methodologies are used for object tracking: (1) point tracking, (2) kernel tracking, and (3) silhouette tracking (Yilmaz, Javed, & Shah, 2006). Structural health monitoring is another object tracking application by which buildings' frequencies are determined from video recordings (Feng & Feng, 2018).

In recent years, machine learning techniques have been developed to fulfill the demand for frequent use of image processing and computer vision applications in the civil engineering field. The most frequent methodology for image processing applications is a kind of neural network algorithm named *Convolutional Neural Networks* (CNN) that uses a framework analogous to the functioning of an animal's neural cortex. LeCun et al. (1989) first applied CNN methodology, and applied it for identifying written numbers. However, after more than a decade, this methodology has become more applicable to the field following the spread of GPU technologies. Additionally, there are various competitions in image recognition for various fields. Traffic sign recognition is one of the most prominent areas in these competitions (Stallkamp, Schlipsing, Salmen, & Igel, 2012). Other than traffic sign recognition, the CNN method is used for several areas in the field of civil engineering such as safety level assessment from site images in construction projects (Ding et al., 2018; Fang, Ding, Luo, & Love, 2018), and crack detection in structural components (Cha, Choi, & Buyukozturk, 2017; Huang, Li, & Zhang, 2018). For assessing the impact of these studies and new trends in the field of construction building technology, a scientometric analysis is a convenient methodology which has been used for various fields. In this study, these various applications are tracked with a scientometric analysis, and frequently-cited publications are investigated through their keywords and research area.

The impact of publications depends on citation statistics, and scientometrics is a quantitative methodology that uses citation statistics of publications and generates some statistical output using bibliometric data in the literature. In the context of scientometrics, the whole body of literature can be modeled as a network structure (Price, 1965). Additionally, each publication is variable input within scientometrical analysis. Analyzing and defining citation mechanisms are the main way in which articles that are joined to one another with citation and co-citation relationships in order to constitute citation graphs in order to assessing their impacts. Additionally, utilizing inner-citation statistics enables publications to be grouped together. Using scientometric analysis, author and publication contributions can be evaluated after these citation statistics are obtained. In the context of construction building technology, the study of Zhao (2017) is a prominent scientometric review regarding BIM research studies. Other studies have been found to be focused on such topics as safety searches (Jin et al., 2019), and green building searches (Zhao, Zuo, Wu, & Huang, 2019). The focus of the scientometric analysis of this particular study is measuring the impact of computer vision applications for construction building technology in which results are given with visual citation maps. In the first section, the concept of scientometrics is briefly introduced. In subsequent sections, the analysis of current publications on computer vision techniques is given. In the last section, the outputs of scientometric analysis are given. In the conclusion, research trends are discussed.

BACKGROUND

Scientometric methodologies were firstly used in the 1960s in order to quantify scientific impacts. Citation statistics of a publication in the timeline is the main part of a scientometric analysis. Based on citation lineage, visual demonstration of the publication's references and their other cited publications is an efficient methodology for briefly gathering a summary amount of knowledge about the publication and both of its references and cited publications. Each node describes a publication that, in the case of two publications citing mutual references, were scientometrically close each other. With these connections, a whole graph can be generated to represent publications in a network format. In this network graph, highly cited publications are depicted with wider nodes than low-cited publications. The node centrality depends on the citation count of the publication and represents the impact of the publication in the literature. Depending on the citation count, the publication is located more centrally in the group of publications. For labeling publications, classifying and clustering are two essential steps. In this case, the *research front* and the *intellectual base* are two definitions for labeling the publication. The research front represents research area of publications that are cited in the publication. The intellectual base is the research area of the publications in the reference list of the publication. The research front includes instantaneous content and knowledge that stays updated for a period of time. The intellectual base is a constituted set of background knowledge that is referenced by the current publication (Persson, 1994).

Publications become promoted as they get cited by other publications. Frequently cited papers became more visible, especially in search engines. So, they have a high probability of being accessed and cited by researchers. This phenomenon is similar to *rich people becoming richer*—called *the Matthew effect* (Merton, 1968). Besides frequently cited publications, authors usually choose to refer to updated publications in order to append newer knowledge upon recent developments. So, it can be said that most of the publications have a limited period of time in which they are cited. After this period ends, the originality of the publication starts to fade. To define this phenomenon, Price (1965) termed it the *immediacy factor*. Burton and Kebler (1960) modeled this publication aging phenomenon with citation half-life periods.

Scientometric analysis requires a huge amount of computational processing. Various software tools have been developed for conducting scientometric analysis. Scientometric software packages require data files that are imported from research databases such as the WoS and Scopus databases. Table 1 shows several scientometric software packages with their features. Our expectations from the use of scientometric software are for the software to plot a visual image that depicts the whole field of literature, facilitating the spotting of important publications and authors in the search area. Scientometric software packages listed in Table 1 are convenient ways to satisfy these requirements. For this particular analysis, CiteSpace has been chosen because of its providing various statistical outputs, and due to the ease in which it can be accessed and used.

Research Methodology

In this study, files that contained bibliometric data about image processing publications for civil engineering were retrieved from the Web of Science (WOS) database. Bibliometric data files were downloaded from the WOS database, which includes titles, authors, abstracts and reference lists of searched publications. By using CiteSpace software, scientometric analysis is made based on these bibliometric data. The main concern about the data files is they should cover the whole literature about image processing studies regarding construction building technology. To gather the bibliometric data files, in the Web of Science

Table 1. List of scientometric software programs

Software	Characteristics
VOSviewer	A Java-based software package that analyzes the literature and develops a visual network display of citations (van Eck and Waltman 2009)
CiteSpace	A Java-based software package that analyzes the literature and develops a visual network display of citations (Chen, 2006).
Bibexcel	Developed for managing scientometric data and visualizing maps. It can cooperate with other software packages such as Excel (Persson et al. 2009).
Gephi	A Java based open-source software package that analyzes the literature and develops a visual network display of citations (Bastian, M., Heymann, S., & Jacomy, M., 2009).

database, the advanced search option is selected. Then, for retrieving the publications, the query is inserted as "TS=("image processing" or "computer vision" or "image-processing" or "computer-vision") AND (SU="Construction & Building Technology" or WC="Engineering, Civil")". The result of the query appears after the search process is performed. Then, the bibliometric data files can be obtained by clicking the "export" button and selecting "other file formats." The file has to include a full record and cited references, and the file format is plain text format. The criterion of retrieving the bibliometric data has to cover the whole body of civil engineering and construction building technology research studies. This search returned 1,653 publications and 32,104 citations.

The downloaded files can be directly imported into CiteSpace software. Then, several types of networks are generated based on these files by using text-mining techniques. The network types are listed in Table 2. Each network type involves its own node and connection type combinations. There are three connection types designed in CiteSpace. The first one involves co-author type connections. In this connection type, depending on the node type, each node defines an author, institutionm or country, and each connection between nodes shows a co-authorship relationship between the authors, institutions, or countries. The second connection type involves co-occurrences where each connection between nodes shows co-occurrences of individuals depending on node types such as categories or keywords. The last connection type involves co-citation type connections which can be defined as each connection representing mutual citations publications. There are a total of nine different networks that can be defined in CiteSpace, as listed in Table 2. Besides, the user can trim the network by defining different time ranges by which to view the literature overview within different periods. In software terminology, these time periods are called *time slices* (Chen, 2006).

One of the important attributes of CiteSpace is its ability to eliminate infrequently cited publications. Table 3 lists selection criteria for the filtering process. The mutual principle of whole criteria is prominent publications are selected that was published in a certain time slice. The first row's criterion, named "Top N" rule, selects most N cited publications from each slice. The second row's criterion, named the "Top N%" rule, selects the top "N percent" of publications from each slice. There, three different statistical parameters are defined for threshold values that filter the publications by their citation counts. "Usage 180" is a newly introduced parameter that uses the download-count from the WOS database for each publication over the last 180 days. After choosing one of these options, the software eliminates some publications relative to their citation performance before visualizing the results, in order to clear away statistically insignificant items. In addition to the filtering process, the merging process unifies items belong to different period in order to build a network large enough to model the whole literature field.

Table 2. Network types in CiteSpace

Node Type	Connection	Description
Author	Co-authorship	Shows which authors collaborated with whom
Institution	Co-authorship	Shows co-authors' affiliations
Country	Co-authorship	Shows co-authors' countries
Term	Co-occurrence	Shows co-occurrences of different terms in mutual resources
Keyword	Co-occurrence	Shows co-occurrences of different keywords in mutual resources
Category	Co-occurrence	Shows co-occurrences of different WOS categories
Cited Reference	Co-citation	Shows citation relationships between publications in reference lists
Cited Author	Co-citation	Shows citation relationships between authors
Cited Journal	Co-citation	Shows citation relationships between journals

Co-Authorship Network

In most cases, several authors collaborate for publications. In the scientometric context, this collaborating can be tracked and viewed. As shown in Table 2, author, institution, and country network types use co-authorship-type connections that show collaborations between authors, institutions, and countries, respectively. In the networks generated by CiteSpace, connection thickness is related in terms of collaborating frequencies. Citation statistics affects node visibility and thickness. Figure 1 shows the author network and collaborations between the authors.

Figure 1 shows the most-collaborated-with authors and their collaborators. It can be seen that some collaborators work as groups. However, some other authors were spread around the graph. The main reason for that was that these authors' collaborators did not get enough citations to be seen in the graph. In the authorship network, there are a total of 249 authors who contributed to the field. From the selection criteria, the "top 30%" of most-cited authors are selected from each time slice. The reason for this is that, as compared to the number of references, there are fewer authors and connections available for building a network. After pruning and merging processes, out of 249 authors, only 5% of authors ended up being plotted and labeled. Table 4 shows the top ten productive authors, along with their affiliations and research areas. *Half-life* is a required period for decreasing annual citation quantity to its *half-value*. Articles that have longer citation half-lives, being cited, would be continued for a longer period of time than others would. These kinds of articles may be considered as being *classical publications*. By contrast, *transient publications* get a high citation count in a short amount of time, yet the number of citations they receive decays after a while (Price 1965).

Table 3. Selection options

Selection Criterion	Description
Top N	Selects the "N" most cited items from within a given slice
Top N%	Selects the top "N" percent of items most frequently cited from within a given slice
Thresholds Values	Citation-counts (c), Co-citation counts (cc) and Co-citation coefficients (ccv)
Usage 180	The items most frequently downloaded in last 180 days.

Figure 1. Co-authorship network for image processing studies

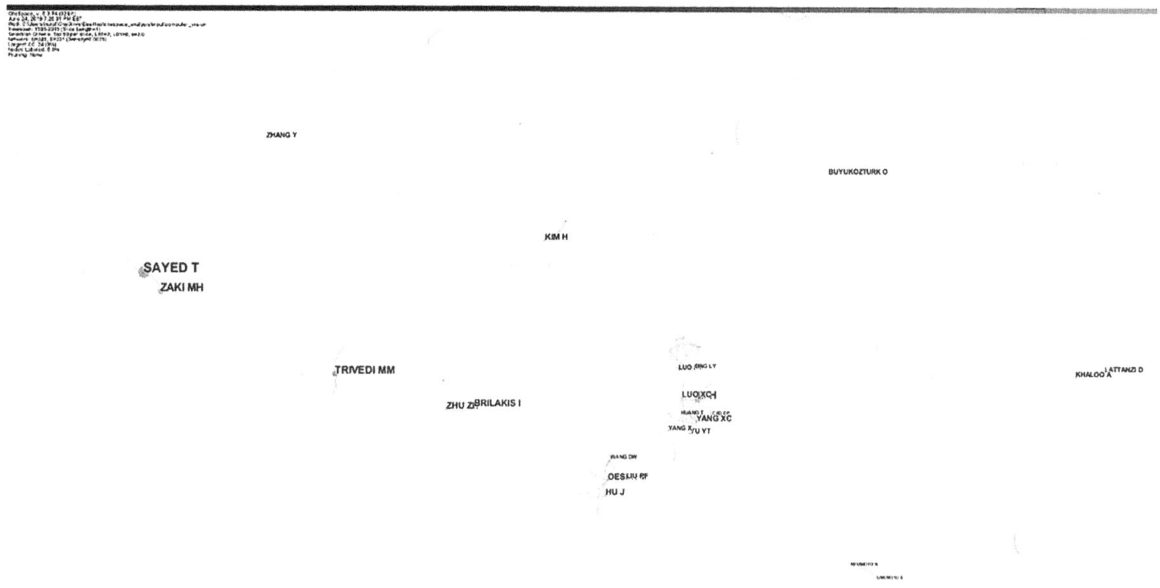

Table 4. The most-collaborated-with authors in field.

Co-authorship	Half-Life	Author	Affiliation	Area of Expertise
26	3	Tarek Sayed	University of British Columbia	Transportation
15	2	Mohamed H. Zaki	University of British Columbia	Transportation
14	1	Heng Li	Hong Kong Polytechnic University	Construction Management
13	3	Mohan Trivedi	University of California	Intelligent Vehicles
9	7	Zhenhua Zhu	Concordia University	Construction Management
8	5	Ioannis Brilakis	University of Cambridge	IT in Construction
7	1	Marcus Oeser	RWTH Aachen University	Pavement and Traffic Engineering
7	3	Hongjo Kim	Yonsei University	Construction Management
7	1	Jing Hu	Southeastern University	Material
6	3	Necati Catbas	University of Central Florida	Structural Identification

Countries and Institutes

As is seen in Table 2, CiteSpace can display the contributions of countries and institutes. Similar to authorship networks, each node is joined to other nodes with co-authorship type connections. For instance, this occurs if there is frequent collaboration by authors from two different countries. This collaboration frequency results in connection thickness between these nodes representing countries, citing statistics effect node sizes. From the bibliometric data, 42 countries contributed to the field. The United States is the most contributing country, with 404 articles. China is second, with 382 articles. The third one is Canada, with 109 articles. In the case of institutions, the most active institution is Southeast University, from China, with 28 articles. The second one is the University of British Columbia, from Canada, with 26 articles. The contribution of institutions and countries can be seen from Figure 2.

Co-Citation Network and Clustering

Besides reference lists, keywords and titles are another such viewpoint for visualizing publications' research areas. In the case of one publication being cited by different publications, the citing publications are called *co-cited publications*. The main point is that the citation statistics define co-citation strength. To view cited references with co-citation connection networks, the "cited reference" selection from Table 2 is chosen. In the case of choosing the "cited reference" as node type in Table 2's eighth row, CiteSpace will display frequently cited references and publications, and will connect them with co-citation relationships. The main drawback of the datafiles is there is no information about cited references' research areas. To overcome this drawback, CiteSpace can build clusters within the network. To infer referenced publications' research areas, CiteSpace utilizes citing publications' research areas that can be acquired directly from bibliometric data files. Figure 2 describes a citation scheme of a publication by several publications that have bibliometric data by using this concept, the cited reference's research areas can be inferred. CiteSpace clusters publications by using nearness factors and, after that, clusters the publications according to usage of the most frequently used phrases included in the cluster.

Figure 2. Contributions of countries and institutions.

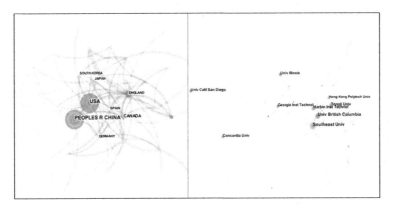

Figure 3. Inferring the topic of a referenced publication

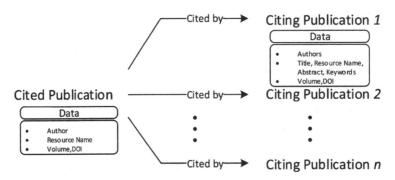

Figure 4. The network of clustered publications

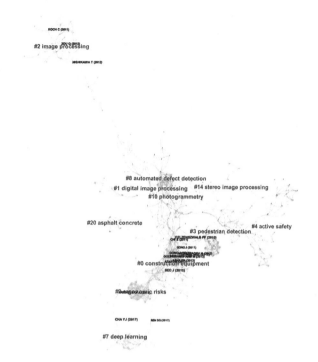

From the alternative options in Table 3, "top N" criterion is chosen. The top 50 most frequently cited publications are selected in order for the software to give such prominent publications that considers priorities in the network-building process. Based on these principles, the publication network is produced. It is shown in Figure 4. Every node represents an individual publication, and the node size depends on the citation count.

As shown in Figure 4, the software builds groups based on publications' having inter-relationships with co-citation relationships. It does so in order that publications can easily be clustered, After that, the clusters gain labels from the most frequently appeared phrases in publications within clusters. There are three different options—abstract terms, indexing terms, and title terms—to be scanned in order to tag the clusters. Figure 4 shows clusters that are labeled with index terms. Sometimes the most frequently used keywords do not represent research backgrounds. For instance, the label of the first cluster—"#construction equipment"—refers to a quite generic phrase rather than to a research field or methodology. The second case label of the cluster—"#1 digital image processing"—indicates the methodology. In fact, the whole obtained methodology of the publications is related to image processing. Cluster numbering is sorted by ascending order, with lower numbers indicating the most important, highest-ranking clusters.

The Network of Co-Occurring Keywords, and Usage Bursts

Co-occurring keywords are keywords that mutually used by different publications. Similar to co-author and co-citation relationships, co-usage frequencies play an important role in determining relationship strength between keywords so that it affects line thickness between nodes in the network plot. By using statistical outputs, users can track trends by monitoring keyword usage. In the results of co-occurring keyword analysis, the overall analysis will firstly be divided into three decades, for presentation of the data. Then, the overall analysis is given with burst values. Bursts are suddenly increasing frequencies of keyword usage. Table 4 is given for determining the changing popularity of keywords. The most frequently used keywords in different time ranges are listed.

As shown in Table 5, generic terms more frequently appeared. By investigating the variation of rankings, it can be inferred that the number of studies about concrete has increased in search. The main reason for that is that the keyword "concrete" has undergone a trend towards its increasingly being used over the last three decades. Besides using frequencies, bursts are important parameters for investigating the variation of the usage frequency of keywords. Table 6 shows keywords that have had top usage bursts in the literature field as a whole.

Table 5. Keyword frequencies for different decades

1991-2001		2001-2011		2011-2019	
Frequency	**Keyword**	**Frequency**	**Keyword**	**Frequency**	**Keyword**
25	image processing	83	image processing	191	image processing
5	system	44	computer vision	150	computer vision
4	transport	26	system	93	system
4	computer vision	17	model	85	model
2	Gp	14	algorithm	65	concrete
2	automation	14	digital image processing	63	tracking
2	segmentation	13	segmentation	53	digital image processing
2	modelling	13	image analysis	44	classification
2	vision	12	classification	43	performance
2	image analysis	12	image	41	image analysis

Table 6. Top keywords with top burst values

Frequency	**Burst**	**Keyword**
278	16.66	image processing
32	5.63	crack detection
59	5.26	digital image processing
17	5.18	pattern recognition
35	5.08	recognition
12	4.83	damage
12	4.83	surface
24	4.5	mechanical property
39	4.18	segmentation
29	3.99	simulation

Figure 5. Co-occurrence of keywords in the research field

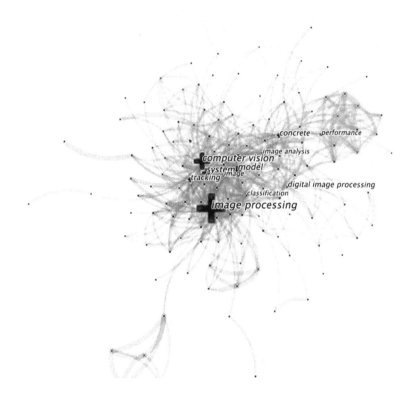

As we focus on research topics, two keywords draw attention to display research trends. These are "pattern recognition" and the other one is "crack detection." The other keywords represent more generic definitions. For instance, the term "pattern recognition" is a subfield of machine learning techniques, and "crack detection" addresses a specific research area. Figure 5 shows the co-occurrence network of keywords drawn from image processing studies in the field of civil engineering.

The Network of Keywords from the Most Downloaded Publications

The Web of Science database has an important feature for acquiring download counts. The database logs the download statistics and inserts them into the publications' data. CiteSpace accesses the download count from bibliometric data files and selects prominent studies, using these download counts. Depending on the type of network, 'unattractive' (uncited) publications or keywords that belong to such publications will not appear in the network plots.

Citation Bursts and Half-lives

Similar to usage bursts, citation bursts are rapid increases in the gaining of citations. Two main occasions cause citation bursts. The first kind of cause is an important discovery that leads to a number of searches that delve into outcomes of the discovery. The second kind of cause is the emergence of newly-appearing opportunities and necessities affecting various related research studies. Kleinberg's study (2002) proposed a detection algorithm for detecting bursts in email streams. Citation half-life is

Table 7. Top cited publications with other parameters

Resource	Keyword	Citation	Burst	Half-Life (Years)
(Cha et al., 2017)	Identification; Architectures; Segmentation; Algorithm; Surfaces; Images	25		1
(Brilakis, Park, & Jog, 2011)	Artificial Intelligence; Automation Imaging Techniques; Automatic Identification Systems; Models; Information Technology	22		6
(Seo et al., 2015)	Construction Safety and Health; Computer Vision; Monitoring	22	6.96	3
(Felzenszwalb, Girshick, McAllester, & Ramanan, 2009)	Object Recognition; Deformable Models; Pictorial Structures; Discriminative Training; Latent SVM.	21		5
(Park & Brilakis, 2012)	Pattern Recognition; Construction Worker; Detection Tracking; Remote Sensing; Image Processing; Computer Vision	21		6
(Ismail, Sayed, & Saunier, 2013)	Camera Calibration; Traffic Monitoring; Traffic Data Collection; Video Analysis; Road User Tracking	20	4.76	2
(Zou, Cao, Li, Mao, & Wang, 2012)	Crack Detection; Edge Detection; Edge Grouping; Tensor Voting; Shadow Removal	20	4.73	5
(Rezazadeh Azar, Dickinson, & McCabe, 2012)	Tracking; Frames; Computing In Civil Engineering; Vehicle Loads; Algorithms, Trucks; Load Factors; Earthmoving	19	3.93	4
(Nishikawa, Yoshida, Sugiyama, & Fujino, 2012)	Genetic Algorithms; Optimization; Network	19	6.74	6
(Chi & Caldas, 2011)	Visual Surveillance; Project-Management; Tracking; System; Productivity; Algorithms	18		6

another parameter that is a period for decreasing annual citation count to its half-value. Publications with a longer half-life are cited longer than other publications with shorter half-lives. Classical or benchmark studies are prone to have longer half-life durations. In getting citation statistics, transient publications are short-lived publications that are frequently cited for a short period of time, and then abruptly decay, start to be cited less frequent (Price, 1965). Table 7 displays the most cited publications with citation burst and half-life values.

CONCLUSION

In the area of construction-building technology, image processing studies have focused attention on "concrete search" and "crack detection." The main reason is the priority that investigators necessarily place on monitoring and measuring structural deformations. Another highlighted output is a cluster Figure 4 termed "deep learning field." From Table 7, the study of Cha, Choi, and Buyukozturk (2017) is related to the "convolutional neural network algorithm." Despite its research focus being quite new to the field, their study has gained the best citation record in its overall area of research. From this analysis, it can be concluded that image processing applications increasingly utilize machine learning techniques.

REFERENCES

Bas, E., Tekalp, A. M., & Salman, F. S. (2007). Automatic vehicle counting from video for traffic flow analysis. In *2007 IEEE Intelligent Vehicles Symposium* (pp. 392–397). 10.1109/IVS.2007.4290146

Bastian, M., Heymann, S., & Jacomy, M. (2009). Gephi: an open source software for exploring and manipulating networks. Icwsm, *8*(2009), 361–362. doi:10.1016/j.aei.2011.01.003

Brilakis, I., Park, M.-W., & Jog, G. (2011). Automated vision tracking of project related entities. *Advanced Engineering Informatics*, *25*(4), 713–724. doi:10.1016/j.aei.2011.01.003

Burton, R. E., & Kebler, R. W. (1960). The "half-life" of some scientific and technical literatures. *American Documentation*, *11*(1), 18–22.

Cha, Y.-J., Choi, W., & Buyukozturk, O. (2017). Deep learning-based crack damage detection using convolutional neural networks. *Computer-Aided Civil and Infrastructure Engineering*, *32*(5), 361–378. doi:10.1111/mice.12263

Chen, C. (2006). CiteSpace II: Detecting and visualizing emerging trends and transient patterns in scientific literature. *Journal of the American Society for Information Science and Technology*, *57*(3), 359–377. doi:10.1002/asi.20317

Chi, S., & Caldas, C. H. (2011). Automated object identification using optical video cameras on construction sites. *Computer-Aided Civil and Infrastructure Engineering*, *26*(5), 368–380.

Ding, L., Fang, W., Luo, H., Love, P. E. D., Zhong, B., & Ouyang, X. (2018). A deep hybrid learning model to detect unsafe behavior: Integrating convolution neural networks and long short-term memory. *Automation in Construction*, *86*, 118–124.

Fang, W., Ding, L., Luo, H., & Love, P. E. D. (2018). Falls from heights: A computer vision-based approach for safety harness detection. *Automation in Construction*, *91*, 53–61. doi:10.1016/j.autcon.2018.02.018

Felzenszwalb, P. F., Girshick, R. B., McAllester, D., & Ramanan, D. (2009). Object detection with discriminatively trained part-based models. *IEEE Transactions on Pattern Analysis and Machine Intelligence*, *32*(9), 1627–1645. doi:10.1109/TPAMI.2009.167 PMID:20634557

Feng, D., & Feng, M. Q. (2018). Computer vision for SHM of civil infrastructure: From dynamic response measurement to damage detection – A review. *Engineering Structures*, *156*, 105–117. doi:10.1016/j.engstruct.2017.11.018

Huang, H., Li, Q., & Zhang, D. (2018). Deep learning-based image recognition for crack and leakage defects of metro shield tunnel. *Tunnelling and Underground Space Technology*, *77*, 166–176.

Ismail, K., Sayed, T., & Saunier, N. (2013). A methodology for precise camera calibration for data collection applications in urban traffic scenes. *Canadian Journal of Civil Engineering*, *40*(1), 57–67. doi:10.1139/cjce-2011-0456

Jin, R., Zou, P. X. W., Piroozfar, P., Wood, H., Yang, Y., Yan, L., & Han, Y. (2019). A science mapping approach-based review of construction safety research. *Safety Science*, *113*, 285–297. doi:10.1016/j.ssci.2018.12.006

Kleinberg, J. (2002). Bursty and hierarchical structure in streams. In *Proceedings of the eighth ACM SIGKDD international conference on Knowledge discovery and data mining* (pp. 91–101). ACM. 10.1145/775047.775061

LeCun, Y., Jackel, L. D., Boser, B., Denker, J. S., Graf, H. P., Guyon, I. ... Hubbard, W. (1989). Handwritten digit recognition: applications of neural net chips and automatic learning. In F. Fogelman, J. Herault, & Y. Burnod (Eds.), Neurocomputing, Algorithms, Architectures and Applications. Les Arcs, France: Springer.

Merton, R. K. (1968). The Matthew effect in science. *Science, 159*(3810), 56–63. doi:10.1126cience.159.3810.56

Murphy, K., Torralba, A., Eaton, D., & Freeman, W. (2006). Object detection and localization using local and global features. In *Toward Category-Level Object Recognition* (pp. 382–400). Springer. doi:10.1007/11957959_20

Nishikawa, T., Yoshida, J., Sugiyama, T., & Fujino, Y. (2012). Concrete crack detection by multiple sequential image filtering. *Computer-Aided Civil and Infrastructure Engineering, 27*(1), 29–47.

Park, M.-W., & Brilakis, I. (2012). Construction worker detection in video frames for initializing vision trackers. *Automation in Construction, 28*, 15–25. doi:10.1016/j.autcon.2012.06.001

Persson, O. (1994). The intellectual base and research fronts of JASIS 1986–1990. *Journal of the American Society for Information Science, 45*(1), 31–38. doi:10.1002/(SICI)1097-4571(199401)45:1<31::AID-ASI4>3.0.CO;2-G

Price, D. J. D. S. (1965). Networks of scientific papers. *Science, 149*(3683), 510–515. doi:10.1126cience.149.3683.510 PMID:14325149

Rezazadeh Azar, E., Dickinson, S., & McCabe, B. (2012). Server-customer interaction tracker: Computer vision–based system to estimate dirt-loading cycles. *Journal of Construction Engineering and Management, 139*(7), 785–794. doi:10.1061/(ASCE)CO.1943-7862.0000652

Seo, J., Han, S., Lee, S., & Kim, H. (2015). Computer vision techniques for construction safety and health monitoring. *Advanced Engineering Informatics, 29*(2), 239–251. doi:10.1016/j.aei.2015.02.001

Stallkamp, J., Schlipsing, M., Salmen, J., & Igel, C. (2012). Man vs. computer: Benchmarking machine learning algorithms for traffic sign recognition. *Neural Networks, 32*, 323–332. PMID:22394690

Van Eck, N., & Waltman, L. (2009). Software survey: VOSviewer, a computer program for bibliometric mapping. Scientometrics, 84(2), 523-538. PMID:22394690

Yamaguchi, T., & Hashimoto, S. (2010). Fast crack detection method for large-size concrete surface images using percolation-based image processing. *Machine Vision and Applications, 21*(5), 797–809. doi:10.100700138-009-0189-8

Yilmaz, A., Javed, O., & Shah, M. (2006). Object tracking: A survey. *ACM Computing Surveys, 38*(4), 13, es. doi:10.1145/1177352.1177355

Zhao, X. (2017). A scientometric review of global BIM research: Analysis and visualization. *Automation in Construction*, *80*, 37–47. doi:10.1016/j.autcon.2017.04.002

Zhao, X., Zuo, J., Wu, G., & Huang, C. (2019). A bibliometric review of green building research 2000–2016. *Architectural Science Review*, *62*(1), 74–88. doi:10.1080/00038628.2018.1485548

Zou, Q., Cao, Y., Li, Q., Mao, Q., & Wang, S. (2012). CrackTree: Automatic crack detection from pavement images. *Pattern Recognition Letters*, *33*(3), 227–238. doi:10.1016/j.patrec.2011.11.004

Chapter 7

High Performance Concrete (HPC) Compressive Strength Prediction With Advanced Machine Learning Methods:
Combinations of Machine Learning Algorithms With Bagging, Rotation Forest, and Additive Regression

Melda Yucel

Istanbul University-Cerrahpaşa, Turkey

Ersin Namlı

ⓘ https://orcid.org/0000-0001-5980-9152

Istanbul University-Cerrahpasa, Turkey

ABSTRACT

In this chapter, the authors realized prediction applications of concrete compressive strength values via generation of various hybrid models, which are based on decision trees as main a prediction method. This was completed by using different artificial intelligence and machine learning techniques. In respect to this aim, the authors presented a literature review. The authors explained the machine learning methods that they used as well as with their developments and structural features. Next, the authors performed various applications to predict concrete compressive strength. Then, the feature selection was applied to a prediction model in order to determine parameters that were primarily important for the compressive strength prediction model. The authors evaluated the success of both models with respect to correctness and precision prediction of values with different error metrics and calculations.

DOI: 10.4018/978-1-7998-0301-0.ch007

INTRODUCTION AND LITERATURE REVIEW

Concrete, which is a structural material frequently used in civil engineering, is composed from water, cement, and various mineral or chemical materials. Engineers use this material to connect structural components to each other. For instance, in a beam and a column, and each component in place during the construction of structures. In this regard, strength and quality of concrete is important for ensuring resistance and sustainability of structures to various factors.

This case causes a considerable problem related to determining concrete strength with precision based on the features of the materials themselves, together with their mixture rates, acting to quality of concrete (Küçük, 2000). However, the process of calculating the compressive strength of concrete is a time-intensive and expensive operation. Altogether, calculating and observing concrete strength takes 28 days (Turkish Standards Institute, 2000). On the other hand, various environmental effects, such as air and gas rates, temperature, humidity, and the properties of the sample tools or techniques used for measurement contribute to the strength of concrete. For this reason, the strength values could deviate from the real results and so, show change.

Nowadays, various advanced methods, which can be useful alternatives to traditional laboratory analyses and test methods, may be utilized in to effectively and rapidly calculate and determine numerical values for structural materials. Considering that these methods are frequently machine learning and artificial intelligence prediction techniques, serve the benefit of preventing loss of time and effort, with cost residuals.

In the first periods when researchers operated machine learning and statistical applications, nondestructive tests were greatly preferred by many researchers, and numerous studies were realized with these techniques. One example of this technique is the ultrasonic pulse velocity method, which may be beneficial to researchers because it uses its own values to predict compressive strength. With this aim, the authors carried out various studies in the literature review that follows:

Hoła and Schabowicz (2005) performed artificial neural networks (ANNs) implementation using the features obtained by means of the nondestructive testing as input variables for prediction of concrete compressive strengths. Next, Kewalramani and Gupta (2006) used multiple regression analysis and ANNs to predict the concrete compression strength of samples belonging to concrete mixtures that have two different sizes and forms in longtime, by used ultrasonic pulse velocities and weight values. In another study, the authors applied multilayer neural network, which is one kind of ANNs technique, to estimate compressive strength of concrete by values of ultrasonic pulse velocity besides concrete mixtures properties (Trtnik, Kavčič & Turk, 2009).

On the other hand, Bilgehan and Turgut (2010) presented an application regard to performing predictions for the compressive strength of concrete based on the ultrasonic pulse velocity method. In this study, the authors generated a great deal of data through the usage of different concrete parts of a variety of ages and concrete rates. The authors did this by taking samples from concrete structures with the aim of training of ANNs. Nevertheless, Atici (2011) performed a study that assessed compressive strength in different curing times of concrete mixtures containing several amounts blast furnace cinder and fly ash by depend on qualities and values. The authors obtained this information via rebound number from a non-destructive test, ultrasonic pulse velocity of additive agents forecasted via ANNs, and multiple regression analysis methods.

In addition to these studies, intelligent systems working with various algorithms have started to come into prominence and use thanks to advancing technology and computational methods. One of these is fuzzy logic methods, and some studies are as follows:

Topcu and Sarıdemir (2008) used fuzzy logic methods and ANNs in forecasting the daily period compressive strength of three different types of concrete with fly ashes inside and high and low-level lime. Next, Cheng, Chou, Roy and Wu (2012) developed a prediction tool by combining three different methods, including fuzzy logic, support vector machine, and a genetic algorithm under the name of "Evolutionary Fuzzy Support Vector Machine Inference Model for Time Series Data" to foresee of the value of compressive strength for high performance concretes (HPC). by depend on an opinion that determining of these concretes' behavior is more difficult opposite to traditional concretes.

Nevertheless, from other respects, methods like artificial neural networks, various kinds of regression analysis, and decision tree models are some of the principal techniques that have the most usage and preferability with regards to artificial intelligence applications, such as prediction, forecasting, recognition, classification by researchers and performers of machine learning. In this direction, Naderpour, Kheyroddin and Amiri (2010) carried out a study that dealt with ANNs. In this study, the authors determined the compressive strengths of concretes covered with fiber-reinforced polymer (FRP) by using the properties of FRPs and concrete sample. Also, the authors composed five separate models to determine HPC compression strength with usage of k-nearest neighbor machine learning method. This was optimized using a type of differential evolution, together with generalized regression neural networks and stepwise regressions (Ahmadi-Nedushan, 2012). An application in which recycled aggregates were used, was actualized by Duan, Kou, and Poon in 2013. In this study, an ANN model operating via back propagation algorithm was established with 14 inputs for estimation of the compressive strength for concrete varieties with recycled aggregate, which were obtained from 168 different mixture data taking place in literature.

Authors executed a prediction application made in 2013 for high performance concrete, and they applied many advanced machine learning techniques and algorithms to determine compressive strength. These techniques are ensemble-structured ANNs, which consisted of using methods such as bagging and gradient boosting. Moreover, authors implemented a wavelet transform operation combined with prediction models for the purpose of increasing of precision of these models (Erdal, Karakurt & Namli, 2013). Furthermore, Akande, Owolabi, Twaha and Olatunji (2014) investigated the support vector machines by comparing them with ANNs from respects that observing of estimation success for concrete compressive strength. Chou, Tsai, Pham and Lu (2014) presented an extent study regarding the compressive strength prediction issue of high-performance concrete. from the authors used several enhanced machine learning algorithms, such as multilayer perceptron (a kind of artificial neural network), support vector machines, classification and regression tree method (CART), and linear regression. In addition to these applications, the authors applied ensemble models for prediction of this mechanic property.

Additionally, another study in which Deshpande, Londhe, and Kulkarni (2014) applied three various methods is the prediction of compression strength 28-day values of concrete consisting of recycled aggregates which was used as additives material. With this aim, specific input parameters, such as different aggregate types, water and cement amounts, and the other material rates were determined to realize the prediction process using two different methods besides ANNs; model tree and non-linear regression.

Nikoo, Zarfam, and Sayahpour (2015) used a different type of ANNs. The authors ran these neural networks, consisting of self-organizing feature maps, to estimate the compressive strength of 173 concrete mixtures. Additionally, the authors turned ANNs models with self-organizing maps to optimized-case design through use of genetic algorithm. Furthermore, in 2016, Chopra, Sharma and Kumar performed a study via ANNs and genetic programming methods by using the data obtained in different curing times consisting from 28, 56, and 91 days. This study served the purpose of developing models for determining concrete compressive strength values.

Furthermore, Chithra, Kumar, Chinnaraju, and Ashmita (2016) used multiple regression analysis with ANNs to predict the compression strength of numerous high-strength concrete samples, including particular rates of different additions materials, such as nano silica and copper slag. In addition, these predictions were directed to curing times consisting of six different days, like 1, 3, 7, and so. Additionally, Behnood, V. Behnood, Gharehveran, and Alyamac (2017) utilized an algorithm that is called M5P model tree to estimate compressive strength for normal and high strength concretes (HPC).

In another study conducted by Akpinar and Khashman in 2017, the authors used ANNs to carry out classification, which is a type of machine learning applications. In this direction, separating the mixture of concretes according to different strength level became possible through an improved model.

ANN models, which possesses a feed-forward structure working via a back propagation algorithm were developed to estimate the compressive strength and concrete tensile strength that have steel-slag aggregate. For modeling of ANN prediction structure in this study, the authors determined various input parameters, including different curing times, water/cement rate, and value that replacement of a type of blast furnace slag with certain amounts of granite (Awoyera, 2018).

Behnood and Golafshani (2018) applied a two-stage work that grey wolf optimization method was benefited for construction of the most proper ANNs model to predict the compressive strength of concrete mixtures that a part of Portland cement was constituted by alter with silica fume material, which comes in use for nature from many respects. However, Bui, Nguyen, Chou, Nguyen-Xuan and Ngo (2018) developed an expert system consisting of combinations of ANNs and metaheuristic firefly algorithm (FA) to estimate tensile and compression strengths of high-performance concretes. Yu, W. Li, J. Li and Nguyen (2018) also developed a prediction model with support vector machines for high performance concretes. In addition to this, the authors optimized the operated technique parameters, which is important on increasing of model performance and success, via cat swarm optimization, which is one of the metaheuristic algorithms.

Also, Yaseen et al. (2018) used extreme learning machine technique besides multivariate adaptive regression spline, M5 tree model, and support vector regression, for determination of compressive strength of concrete adding foamed. Similarly, Al-Shamiri, Kim, Yuan and Yoon (2019) realized the prediction of concrete compressive strength for concrete with high-strength quality by using a newer method compared to other methods, called extreme learning machine. For this purpose, mentioned method is benefited respect to training of a model, which was developed with the usage of many experiment data, by artificial neural networks.

In this study, the authors improved hybrid prediction models with one of the decision tree models, which consisted of Random Forest and ANNs as base learners. These were combined with different advanced machine learning techniques, namely Additive Regression, Rotation Forest, and Bagging, which are ensemble learning algorithms, with the aim of determining the HPC compressive strength for concretes. In addition, following this process, the authors performed feature selection application to determine mainly effective parameters on a model developed for prediction of values of concrete compressive strength.

Also, the authors calculated various error metrics, composing mean absolute error (MAE), root mean squared error (RMSE), relative absolute error (RAE) and root relative squared error (RRSE), and coefficient of correlation (R) to evaluate prediction success belonging to first main models and next models, which are performed for the optimization of features.

MATERIAL AND METHODS

Base Algorithms

Rotation Forest

Rotation Forest (RotF) is a method that applies a principle that is based on generating ensemble classifiers through decision trees, which are trained separately (Kuncheva & Rodríguez, 2007).

The Rotation Forest algorithm operates in a similar way to random forest in that the classification (or regression) process is realized by using more one tree. In this direction, Bootstrap algorithm, which is one of ensemble-based classification methods, is used as a base classifier method with the aim of training each of the data groups. On the other hand, in this method, analysis of principal component (PCA) determines attributes belonging to Bootstrap data groups, which is generated in order to train each decision tree. In the end of the process, the user determines the main/principal components belonging to K attribute parts, and the user generates a new attribute set by combining each of these components obtained in each attribute group for the data groups (Kılınç, Borandağ, Yücalar, Özçift & Bozyiğit, 2015; Kuncheva & Rodríguez, 2007). Ensuring difference is made possible by performing the process separately for all decision trees.

In this paper, this new attribute set is constructed with nonlinearly, arranged from the relationships between attributes of PCA technique, which was proposed by Karl Pearson in 1901 (Rokach & Maimon, 2015).

Figure 1 shows the pseudo code of rotation forest algorithm, which is explained by the operating logic of this method. In Figure 1, as expressed that in previous, initially, attributes are split for base classifiers, namely decision trees of sub data groups ($S_{i,j}$) for training. This division is made randomly with M number, which is the same in each attribute subset ($F_{i,j}$). Also, sample quantity within each data groups is composed from three-quarters of $S_{i,j}$, and this portion of samples is called as *'Bootstrap sample'* ($S_{i,j}'$). Following this step, PCA analysis is applied to all $S_{i,j}'$ samples, and the main classifier is built after in various steps.

Bagging

Bagging (bootstrap aggregating) is one of the best-known ensemble-based machine learning algorithms. In order to form an aggregated predictor, bagging method produces multiple predictor variants (Breiman, 1996). The bagging algorithm improves the prediction accuracy of model by operating a training data set before base learner's prediction process (Dietterich, 2000).

Bagging algorithm makes the selection of subsamples without any replacement for training data. After that, the subsample is used instead of original data set for fitting base classifier or predictor, which finally builds the model for next iteration. Nevertheless, bagging ensemble method, which is proposed for regression models, was clarified via several equations that follow (Aydogmus et al., (2015); Bühlmann, & Yu, 2002).

Figure 1. Rotation forest pseudo code
(Rokach & Maimon, 2015)

Rotation Forest

Require: I (a base inducer), S (the original training set), T (number of iterations), K (number of subsets),

1: **for** $i = 1$ to T **do**
2:   Split the feature set into K subsets: $F_{i,j}$ (for $j=1..K$)
3:   **for** $j = 1$ to K **do**
4:     Let $S_{i,j}$ be the dataset S for the features in $F_{i,j}$
5:     Eliminate from $S_{i,j}$ a random subset of classes
6:     Select a bootstrap sample from $S_{i,j}$ of size 75% of the number of objects in $S_{i,j}$. Denote the new set by $S'_{i,j}$
7:     Apply PCA on $S'_{i,j}$ to obtain the coefficients in a matrix $C_{i,j}$
8:   **end for**
9:   Arrange the $C_{i,j}$, for $j = 1$ to K in a rotation matrix R_i as in the equation:

$$R_i = \begin{bmatrix} a_{i,1}^{(1)}, a_{i,1}^{(2)}, \dots, a_{i,1}^{(M_1)} & [0] & \dots & [0] \\ [0] & a_{i,2}^{(1)}, a_{i,2}^{(2)}, \dots, a_{i,2}^{(M_2)} \dots & & [0] \\ \dots & \dots & \dots & \dots \\ [0] & [0] & \dots a_{i,k}^{(1)}, a_{i,k}^{(2)}, \dots, a_{i,k}^{(Mk)} \end{bmatrix}$$

10:   Construct R_i^a by rearranging the columns of R_i so as to match the order of features in F
11: **end for**
12: Build classifier M_i using (SR_i^a, X) as the training set

In these equations, D is a training set that includes data couples composed from (X_i, Y_i) as $i=1,2,\dots,n$. X_i components express the multi-dimensional predictor variables, and Y_i is a realization of a variable with a real value. A predictor variable is given in Equation (1) and denoted by Equation (2).

$$E(Y|X=x) = f(x). \tag{1}$$

$$C_n(x) = h_n(D_1,\dots,D_n)(x) \tag{2}$$

Figure 2. Bagging algorithm structure
(Erdal et al., 2013)

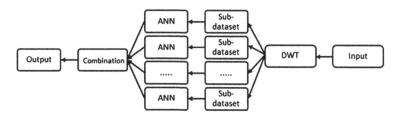

Theoretically, bagging is described in this way: bagging starts with the bootstrapped sample construction process that follows:

$$D_i^* = \left(Y_i^*, X_i^* \right) \tag{3}$$

In accordance with the experimental distribution of the pairs $D_i=(X_i, Y_i)$, where ($i=1,2,...,n$). After that, prediction of the bootstrapped predictors, defined via Equation (4), is carried out by using the principle of plug-in:

$$C_n^* \left(x \right) = h_n \left(D_i^*, ..., D_n^* \right) \left(x \right) \tag{4}$$

Finally, the bagged predictor can be obtained with Equation (5).

$$C_{n;B} \left(x \right) = E^* \left[, D_n^* \left(x \right) \right] \tag{5}$$

Additive Regression

On one hand, regression is a method that is used to observe the effect of one or more independent variables on a dependent variable and determine the statistical relationship between these variables. On the other hand, Additive Regression (*AddReg*)is an amplified version of regression method via any algorithm. In this regard, besides that basically regression analysis is performed, this analysis is carried out by applying a learning algorithm to each data set.

Generally, this method expresses that the way of producing any prediction is by collocating the contributions obtained through all off the other separate models. While most of learning algorithms cannot generate the base models on their own, they try to generate a main model group, which optimizes predicted performance according to specific standard by completing each other together with a regression model (Witten, Hall, & Frank, 2011).

Learner Algorithms

Artificial Neural Networks

The method of ANN is an algorithm inspired by the human nervous system. ANNs are black box operations that are based on nonlinear basis functions (Malinov, Sha, & McKeown, 2001). ANN algorithms have wide application areas due to the ability of solving highly non-linear complex problems (Seyhan, Tayfur, Karakurt, & Tanoglu, 2005). ANNs models have ability, which is made possible that be able to learn the connection of input parameters with the output or target parameter because they benefit from mathematical training processes (Dahou, Sbartaï, Castel, & Ghomari, 2009).

Back-propagation is explained as a technique that calculates an error between actual values and predicted values and propagates the error information back through the network to each node in each layer.

Error back-propagation method includes two significant parameters for learning the target variable or variables. These are composed of learning rate and momentum. Learning rate expresses the quantity of comprehension by network – in other words changing or updating of weight values according to observed errors. Also, momentum value means that weights wish to perform gradient in a direction, where weight values are changed.

Conventional backpropagation is formulated in the equation that follows (Erdal et al., 2013). Output value, which belongs to any l^{th} neuron exists in the nth layer and can be calculated with Equation (6).

$$y_l^n \left(t \right) = \phi \left[\sum_{j=1}^{p} w_{lj}^n \left(t \right) y_j^{n-1} \left(t \right) + \Psi_l^n \right] \tag{6}$$

In Equation (6), ϕ.is the activation function used for the ultimate summation obtained with weight values for inputs. Also, w_{lj}^n indicates the weight values for connections between neurons in layers, y_j^{n-1} is the input value for any j neuron within the previous layer, t is value of iteration or time indicator, and $\Psi_l^n = w_{l_0}^n \left(t \right)$ defines the bias value for related l neuron. Furthermore, value of synaptic connection weight ($w_{ji}^n \left(t \right)$ can be calculated via Equation (7), where $\Delta w_{ji}^n \left(t \right)$ expresses the differentiation of weight value compared to the previous time.

$$w_{ji}^n \left(t+1 \right) = w_{ji}^n \left(t \right) + \Delta w_{ji}^n \left(t \right) \tag{7}$$

In Figure 3, an ANN structure generates from three layers, which correspond to the input layer, the hidden layer, and the output layer respectively, is shown representatively. Hidden layers can become as more from indicating in this presentation. In this Figure, x_m is expressed the data given to each ith input nodes by i=1, 2, …, m.

Random Forest

On one hand, each node within a forest structure is parted according to the best split among all model parameters. On the other hand, each node existing in a random forest is split using the best predictor among a subset of predictor variables randomly chosen at that node. At the same time, RF also is expressed as a different model of Bagging ensemble algorithm. The intercorrelation of each independent tree classifiers in the forest ensures that researchers find out the error rate of the random forest base learner. Thus, accuracy for performance belonging to the RF model arises by based on each tree learner (Khoshgoftaar, Golawala, & Van Hulse, 2007; Ozturk, Namli, & Erdal, 2016).

In regression problems, RF is defined as non-parametric regression approximation. This process forms of a set of M trees (Adusumilli, Bhatt, Wang, Bhattacharya, & Devabhaktuni, 2013).

The following expression as p-dimension input vector, which builds a forest, is $\{T_1(X), T_2(X), …, T_M(X)\}$ where, $(X_1, X_2, …, X_p\}$.

The authors present M outputs related to each tree produced by ensemble algorithm via Equation (8).

Figure 3. Structure of artificial neural networks (ANNs)

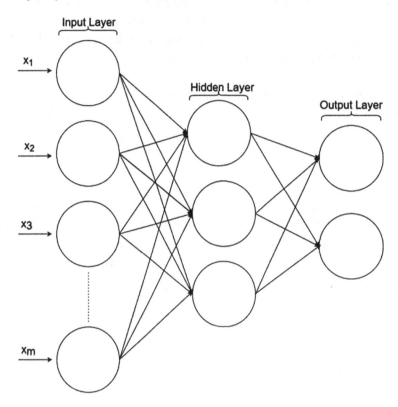

$$\hat{Y}_1 = T_1\left(X\right),...,\hat{Y}_M = T_M\left(X\right) \tag{8}$$

where, $\hat{Y}_m, m = 1,.., M$ is the mth tree output. Also, the average value of predictions determines the output value.

Feature Selection

Researchers apply feature, or attribute or variable selection application, with the purpose of observing and determining the essential and significant components that are effective over any data model, and owing to this, reducing the load of input space (Rokach & Maimon, 2015). Actually, the aim of this technique is improving prediction success and decreasing of error that occurs according to the real values of data samples for any machine-learning model. However, this cannot be possible every time.

Moreover, researchers regard the principal subject in terms of machine learning technology as selection of features. The reason for this is to ensure usage of less quantity during learned of data, to improve learning possibility of more intense things, and to complete the operation process in less time by the feature selection technique (Hall, 1999). In addition, for the selection of features, numerous different technique and algorithms can be used in machine learning applications, such as prediction, classification, pattern recognition, etc.

Backward elimination by Marill and Green, best-first search by Winston, and genetic algorithms by Goldberg, nearest-neighbor learning by Kibler with Aha, and Cardie, and correlation-based feature selection method by Hall (which the authors utilized in this study for feature selection), are among some of these techniques (Witten & Frank, 2005).

Correlation-based feature selection method (CFS; seen as '*Cfs subset evaluator*' in Weka software) is a technique used in statistical relationships between attributes. For this reason, the technique determines the effective features by some select features according to relations between each, other besides relations according to the output or target features (Cichosz, 2015). This case can be considered as correlation, and so, this technique may become a correlation-based selector or evaluator.

On the other hand, the CFS evaluator has many advantages for selecting principal features, such as operating easily and rapidly, ability of clear the not effective features, and so prediction can be improved many times. In this regard, the technique uses a heuristic function related with correlation and realizes of rank the subsets of features. The evaluation function for CFS is given in Equation (9) (Hall, 1999).

$$M_s = \frac{k\overline{r_{cf}}}{\sqrt{k + k\left(k-1\right)\overline{r_{ff}}}} \tag{9}$$

In this equation, M_s indicates the heuristic value belonging to any S subset of features, including k features. Also, $\overline{r_{cf}}$ is the average value of correlations existing among each feature with output class, and $\overline{r_{ff}}$ is the average of features' correlations according to each other.

CALCULATIONS AND EMPIRICAL RESULTS

Dataset Description

In this study, the authors obtained a dataset containing various parameters acting to measure concrete compressive strength. This dataset includes defined information obtained from performed experiments at the University of California (Yeh, 1998a, 1998b). In addition, these samples amount to 1030, and the samples consisted of normal Portland cement, which has various additives and was cured via normal conditions. Also, these numerical observations for model samples were gained from research laboratories taking place in a different university.

Table 1 represents the dataset and its various properties, and Figures 4-12 show the number of each feature label for samples within dataset.

Cross Validation

Cross validation is a technique applied for observation by collecting the data samples inside independent groups (namely, fold) for developing prediction performance of any machine learning algorithm to design a model.

Table 1. Properties of inputs and output in data set

Features	Unit	Ultimate Values	
		Min	Max
Cement	kg/m^3	102.0	540.0
Blast-furnace slag	kg/m^3	11.0	359.4
Fly ash	kg/m^3	24.5	200.1
Water	kg/m^3	121.8	247.0
Superplasticizer	kg/m^3	1.7	32.2
Coarse aggregate	kg/m^3	801.0	1145.0
Fine aggregate	kg/m^3	594.0	992.6
Age of testing	Day	1.0	365.0
Concrete compressive strength	MPa	2.3	82.6

Figure 4. Frequency of age input feature for data samples

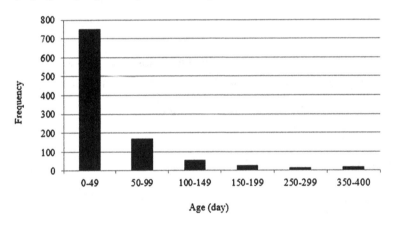

Figure 5. Frequency of blast furnace input feature for data samples

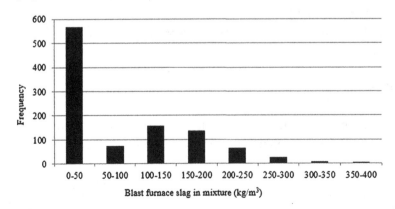

Figure 6. Frequency of fly ash input feature for data samples

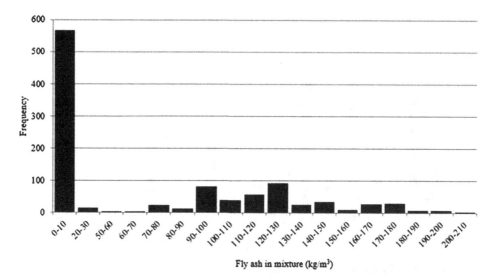

In this respect, this method applies an approach that is related to dividing samples into two separate groups consisting of training, in other words learning, and testing part, which is preferred for validating of model. Thus, these groups should be sequenced consecutively. On the other hand, k folds exist in classic form of this method (Refaeilzadeh, Tang, & Liu, 2009).

Figure 13 represents the operation logic of k-fold cross validation method. As mentioned before, this data model is first split as training and testing data, and then each testing fold is evaluated by means of learning and modelling of all training folds (as (k-1) fold). As a result of realizing all cross-validation operations, obtained performance results are collected in the final prediction value for operated algorithm.

However, in this study, the authors preferred cross-validation, as k is equal to 10-fold to detect of prediction success and performances of proposed machine and ensemble machine learning models. In this study, the dataset, which was randomized initially, is composed of 10 separate folds, and their single fold is a test and the remaining 9 folds are training folds. The authors carry out calculation of error values belonging to each cross-validation set, and the following total error value can be obtained after this sequential process.

Figure 7. Frequency of superplasticizer input feature for data samples

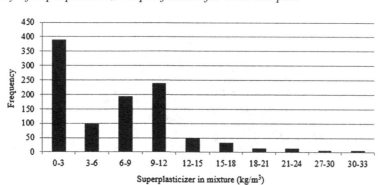

Figure 8. Frequency of cement input feature for data sample

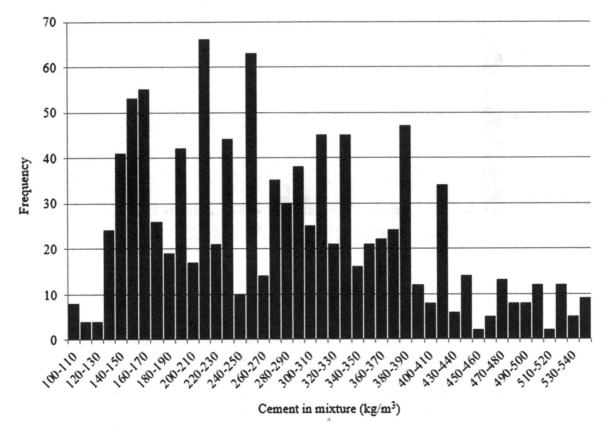

Figure 9. Frequency of water input feature for data samples

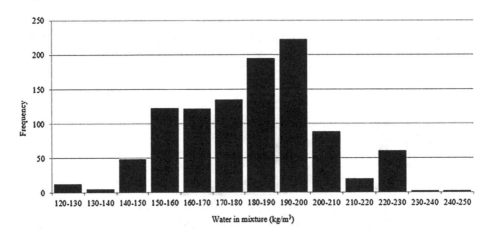

Figure 10. Frequency of coarse aggregate input feature for data samples

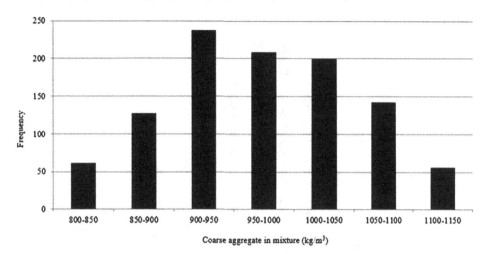

Performance Evaluation Metrics

In Table 2, various error evaluation metrics are composed from mean absolute error, relative absolute error, root mean squared error, and root relative squared error, which are considered generally as basic observations and are frequently used. These are given and intended for evaluating of success of machine-learning prediction models.

Table 3 represents formulations of metrics used in this study.

On the other hand, a different measurement was realized for evaluating error as commonly. This method is called Synthesis Index (SI) calculation. In this respect, the authors used ΣSI as a comprehensive performance measure, and its equation is expressed via Equation (10).

Figure 11. Frequency of fine aggregate input feature for data samples

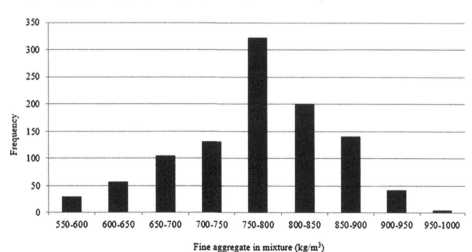

Figure 12. Frequency of concrete compressive strength output feature for data samples

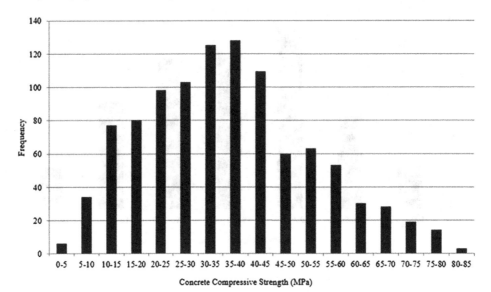

Figure 13. k-fold cross validation method
(Chou, Lin, Pham, & Shao,2015)

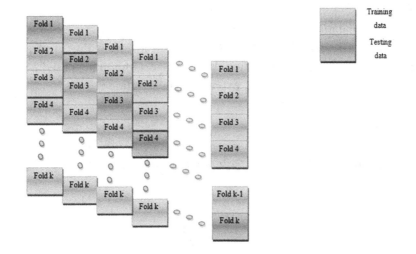

Table 2. Error evaluation metrics

Abbreviations	Definitions
R	Correlation Coefficient
MAE	Mean Absolute Error
RAE	Relative Absolute Error
RMSE	Root Mean Squared Error
RRSE	Root Relative Squared Error

Table 3. Formulation of evaluation metrics

Evaluation Metrics	Equations
Correlation Coefficient (R)	$$\frac{n\sum y \cdot y' - \left(\sum y\right)\left(\sum y'\right)}{\sqrt{n\left(\sum y^2\right) - \left(\sum y\right)^2} - \sqrt{n\left(\sum y'^2\right) - \left(\sum y'\right)^2}} \cdot$$
Mean Absolute Error (MAE)	$$\frac{1}{n}\sum_{i=1}^{n}\left|y - y'\right| \cdot$$
Root Absolute Error (RAE)	$$\frac{\sum_{i=1}^{n}\left|y' - y\right|}{\sum_{i=1}^{n}\left|\hat{y} - y\right|}$$
Root Mean Squared Error (RMSE)	$$\sqrt{\frac{1}{n}\sum_{i=1}^{n}\left(y_i - y_i'\right)^2}$$
Root Relative Squared Error (RRSE)	$$\sqrt{\frac{\sum_{i=1}^{n}\left(y' - y\right)^2}{\sum_{i=1}^{n}\left(\hat{y} - y\right)^2}}$$

$$\Sigma SI = (1\text{-}ErrorSI) + PerformanceSI \tag{10}$$

Where ErrorSI (E-SI) and PerformanceSI (P-SI) are calculated by the following formula as in Equation (11) (Huang et al., 2016):

$$s1 = \frac{1}{m}\sum_{i=1}^{m}\left(\frac{P_i - P_{min,i}}{P_{max,i} - P_{min,i}}\right) \tag{11}$$

Discussion and Results

According to these results, the authors found AddReg-RF hybrid model to be superior to other methods. One of the causes was that the correlation coefficient (R) belonging to mentioned hybrid model is the best value with the point of prediction performance, according to other ones. In addition to this performance, the authors observed the least values in terms of all of the error evaluation metrics towards this hybrid model.

After the feature selection process, which was used to determine the fundamental effective parameters for the concrete compressive strength prediction model, the authors observed that the results were close to the results obtained with eight features existing in the initial model. However, observed values for prediction performance were a decreased to a lesser degree compared to the initial model. On the other hand, the selected features are cement, water, superplasticizer, and age.

Table 6 shows the percentage change of performance results (error metrics) between the first model and after feature selection process. The percentage change in R values are in an acceptable range.

In Table 7 and Table 8, the authors combined all performance metrics with error SI method, and the authors ranked the results for readers to easily interpret them.

Table 4. Performance results of machine and ensemble machine learning models

	RF	ANN	Bagging-RF	Bagging-ANN	AddReg-RF	AddReg-ANN	RotF-RF	RotF-ANN
R	0.9630	0.9023	0.9598	0.9149	0.9713	0.9187	0.9464	0.9165
MAE	3.1963	5.5957	3.5096	5.2329	2.4945	5.1139	4.0526	5.1697
RMSE	4.6031	7.2950	4.8649	6.7596	4.0025	6.6883	5.6928	6.8207
RAE	23.7152	41.5172	26.0397	38.8257	18.5076	37.9422	30.0726	37.3617
RRSE	27.5379	43.6426	29.104	40.439	23.9449	40.013	34.0613	40.8102

Table 5. Performance results of machine and ensemble machine learning models applied selection of features

	RF	ANN	Bagging-RF	Bagging-ANN	AddReg-RF	AddReg-ANN	RotF-RF	RotF-ANN
R	0.9259	0.8501	0.9229	0.8633	0.9444	0.8573	0.9111	0.8662
MAE	4.5607	6.9468	4.7796	6.5228	3.6391	6.7875	5.2162	6.4578
RMSE	6.3130	9.0045	6.4429	8.4303	5.4946	8.7302	6.9190	8.4017
RAE	33.8425	51.5491	35.4670	48.4025	27.0039	50.3671	38.7069	47.9203
RRSE	37.7721	53.8767	38.5499	50.4410	32.8754	52.2352	41.3982	50.2696

Table 6. Percentage change of performance results (error metrics) between first model and after feature selection process

	RF	ANN	Bagging-RF	Bagging-ANN	AddReg-RF	AddReg-ANN	RotF-RF	RotF-ANN
R	-3.85	-5.79	-3.84	-5.64	-2.77	-6.68	-3.73	-5.49
MAE	42.69	24.15	36.19	24.65	45.88	32.73	28.71	24.92
RMSE	37.15	23.43	32.44	24.72	37.28	30.53	21.54	23.18
RAE	42.70	24.16	36.20	24.67	45.91	32.75	28.71	28.26
RRSE	37.16	23.45	32.46	24.73	37.30	30.55	21.54	23.18

CONCLUSION

As the authors mentioned before, researchers presented various methods and models for prediction of HPC compressive strength. Machine learning-based studies showed that ML is one of most influential methods in this area. The construction sector is an industry where international competition is very intense. To survive in this intense and ruthless competitive environment, managing time efficiently will be a great advantage. In this case, waiting for the results of concrete compressive strength for days will cause unnecessary time loss, increase construction completion time, and increase costs.

The authors have proposed an advanced ML based prediction model of HPC in this study. The prediction model could be used as a decision support system that could eliminate the time to wait for laboratory test results. According to sudden solutions of ML model execution, the process of construction never stops.

In this study, the authors explained eight different ML and EML techniques, composed from RF and ANN algorithms, which were used as base learners for prediction performance comparison of Rotation Forest, Additive Regression, and Bagging EML methods, besides only themselves. In addition, the authors applied a feature selection process to the model to decrease calculation time and to make accurate predictions with less information. The results obtained showed that there are acceptable declines in R values. Finally, the results also demonstrate that the Additive Regression method with Random Forest base learner is the most accurate model in predicting HPC compressive strength, according to evaluation metrics.

Table 7. SI results of machine and ensemble machine learning models

	1-R	**MAE**	**RMSE**	**RAE**	**RRSE**	**SI Average Value**	**Rank**
RF	0.12	0.23	0.18	0.23	0.18	0.19	2
ANN	1.00	1.00	1.00	1.00	1.00	1.00	8
Bagging-RF	0.17	0.33	0.26	0.33	0.26	0.27	3
Bagging-ANN	0.82	0.88	0.84	0.88	0.84	0.85	7
AddReg-RF	0.00	0.00	0.00	0.00	0.00	0.00	1
AddReg-ANN	0.76	0.84	0.82	0.84	0.82	0.82	5
RotF-RF	0.36	0.50	0.51	0.50	0.51	0.48	4
RotF-ANN	0.79	0.86	0.86	0.82	0.86	0.84	6

Table 8. SI results of machine and ensemble machine learning models after feature selection process

	1-R	MAE	RMSE	RAE	RRSE	SI Average Value	Rank
RF	0.20	0.28	0.23	0.28	0.23	0.24	2
ANN	1.00	1.00	1.00	1.00	1.00	1.00	8
Bagging-RF	0.23	0.34	0.27	0.34	0.27	0.29	3
Bagging-ANN	0.86	0.87	0.84	0.87	0.84	0.86	6
AddReg-RF	0.00	0.00	0.00	0.00	0.00	0.00	1
AddReg-ANN	0.92	0.95	0.92	0.95	0.92	0.93	7
RotF-RF	0.35	0.48	0.41	0.48	0.41	0.42	4
RotF-ANN	0.83	0.85	0.83	0.85	0.83	0.84	5

REFERENCES

Adusumilli, S., Bhatt, D., Wang, H., Bhattacharya, P., & Devabhaktuni, V. (2013). A low-cost INS/GPS integration methodology based on random forest regression. *Expert Systems with Applications, 40*(11), 4653–4659. doi:10.1016/j.eswa.2013.02.002

Ahmadi-Nedushan, B. (2012). An optimized instance-based learning algorithm for estimation of compressive strength of concrete. *Engineering Applications of Artificial Intelligence, 25*(5), 1073–1081. doi:10.1016/j.engappai.2012.01.012

Akande, K. O., Owolabi, T. O., Twaha, S., & Olatunji, S. O. (2014). Performance comparison of SVM and ANN in predicting compressive strength of concrete. *IOSR Journal of Computer Engineering, 16*(5), 88–94. doi:10.9790/0661-16518894

Akpinar, P., & Khashman, A. (2017). Intelligent classification system for concrete compressive strength. *Procedia Computer Science, 120*, 712–718. doi:10.1016/j.procs.2017.11.300

Al-Shamiri, A. K., Kim, J. H., Yuan, T. F., & Yoon, Y. S. (2019). Modeling the compressive strength of high-strength concrete: An extreme learning approach. *Construction & Building Materials, 208*, 204–219. doi:10.1016/j.conbuildmat.2019.02.165

Atici, U. (2011). Prediction of the strength of mineral admixture concrete using multivariable regression analysis and an artificial neural network. *Expert Systems with Applications, 38*(8), 9609–9618. doi:10.1016/j.eswa.2011.01.156

Awoyera, P. O. (2018). Predictive models for determination of compressive and split-tensile strengths of steel slag aggregate concrete. *Materials Research Innovations, 22*(5), 287–293. doi:10.1080/14328 917.2017.1317394

Aydogmus, H. Y., Erdal, H. İ., Karakurt, O., Namli, E., Turkan, Y. S., & Erdal, H. (2015). A comparative assessment of bagging ensemble models for modeling concrete slump flow. *Computers and Concrete, 16*(5), 741–757. doi:10.12989/cac.2015.16.5.741

Behnood, A., Behnood, V., Gharehveran, M. M., & Alyamac, K. E. (2017). Prediction of the compressive strength of normal and high-performance concretes using M5P model tree algorithm. *Construction & Building Materials, 142*, 199–207. doi:10.1016/j.conbuildmat.2017.03.061

Behnood, A., & Golafshani, E. M. (2018). Predicting the compressive strength of silica fume concrete using hybrid artificial neural network with multi-objective grey wolves. *Journal of Cleaner Production, 202*, 54–64.

Bilgehan, M., & Turgut, P. (2010). Artificial neural network approach to predict compressive strength of concrete through ultrasonic pulse velocity. *Research in Nondestructive Evaluation, 21*(1), 1–17. doi:10.1080/09349840903122042

Breiman, L. (1996). Bagging predictors. *Machine Learning, 24*(2), 123–140. doi:10.1007/BF00058655

Bühlmann, P., & Yu, B. (2002). Analyzing bagging. *Annals of Statistics, 30*(4), 927–961. doi:10.1214/aos/1031689014

Bui, D. K., Nguyen, T., Chou, J. S., Nguyen-Xuan, H., & Ngo, T. D. (2018). A modified firefly algorithm-artificial neural network expert system for predicting compressive and tensile strength of high-performance concrete. *Construction & Building Materials, 180*, 320–333. doi:10.1016/j.conbuildmat.2018.05.201

Cheng, M. Y., Chou, J. S., Roy, A. F., & Wu, Y. W. (2012). High-performance concrete compressive strength prediction using time-weighted evolutionary fuzzy support vector machines inference model. *Automation in Construction, 28*, 106–115. doi:10.1016/j.autcon.2012.07.004

Chithra, S., Kumar, S. S., Chinnaraju, K., & Ashmita, F. A. (2016). A comparative study on the compressive strength prediction models for high performance concrete containing nano silica and copper slag using regression analysis and artificial neural networks. *Construction & Building Materials, 114*, 528–535. doi:10.1016/j.conbuildmat.2016.03.214

Chopra, P., Sharma, R. K., & Kumar, M. (2016). Prediction of compressive strength of concrete using artificial neural network and genetic programmings. *Advances in Materials Science and Engineering, 2016*, 1–10. doi:10.1155/2016/7648467

Chou, J. S., Lin, C. W., Pham, A. D., & Shao, J. Y. (2015). Optimized artificial intelligence models for predicting project award price. *Automation in Construction, 54*, 106–115. doi:10.1016/j.autcon.2015.02.006

Chou, J. S., Tsai, C. F., Pham, A. D., & Lu, Y. H. (2014). Machine learning in concrete strength simulations: Multi-nation data analytics. *Construction & Building Materials, 73*, 771–780. doi:10.1016/j.conbuildmat.2014.09.054

Cichosz, P. (2015). *Data mining algorithms: explained using R.* Chichester, UK: John Wiley & Sons. doi:10.1002/9781118950951

Dahou, Z., Sbartaï, Z. M., Castel, A., & Ghomari, F. (2009). Artificial neural network model for steel–concrete bond prediction. *Engineering Structures, 31*(8), 1724–1733. doi:10.1016/j.engstruct.2009.02.010

Deshpande, N., Londhe, S., & Kulkarni, S. (2014). Modeling compressive strength of recycled aggregate concrete by artificial neural network, model tree and non-linear regression. *International Journal of Sustainable Built Environment, 3*(2), 187–198.

Dietterich, T. G. (2000). An experimental comparison of three methods for constructing ensembles of decision trees: Bagging, boosting, and randomization. *Machine Learning, 40*(2), 139–157. doi:10.1023/A:1007607513941

Duan, Z. H., Kou, S. C., & Poon, C. S. (2013). Prediction of compressive strength of recycled aggregate concrete using artificial neural networks. *Construction & Building Materials, 40*, 1200–1206. doi:10.1016/j.conbuildmat.2012.04.063

Erdal, H. I., Karakurt, O., & Namli, E. (2013). High performance concrete compressive strength forecasting using ensemble models based on discrete wavelet transform. *Engineering Applications of Artificial Intelligence, 26*(4), 1246–1254. doi:10.1016/j.engappai.2012.10.014

Hall, M. A. (1999). *Correlation-based feature selection for machine learning* (Doctoral dissertation). Retrieved from https://www.cs.waikato.ac.nz/~mhall/thesis.pdf

Hoła, J., & Schabowicz, K. (2005). Application of artificial neural networks to determine concrete compressive strength based on non-destructive tests. *Journal of Civil Engineering and Management, 11*(1), 23–32. doi:10.3846/13923730.2005.9636329

Huang, S., Wang, B., Qiu, J., Yao, J., Wang, G., & Yu, G. (2016). Parallel ensemble of online sequential extreme learning machine based on MapReduce. *Neurocomputing, 174*, 352–367.

Kewalramani, M. A., & Gupta, R. (2006). Concrete compressive strength prediction using ultrasonic pulse velocity through artificial neural networks. *Automation in Construction, 15*(3), 374–379. doi:10.1016/j.autcon.2005.07.003

Khoshgoftaar, T. M., Golawala, M., & Van Hulse, J. (2007). An empirical study of learning from imbalanced data using random forest. In *Proceedings of 19th IEEE International Conference on Tools with Artificial Intelligence (ICTAI 2007)*. Patras, Greece: IEEE. 10.1109/ICTAI.2007.46

Kılınç, D., Borandağ, E., Yücalar, F., Özçift, A., & Bozyiğit, F. (2015). Yazılım hata kestiriminde kolektif sınıflandırma modellerinin etkisi. In *Proceedings of IX. Ulusal Yazılım Mühendisliği Sempozyumu*. Bornava-İzmir.

Küçük, B. (2000). Factors providing the strength and durability of concrete. *Pamukkale Üniversitesi Mühendislik Bilimleri Dergisi, 6*(1), 79–85.

Kuncheva, L. I., & Rodríguez, J. J. (2007). An experimental study on rotation forest ensembles. In M. Haindl, J. Kittler, & F. Roli (Eds.), *Multiple classifier systems* (pp. 459–468). Heidelberg, Germany: Springer-Verlag. doi:10.1007/978-3-540-72523-7_46

Malinov, S., Sha, W., & McKeown, J. J. (2001). Modelling the correlation between processing parameters and properties in titanium alloys using artificial neural network. *Computational Materials Science, 21*(3), 375–394. doi:10.1016/S0927-0256(01)00160-4

Naderpour, H., Kheyroddin, A., & Amiri, G. G. (2010). Prediction of FRP-confined compressive strength of concrete using artificial neural networks. *Composite Structures, 92*(12), 2817–2829.

Nikoo, M., Zarfam, P., & Sayahpour, H. (2015). Determination of compressive strength of concrete using self organization feature map (SOFM). *Engineering with Computers, 31*(1), 113–121.

Ozturk, H., Namli, E., & Erdal, H. I. (2016). Modelling sovereign credit ratings: The accuracy of models in a heterogeneous sample. *Economic Modelling, 54*, 469–478.

Refaeilzadeh, P., Tang, L., & Liu, H. (2009). Cross-validation. In L. Liu & M. T. Özsu (Eds.), *Encyclopedia of database systems* (pp. 24–154). Boston, MA: Springer.

Rokach, L., & Maimon, O. (2015). *Data mining with decision trees theory and applications* (2nd ed., Vol. 81). Singapore: World Scientific.

Seyhan, A. T., Tayfur, G., Karakurt, M., & Tanoglu, M. (2005). Artificial neural network (ANN) prediction of compressive strength of VARTM processed polymer composites. *Computational Materials Science, 34*(1), 99–105. doi:10.1016/j.commatsci.2004.11.001

Topcu, I. B., & Sarıdemir, M. (2008). Prediction of compressive strength of concrete containing fly ash using artificial neural networks and fuzzy logic. *Computational Materials Science, 41*(3), 305–311.

Trtnik, G., Kavčič, F., & Turk, G. (2009). Prediction of concrete strength using ultrasonic pulse velocity and artificial neural networks. *Ultrasonics, 49*(1), 53–60. doi:10.1016/j.ultras.2008.05.001 PMID:18589471

Turkish Standards Institute (TSE). (2000). *Requirements for design and construction of reinforced concrete structures (ICS 91.080.40)*. Ankara, Turkey: Turkish Republic Ministry of Public Works and Settlement.

Wang, Q. (2007). Artificial neural networks as cost engineering methods in a collaborative manufacturing environment. *International Journal of Production Economics, 109*(1-2), 53–64. doi:10.1016/j.ijpe.2006.11.006

Witten, I. H., & Frank, E. (2005). *Data Mining: Practical machine learning tools and techniques* (2nd ed.). San Francisco, CA: Morgan Kaufmann.

Witten, I. H., Hall, M. A., & Frank, E. (2011). *Data mining: practical machine learning tools and techniques*. Burlington, MA: Morgan Kaufmann Series.

Yaseen, Z. M., Deo, R. C., Hilal, A., Abd, A. M., Bueno, L. C., Salcedo-Sanz, S., & Nehdi, M. L. (2018). Predicting compressive strength of lightweight foamed concrete using extreme learning machine model. *Advances in Engineering Software, 115*, 112–125. doi:10.1016/j.advengsoft.2017.09.004

Yeh, I. C. (1998a). Modeling of strength of high-performance concrete using artificial neural networks. *Cement and Concrete Research, 28*(12), 1797–1808. doi:10.1016/S0008-8846(98)00165-3

Yeh, I. C. (1998b). Modeling concrete strength with augment-neuron networks. *Journal of Materials in Civil Engineering*, *10*(4), 263–268. doi:10.1061/(ASCE)0899-1561(1998)10:4(263)

Yu, Y., Li, W., Li, J., & Nguyen, T. N. (2018). A novel optimised self-learning method for compressive strength prediction of high-performance concrete. *Construction & Building Materials*, *184*, 229–247.

Chapter 8
Artificial Intelligence Towards Water Conservation:
Approaches, Challenges, and Opportunities

Ravi Sharma

Symbiosis Institute of International Business, Symbiosis International University, India

Vivek Gundraniya

Symbiosis Institute of International Business, Symbiosis International University, India

ABSTRACT

Innovation and technology are trending in the industry 4.0 revolution, and dealing with environmental issues is no exception. The articulation of artificial intelligence (AI) and its application to the green economy, climate change, and sustainable development are becoming mainstream. Water is a resource that has direct and indirect interconnectedness with climate change, development, and sustainability goals. In recent decades, several national and international studies revealed the application of AI and algorithm-based studies for integrated water management resources and decision-making systems. This chapter identifies major approaches used for water conservation and management. On the basis of a literature review, the authors will outline types of approaches implemented through the years and offer instances of the ways different approaches selected for water conservation and management studies are relevant to the context.

INTRODUCTION

Artificial intelligence (AI) is not a new concept in the modern era of technology. AI was introduced in ancient time, but the word "artificial intelligence" was not coined until 1956 when Dartmouth College introduced the word Artificial Intelligence. Scientists introduced this word to describe the human mind thinking as the "Symbolic System" (Buchanan, 2005). Artificial intelligence can be captured as a brain-mimicry. AI refers to computer systems that "can sense their environment, think, learn, and act in

DOI: 10.4018/978-1-7998-0301-0.ch008

response to what they sense and their programme objectives," according to a World Economic Forum report published in 2018. As the fourth Industrial Revolution gains momentum, technological innovations become increasingly connected and accessible. AI has a progressive role as both knowledge and non- knowledge-based activities towards creating a positive impact while delving into urgent environmental challenges. AI together with the big data sources, information accessibility, recent advances in AI algorithms, advancements in hardware, and innovative technologies have resulted in AI to propelled out of the lab and into daily based life and industry sectors, including the energy and environment sectors (World Economic Forum, 2018). The United Nations Sustainable Development Goals (SDGs) provides another motivation for the implementation of innovative technologies for the challenges facing the environment and humanity. In the current and rapidly changing environmental issues, researchers see the role of AI as a game changer for environmental predictions and decision making. According to Coumou and Rahmstorf (2012), with climate change impacts worsening in the coming decades, as believed by many scientists, they result in increased extreme weather events and disasters. Along with climate change issues, the loss of biodiversity, conservation efforts, ocean health, clean air, and water security are important critical challenges that require transformative actions in the next decade.

According to the United Nations, the world population will increase to 8.5 billion by 2030, and demographers expect India to have the largest population (United Nations Department of Economic and Social Affairs, 2015). With this population size amalgamated with changing weather extremes and events, countries across the world will face water scarcity. The inefficient mechanism for water resource management is another reason contributing to water scarcity. The agriculture sector is another sector where the demand for water has increased. This sector is accountable for about 70% of the total global water withdrawal (FAO, 2017) and with the increase in population, the demand for water and efforts for conservation of resources will also show a trend by 2050. With the growing demand on water, there is a dire need for the implementation of innovative technologies that contribute to water conservation efforts. AI could play a crucial role in solving this issue. Managing water resources, designing efficient supply systems, optimizing present water resources, and planning infrastructure of water resources and safe water mechanisms can be efficiently addressed by the use of AI technologies in water resource management (Lin et al., 2018). Scientists have introduced AI in the water management field for reducing wastage of valuable resource. Researchers have reported the breakthrough cases advocating AI usage to effectively reduce water wastage. Various researchers including Tsanis et al. (2008), Yesilnacar et al. (2008), and Sahoo et al. (2006) used the effective, successful AI as a counterfeit mechanism to traditional hydrological approaches, and Kia et al. (2012), Sahay and Srivastava (2014) used for flood monitoring and management of water.

The latest report predicted that the market of the AI would grow from the $21.46 billion that it generated in 2018 to $190.61 billion in 2025, which is a CAGR of 36.62%. The major reason for the growing AI market would be big data, increasing demand of the cloud-based system, advancement of AI-enabled analytical systems (Markets and Markets, 2018). India has recorded many developments in the water management because of the increased trend in demand. AI is also one of the areas in technological innovation in India that researchers have explored for water management and developed algorithms for water quality (Kulshreshtha and Shanmugam, 2018; Chatterjee et al., 2017; Visalakshi and Radha, 2017). The Ministry of Water is also going forward to collaborate with Google platform for developing an AI model for weather forecasts and a flood warning system for the users. In 2017, Microsoft Corporation announced the AI for Earth program for the organizations and people, which aids them in resolving environmental challenges through the power of AI. This program includes applications to climate change, biodiversity, water, and agriculture. These efforts show the trend of AI supporting the conservation of biodiversity and natural resources in India.

In recent decades, several studies at the national and international levels have revealed the application of AI and algorithm-based studies for integrated water management resources and conservation-related decision-making systems. During the current review, the researcher attempts to identify the major approaches used for water conservation and management. On the basis of the literature review, the author will attempt to outline types of approaches implemented through the years. Furthermore, the author offers instances of different approaches selected for water conservation and management studies relevant to the context. Finally, this review study finds that the Artificial Neural Network (ANN) approach is common globally and also specific the context in India towards the water domain studies. The scope of the study is an attempt to review these broad concerning water management and conservation of water resources. In the present study, the authors reviewed published papers from Scopus online database journals to date which used artificial intelligence as a tool in water conservation and management. The author narrates the objectives of the review study in the following research questions: (1) Which methodology or approach is commonly used in studies for water conservation and management? (2) What is the trend of the AI method, and what are the application areas in the field of water conservation and management? (3) What are the benefits, complexities, and water conservation requirements included in studies?

The advantage of such a review of literature on AI and water management is to help in studying the trends and heterogeneity in the scope of results. This study also gathers evaluates multiple studies, and it extrapolates results by providing the future scope of development of AI in water management. Overall, the systematic study of AI and water is a valuable research alternative that will attempt to provide, (a) an opportunity through identification of gaps, (b) divergent views on the use of technology as an approach for water conservation and management, (c) flexibility in analyses of the concept of AI and water management with different scopes and purposes, (d) further unification of conceptual and empirical criteria between the technology and sector to understand the innovations among a certain group to solve the major environmental problems nationally or globally, and (e) a useful tool for addressing challenges and barriers to the integration of AI in solving environmental issues that are important to identify the future concerns for researchers.

METHODOLOGY

AI Methodologies and Concept

AI methodologies generally emphasize the application of logic and the formal way of representation of data gathered. Other times, data may be generated through an application-oriented bottom-up approach based on experimental and investigative approaches. Based on these two approaches, the AI methodologies generally consist of broad two categories for research: (a) Conventional AI, and (b) Computational AI. A further elaborated application of Conventional AI is programming, which is attempted by machine learning and emphasizes statistical analysis for calculating probable outcomes. The key areas include:

- **Behavior-based AI:** Behavior-based AI decomposes intelligence into autonomous modules attempted by Behavior-Based AI (BBAI). This area is popular in the field of robotics.

- **Bayesian networks:** Bayesian networks based upon probability theory. The problem domain is represented as a network in graph form that users can understand simply.
- **Expert systems:** Expert systems are programmed to capture the expertise of humans in a particular field, which resembles working like a human expert. This is the most successful method amongst conventional AI.
- **Case-Based Reasoning (CBR):** CBR is used to solve problems based on the historical problem and solution data.

On the other hand, computational intelligence (CI) depends upon complex design algorithms (heuristics) and computation. Whereby using iterative learning, intelligence can be developed by software. The key areas of CI are (a) Neural Networks, in which difficult problems can be solved by the neural network. Here, many nodes are used-based and developed by simulations with the human neuron system; (b) Fuzzy Logic, which refers to reasoning depending on probability. Zero and one are the values used in the binary system, which provides a broader answer in value. Sometime, a hybrid model is also used to find solutions using the two methods described below.

Research Methodology

The present study consists of extensive research on the previous year's papers to assess and analyze their respective applications of different AI approaches in the field of water management and conservation. For this reason, the authors used the "Preferred Reporting Items for Systematic Reviews and Meta-Analyses (PRISMA)" as suggested by Moher et. al. (2009), as a preferred research methodology. The purpose of this methodology is to identify the water conservation area, available AI techniques, and its application. This approach will result in existing approaches and applications for findings based on the assessment and evidence from the studies. When implemented, the meta-analysis will provide a useful path for executing the validated outcomes based on the screened resources. To review the study, the authors processed Scopus database, extracted data, and summarized the relevant papers. The authors used the keyword syntax "Artificial Intelligence" and "Water" to identify the scientific journal articles. Figure 1 illustrates the analysis and procedure of systematic review. Under the exclusion criterion, the authors excluded in-press articles, non-English articles, books, book chapters, conference proceedings, notes, survey reports, reports, and trade journal articles while screening articles. The authors finally sorted the articles based on the relevancy to the research objective. Results revealed six papers related to the AI application using ANN approaches in the field of water management and conservation. In searching for topics specific to the Indian context, the authors conducted a literature review for the same and the discuss important tools and methodologies in the sections presented in this study.

Results and Discussion

Using the PRISMA approach and screening tools as the authors described in the sections above, in context specific to India, the authors retrieved a total of six studies that discussed the application and importance of AI in water management in different application areas. Researchers have reported the application of AI techniques for predictive modelling of scouring phenomenon and sediment transport in rivers as a very useful tool (Haghiabi, 2017; Parsaie et al., 2017; Parsaie et al., 2019). The predominant studies found to be in the water management application area (83%). Researchers widely used the ANN

Figure 1. Classification and screening of papers based on artificial intelligence and water conservation (As per PRISMA Approach)

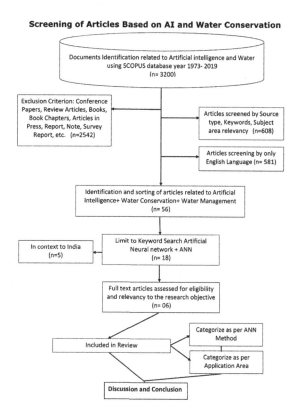

prediction model in the diverse application areas. The major areas of application covered are in the field of water demand, water resource management, water quality, and trapping efficiency using the ANN prediction models. Integrated water management (17%) is another application area where researchers used the AI methodology in ANN non-predictive only in India. Furthermore, researchers successfully reported the AI technique in water quality modelling across the globe (Haghiabi et al., 2018) as well as the prediction of flow discharge in streams (Parsaie and Haghiabi, 2017; Parsaie et al., 2017)

The researchers used the decision support system (DSS) for the framework. This system consists of water balancing terms, water demand-supply, reuse options, basic details, stormwater management, and wastewater management to develop a system using these data for the urban water balancing study conducted for Varanasi city, India (Maurya et al., 2018). The study estimated that the annual water balance is very low, and the actual demand is increasing by 37%, which advocates for an urgent water management intervention to help improve the water balance condition in the area. Maurya et al. (2018) suggested to increase the reuse capacity and improve the water balance of the urban city with the three scenarios that used the data of ground extraction, water balance, annual stormwater, and untreated wastewater. Researchers designed this framework to evaluate the water balance from the reuse to disposal using DSS, but the data that is used in the system is already available from the different departments. To some extent, this acts as a limitation. Instead, the use of real-time data in the framework will provide more accurate and informed results for the water balance and overall management decisions.

In another study, researchers implemented a more advanced approach towards the water management of the Barmer region of Rajasthan State in India. Water quality management through the data-driven intelligence system to understand the parameters of the water quality for the public sector and sustainable usage of groundwater using the spatiotemporal related real-time data has resulted in a meaningful and logical application towards groundwater use (Sinha et al., 2018). Similarly, in a study of groundwater levels and management, researchers implemented prediction modeling using different algorithms and architecture for Tirupati, India, which has a dry climate with an annual rainfall of 700 mm. Researchers used ANN models, such as Radial basis function network, forward neural network, Elman type recurrent neural network, Input delay neural network and the14 different types of algorithms involving data and input selection from the observed lake in Tirupati for the prediction of the groundwater levels (Sujatha and Kumar, 2011). The study is important in terms of advocating the reliability of algorithmic forward neural network technique for its accuracy and application.

Considering the global scenarios, using the application of AI for water specific areas shows a more dominant use of different approaches, including both predictive and non-predictive ANN modeling (Table 1). Researchers use the ANN prediction models in different application areas, such as water quality, water demand, water management, and trap efficiency of the dam. Global studies also reveal some integration of prediction methods with another approach resulting in a more accurate presentation of results for informed decisions. The non-prediction model of the ANN applications shows simulation-based tools, which give an optimal solution for water conservation in arid coastal environments that is helpful for crop and water consumption (Grundmann et al., 2013).

Table 2 summarizes the benefits, complexity, and water conservation requirement parameter of the published studies. The authors state few of them below:

1. **Benefits:** The benefits of the ANN prediction model are higher than the non- prediction model. Researchers can use the prediction models for strategies such as water restriction programs, save water systems, sustainable water consumption, etc. (Adamowski and Karapataki, 2010). Such models are also very useful approaches for problem solving and decision-making for other engineering problems like quality (Raheli et al., 2017), for other flood-prone zones, for uncertainty of the problems (Chiang et al., 2018), for controlling sediment by trap efficiency prediction (Parsaie et al, 2018), and for controlling soil moisture content with cost-effective model (Tsang and Jim, 2016). On the other hand, researchers can use the non-prediction model for any river in which individuals can manage the water quality and quantity. This model gives results in a few hours rather than months (Grundmann et al., 2013).

2. **Complexity:** Complexity in the ANN prediction model as well as the non-predictive modeling both involves parameter availability and selection, uncertainty in results, the problem of the constraints, and algorithm functionality.

3. **Water Conservation Requirement:** ANN prediction and non-prediction methods both show that researchers can use these technologies are information to answer where what, how, and when ques-

Table 1. Description of studies about ANN methods, application area, study purpose, and outcomes

Authors	ANN Methods	Application Area	Study Purpose	Outcomes
Adamowski and Karapataki, 2010	Prediction	Water demand	Purpose of this study is too accurate short-term water demand forecasting using different technology comparisons such as multilayer linear regression and ANN	Results show that the prediction of weekly water demand by Levenberg Marquardt ANN method was more accurate than other methods
Raheli et al., 2017	Prediction	Water Quality	study purpose of this paper is to proposed forecasting model for monthly water quality by an integrated approach using Firefly algorithm(FFA) and multilayer perception(MLP)	Results show that MLP-FFA perform better than MLP model for forecasting of water quality
Grundmann et al., 2013	Non-prediction	Integrated water management	this paper present a new integrated water management tool based on simulation in arid coastal environments for water management	Results show that the integrated simulation approach of SCWPF and ANN provide optimal results in groundwater abstraction and cropping patterns
Chiang et al., 2018	Prediction	water resources management	Purpose of the study is Integration of two technique (ensemble technique and artificial neural networks) of used to reducing uncertainty in the streamflow prediction model.	Results show reducing uncertainty and show that different strategies are consistent and insensitive to the hydrological and physiographical condition.
Parsaie et al., 2018	Prediction	Trap efficiency	Purpose of the study is investigated the trap efficiency (TE) of retention dams by using AI techniques with laboratory experiments.	Results observed vf/vs parameter and its shows that TE is increasing if mean diameter grain size and specific gravity of sediment increase.
Tsang and Jim, 2016	Prediction	Water management	This study employs artificial intelligence algorithms composed of an artificial neural network and fuzzy logic, using weather data to simulate soil moisture changes to develop an optimal irrigation strategy.	Results by using The simulation model shows that improving plant coverage reduce the 20% of water use and maintain the soil moisture content from 0.13 to 0.22 m3/m3

tions about the water. These models can also answer which optimal solution should be provided, so that researchers can make informed decisions for water conservation.

Table 2. Description of the benefits, complexity and water conservation requirement of included studies

Authors	Benefits	Complexity	Water Conservation Requirement
Adamowski and Karapataki, 2010	Resolve the problem of drought in Cyprus by imposing water use restrictions, implementing water-demand reduction programs, optimizing water supply systems.	climatic variables not used in this research include evaporation, wind speed and relative humidity that could be a major factor for an accurate result	water conservation strategies are required after the prediction of water demand which can solve different problems
Raheli et al., 2017	Water quality management, aquatic life well-being and the overall healthcare planning of river systems. It applies to other engineering problems also.	After using lots of data, Uncertainty still there in the model	It is helped to find out water conservation requirements in the river or not by water quality prediction.
Grundmann et al., 2013	multi-criteria optimization within a reasonable time (a couple of hours instead of months) it can be used to manage both water quality and quantity	Further analysis requires for more optimal solution by using appropriate constraints and algorithms.	Best water conservation technique can be implemented using a simulation-based optimal solution of water quality and water problem
Chiang et al., 2018	Reducing uncertainty in the prediction model. It can be used for the different flood-prone zone.	Only hydrological or physiographical condition considers checking sensitivity. The accurate result requires more conditions.	By using this model for a different area, we can find where water conservation is required.
Parsaie et al., 2018	controlling sediment transport	there can be other parameters which are efficient for TE mapping and prediction of dam	By using this we can find TE of the dam and where to put the retention dam so, we can conserve more water
Tsang and Jim, 2016	Reducing the water consumption and maintaining the soil moisture content and Cost-effective than available systems in the market	It used for Green roof irrigation. It cannot be used for big farmland where more data and more money require	water conservation by the simulation model which helped in a reduction in the water consumption

CONCLUSION

Artificial intelligence is a very broad term in the application of the area globally involving many different approaches and tools. Artificial intelligence has now become ubiquitous from the perspective of environmental and regulatory stakeholders while dealing with environmental issues and problems, like resource scarcity. As the authors found in the literature review, the ANN techniques involve the prediction and non-prediction modeling approaches. For this study, the parameters such as the application area, study purpose, outcomes, benefits, complexity, and water conservation requirement were considered. The outcomes suggest that researchers have widely used the ANN approach globally as well as in the Indian context for water resource management and conservation-related practices within the broad application areas that range from demand-supply to quality measurement and solving other issues related to it. The overall advantage of using the AI technique is in its robust mechanism of dealing with the data (both primary and secondary data). The studies reveal that the application and ability of AI integration with other environmental techniques (like GIS, statistical forecasting tools, etc.) makes it more reliable and adds credibility to the forecasting and decision-making process amongst the managers. The study also pointed out that challenges faced in using the AI in the environmental sector overall. The data quality, availability, and uncertainty are few factors that play a crucial role in the implementation of these techniques and approaches.

The complexity is only limited to the condition inclusion for the model in both predictive and non-predictive modeling. The advantage and the opportunity with the application of using a non-prediction model is in its ability to deliver the results in a very short time duration with little complexity, but on another part, this compromises the accuracy and uncertainty. Therefore, researchers can view more algorithm and models-based infusions in the environmental area as an opportunity in the field of the AI application of energy and environment field, if the objective of better results and obtaining an optimal solution for the model is to be accomplished. Within the limitations of available data related to hydrological issues like water balance, water level, quality of water, floods, annual precipitation, and the AI techniques such as ANN can provide a credible and reliable solution in a less cost-effective manner compared to big data mechanisms used in other analytical studies. Many company sources have reported that the software programs inculcated with the neural networks play a strategical step towards the water dynamics, operations, and management.

Overall, if the data availability and data authenticity is ensured, the AI has the capability of dealing with the current changing environmental issues for proper planning and management of those scarce resources. The application of AI when dealing with environmental issues has limitations, but these limitations can present opportunities and recommendations for future studies. The results show that evident and summarized the benefits, disadvantages, and water conservation requirement show that prediction and non-predictive modeling in AI both provide the best result with fewer constraints for obtaining optimal solutions. The limitation of this study is its scope selection used for screening the studies. The authors did not include the exclusion criteria like reports, book chapters and trade studies in the study. Therefore, to further expand the study and obtain more concrete conclusions, researchers could widen the horizon of screening and inclusion criteria. Additionally, future studies can be extended based on different, sectors, industries, etc. The study also reveals that there are no common areas for any studies, which shows that there is continuous research trending in the AI field with different approaches to modeling.

The study recommends that there are larger scopes and requirements of applications in this emerging field of technology. With the industry 4.0 revolution with new innovative and emerging technologies, AI could be a useful approach for the stakeholders to deal with environmental problems. Researchers will appreciate the scope of research in the field in the future, as many countries are investing and collaborating with companies towards sustainable growth, as per the SDGs requirement.

REFERENCES

Adamowski, J., & Karapataki, C. (2010). Comparison of multivariate regression and artificial neural networks for peak urban water-demand forecasting: Evaluation of different ANN learning algorithms. *Journal of Hydrologic Engineering*, 15(10), 729–743. doi:10.1061/(ASCE)HE.1943-5584.0000245

Buchanan, B. G. (2005). A (very) brief history of artificial intelligence. *AI Magazine*, 26(4), 53–53.

Chatterjee, S., Sarkar, S., Dey, N., Sen, S., Goto, T., & Debnath, N. C. (2017). Water quality prediction: multi-objective genetic algorithm coupled artificial neural network-based approach. In *July 2017 IEEE 15th International Conference on Industrial Informatics (INDIN)* (pp. 963-968). IEEE. 10.1109/INDIN.2017.8104902

Chiang, Y. M., Hao, R. N., Zhang, J. Q., Lin, Y. T., & Tsai, W. P. (2018). Identifying the sensitivity of ensemble streamflow prediction by artificial intelligence. *Water (Basel)*, *10*(10), 1341. doi:10.3390/w10101341

Coumou, D., & Rahmstorf, S. (2012). A decade of weather extremes. *Nature Climate Change*, *2*(7), 491–496. doi:10.1038/nclimate1452

FAO. (2017). *Water for sustainable food and agriculture. a report produced for the G20 Presidency of Germany. Food and Agriculture Organization of United Nations*. Rome, Italy: FAO.

Grundmann, J., Schütze, N., & Lennartz, F. (2013). Sustainable management of a coupled groundwater–agriculture hydrosystem using multi-criteria simulation-based optimization. *Water Science and Technology*, *67*(3), 689–698. doi:10.2166/wst.2012.602 PMID:23202577

Haghiabi, A. H. (2017). Estimation of scour downstream of a ski-jump bucket using the multivariate adaptive regression splines. *Scientia Iranica*, *24*(4), 1789–1801. doi:10.24200ci.2017.4270

Haghiabi, A. H., Nasrolahi, A. H., & Parsaie, A. (2018). Water quality prediction using machine learning methods. *Water Quality Research Journal*, *53*(1), 3–13. doi:10.2166/wqrj.2018.025

Kia, M. B., Pirasteh, S., Pradhan, B., Mahmud, A. R., Sulaiman, W. N. A., & Moradi, A. (2012). An artificial neural network model for flood simulation using GIS: Johor River Basin, Malaysia. *Environmental Earth Sciences*, *67*(1), 251–264.

Kulshreshtha, A., & Shanmugam, P. (2018). Assessment of trophic state and water quality of coastal-inland lakes based on fuzzy inference system. *Journal of Great Lakes Research*, *44*(5), 1010–1025. doi:10.1016/j.jglr.2018.07.015

Lin, Y. P., Petway, J., Lien, W. Y., & Settele, J. (2018). Blockchain with artificial intelligence to efficiently manage water use under climate change.

MarketsandMarkets. (2018). *Artificial Intelligence Market worth 190.61 Billion USD by 2025*. Available at https://www.marketsandmarkets.com/PressReleases/artificial-intelligence.asp

Maurya, S. P., Ohri, A., & Singh, P. K. (2018). Evaluation of urban water balance using decision support system of Varanasi City, India. *Nature Environment and Pollution Technology*, *17*(4), 1219–1225.

Moher, D., Liberati, A., Tetzlaff, J., & Altman, D. G.Prisma Group. (2009). Preferred reporting items for systematic reviews and meta-analyses: The PRISMA statement. *PLoS Medicine*, *6*(7). PMID:19621072

Parsaie, A., Azamathulla, H. M., & Haghiabi, A. H. (2017). Physical and numerical modeling of performance of detention dams. *Journal of Hydrology (Amsterdam)*. doi:10.1016/j.jhydrol.2017.01.018

Parsaie, A., Ememgholizadeh, S., Haghiabi, A. H., & Moradinejad, A. (2018). Investigation of trap efficiency of retention dams. *Water Science and Technology: Water Supply*, *18*(2), 450–459. doi:10.2166/ws.2017.109

Parsaie, A., & Haghiabi, A. H. (2017). Mathematical expression of discharge capacity of compound open channels using MARS technique. *Journal of Earth System Science*, *126*(2), 20. doi:10.100712040-017-0807-1

Parsaie, A., Haghiabi, A. H., & Moradinejad, A. (2019). Prediction of scour depth below river pipeline using support vector machine. *KSCE Journal of Civil Engineering, 23*(6), 2503–2513. doi:10.100712205-019-1327-0

Parsaie, A., Yonesi, H., & Najafian, S. (2017). Prediction of flow discharge in compound open channels using adaptive neuro fuzzy inference system method. *Flow Measurement and Instrumentation, 54,* 288–297. doi:10.1016/j.flowmeasinst.2016.08.013

Raheli, B. A.-S., Aalami, M. T., El-Shafie, A., Ghorbani, M. A., & Deo, R. C. (2017). Uncertainty assessment of the multilayer perceptron (MLP) neural network model with implementation of the novel hybrid MLP-FFA method for prediction of biochemical oxygen demand and dissolved oxygen: A case study of Langat River. *Environmental Earth Sciences, 76*(14), 1–16. doi:10.100712665-017-6842-z

Sahay, R. R., & Srivastava, A. (2014). Predicting monsoon floods in rivers embedding wavelet transform, genetic algorithm and neural network. *Water Resources Management, 28*(2), 301–317. doi:10.100711269-013-0446-5

Sahoo, G. B., Ray, C., Mehnert, E., & Keefer, D. A. (2006). Application of artificial neural networks to assess pesticide contamination in shallow groundwater. *The Science of the Total Environment, 367*(1), 234–251. doi:10.1016/j.scitotenv.2005.12.011 PMID:16460784

Sinha, K., Srivastava, D. K., & Bhatnagar, R. (2018). Water quality management through data driven intelligence system in Barmer Region, Rajasthan. *Procedia Computer Science, 132,* 314–322. doi:10.1016/j.procs.2018.05.183

Sujatha, P. & Kumar, G. P. (2011). Prediction of groundwater levels using different artificial neural network architectures and algorithms. In *Proceedings on the International Conference on Artificial Intelligence (ICAI)* (p. 1). The Steering Committee of The World Congress in Computer Science, Computer Engineering and Applied Computing (WorldComp).

Tsang, S. W., & Jim, C. Y. (2016). Applying artificial intelligence modeling to optimize green roof irrigation. *Energy and Building, 127,* 360–369.

Tsanis, I. K., Coulibaly, P., & Daliakopoulos, I. N. (2008). Improving groundwater level forecasting with a feed forward neural network and linearly regressed projected precipitation. *Journal of Hydroinformatics, 10*(4), 317–330.

United Nations Department of Economics and Social Affairs. (2015). *World population projected to reach 9.7 billion by 2050.* Available at http://www.un.org/en/development/desa/news/population/2015-report.html

Visalakshi, S., & Radha, V. (2017). Integrated framework to identify the presence of contamination in drinking water. In *2017 IEEE International Conference on Computational Intelligence and Computing Research (ICCIC)* (pp. 1-5). IEEE.

World Economic Forum. (2018, January). *Harnessing artificial intelligence for the Earth.* Fourth Industrial Revolution for the Earth. In collaboration with PwC and Stanford Woods Institute for the Environment.

Yesilnacar, M. I., Sahinkaya, E., Naz, M., & Ozkaya, B. (2008). Neural network prediction of nitrate in groundwater of Harran Plain, Turkey. *Environmental Geology, 56*(1), 19–25. doi:10.100700254-007-1136-5

Chapter 9
Analysis of Ground Water Quality Using Statistical Techniques:
A Case Study of Madurai City

Keerthy K.
Thiagarajar College of Engineering, India

Sheik Abdullah A.
(iD) https://orcid.org/0000-0001-8707-9927
Thiagarajar College of Engineering, India

Chandran S.
Thiagarajar College of Engineering, India

ABSTRACT

Urbanization, industrialization, and increase in population lead to depletion of groundwater and also deteriorate its quality. Madurai city is one of the oldest cities in India. In this study, the authors assessed the quality of groundwater using various statistical techniques. The researchers collected groundwater samples from 11 bore wells and 5 dug wells in the post-monsoon season, in 2002, and analyzed the samples for physicochemical characterization in the laboratory. They analyzed around 17 physicochemical parameters for all the samples. The aim of the descriptive statistical analysis was understanding the correlation between each parameter. Then, the authors carried out cluster analysis to identify the most affected bore well and dug well in Madurai city.

DOI: 10.4018/978-1-7998-0301-0.ch009

INTRODUCTION

Groundwater makes up to 23% of fresh water in the world. Also, it is easily accessible to all. More than 85% of groundwater is used in rural areas. A total of 50% of groundwater is used for urban drinking needs. Increasing industrialization and population growth lead to groundwater pollution, depletion, contamination, and saltwater intrusion. Level of groundwater table also keep on reducing. Improper solid waste also influences the quality of groundwater. Statistical tools are very helpful to pre-process the data and analyze the relationship between parameters. Patil studied the quality of groundwater using Pearson correlation (Urxqgzdwhu, n.d.). Kim J., Kim, R., Lee, Cheong, Yum, & Chang (2005) investigated the quality of groundwater and identified the factors which influence it using multivariate statistical analysis. Anwar (2014) studied the quality of groundwater using a correlation matrix to analyze the correlation between quality parameters. Jamuna (2018) explored groundwater quality parameters of erosion using multivariate statistical methods. Shahid and Amba (2018) assessed the quality of groundwater of Bangalore city using principal component analysis, cluster analysis, factor analysis, and correlation analysis (Shahid, & Amba, 2018). Sadat-Noori (2013) investigated the quality of the Saveh-Noobaran aquifer to identify the places with the best water quality for drinking using a geographic information system and a water quality index. Thivya (2013) studied the quality of groundwater in different rock-formed aquifers, which resulted suitable for drinking, and for domestic and agriculture usage. Elango and Subramani (2005) studied the spatial distribution of hydrogeochemical constituents of groundwater related to its suitability for agriculture and domestic use.

The objective of this study is to identify the most affected areas in relation to groundwater quality and to determine the water quality parameters which influence water quality for drinking, by using multivariate statistical analysis. Multivariate statistical analysis involves correlation analysis and factor analysis. The authors used Statistical Package for the Social Sciences (SPSS) software for multivariate statistical analysis, and the tool Grapher to plot the piper diagram which shows the cation and anion proportion in groundwater.

Study Area

Madurai is one of the oldest and most holistic cities in Tamilnadu. Madurai is the second largest corporation city in the area and the third largest city by population in Tamil Nadu. It is located on the banks of the Vaigai river. The city is divided into four zones as North, South, East, and West. Madurai is located at 9.93°N 78.12°E. It has an average elevation of 101 meters. The city of Madurai lies on the flat and fertile plain of the Vaigai river, which runs in the northwest-southeast direction through the city dividing it almost into two equal halves.

The district is predominantly underlaid with crystalline formations, and alluvium is found along the courses of the river. Groundwater occurs under phreatic conditions in weathered residuum and interconnected shallow fractures and under semiconfined to confined conditions in deeper fractures. The depth of water in the district varies from 3.13 to 7.66 m bgl during the premonsoon period (May 2006) and 1.86 to 5.74 m bgl during the postmonsoon period. Around Madurai, the authors identified 15 sampling locations to study the quality of groundwater characteristics. They collected the samples from 10 bore wells and 5 dug wells. At the time of sampling, latitude, longitude, date, and time were noted. Figure 2 shows sampling locations. All the groundwater samples were taken to the laboratory and analyzed against the following water quality parameters:

Figure 1. Study area: Madurai city

- **pH:** pH is one of the important parameters in water quality. It denotes the concentration of hydrogen ion in the water sample. As per 10500-2012, the permissible lime of pH is 6.5-8.5. The pH is measured using a digital pH meter.

- **Conductivity:** Conductivity reflects the ability of water to conduct electricity. It depends on the amount of ions dissolved in the water sample. It is denoted by mhos and Siemens.

- **Chlorides:** Presence of chlorides shows the amount of salts of calcium, sodium, and potassium. As per 10500-2012, the permissible limit of chlorides was 250mg/l. Presence of chloride will give salty taste to the water.

- **Fluorides:** Sources of fluoride in groundwater are soil and rock minerals. Arability of fluoride in groundwater is more than in surface water. Permissible limits of fluorides were 1.5mg/l; if it is exceeded, it causes dental fluorosis.

- **Total Dissolved Solids (TDS):** Total dissolved solids in groundwater denote the presence of solids in dissolved form. Conductivity and TDS have correlation. The conductivity factor is 0.67. The permissible limit of TDS in drinking water is 500mg/L. Dissolved organic chemicals reduce the dissolved oxygen level, which reduces the water quality.

- **Hardness:** Hardness is an important parameter which represents the presence of calcium and magnesium ions. Hardness is classified into two types, namely temporary and permanent hardness. Hard water induces the scaling in the boiler and more soap consumption. Neither soft water is good for health, because human body requires some amount of calcium and magnesium minerals. The permissible limit of hardness in groundwater is 75 -115mg/L.

Results and Discussion

The researchers collected groundwater samples from 15 wells, in the year of 2002. They analyzed these collected samples as per the methods prescribed by the American Public Health (2005). The water quality parameters the researchers analyzed were TDS, NO2+NO3, Ca, Mg, K, Cl, SO4, CO3, HCO3, Fluoride, pH, EC, Hardness, Sodium Adsorption Ratio (SAR). They compared the sample results with drinking water quality standards (10500-2012). All these samples ponds cover the entire area of Madurai city. The researchers used statistical techniques to identify the correlation between the water quality parameters and perform multivariate statistical analysis. The main tools they used to interpret the groundwater quality parameters were Weka, Rapidminer, SPSS, and Grapher (Abdullah, Selvakumar, & Abirami, 2017).

Table 1 provides sampling location details. Table 2 shows the results of 15 water quality parameters of 15 different samples. Most of the parameters exceed their permissible limits.

Table 3 shows the statistical analysis results of groundwater quality parameters, which highlight the variation between mean and standard deviation.

Figure 2. Sampling locations in the study area

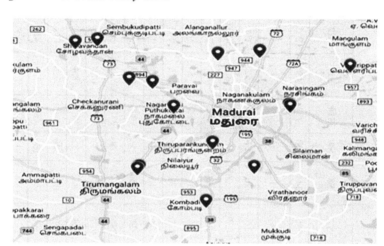

Table 1. Sampling locations

Well No	Well Type	Block	Village	Latitude	Longitude
21027D	Bore Well	Vadipatti	Thiruvedagam	10°01'10"	77°58'20"
21013D	Bore Well	Alanganallur	Manianji	10°00'15"	78°06'09"
21036D	Bore Well	Madurai West	Koolapandi	10°00'30"	78°08'50"
21010D	Bore Well	Madurai East	Karupayoorani	09°55'55"	78°10'40"
21001D	Bore Well	Thiruparankundram	Thiruparankundram	09°52'40"	78°05'50"
21035D	Bore Well	Madurai West	Madurai North	09°50'35"	78°08'30"
21005D	Bore Well	Thiruparankundram	Eliyarpatti	09°47'40"	78°05'25"
21009D	Bore Well	Thirumangalam	Kappalur	09°50'35"	78°01'07"
21014D	Bore Well	Thiruparankundram	Kilakuilkudi	09°55'42"	78°03'18"
21008D	Bore Well	Thiruparankundram	Melamathur	09°58'05"	78°00'28"
83014B	Dug Well	Vadipatti	Mullipallam	10°01'10"	77°57'02"
83025	Dug Well	Thiruparankundram	Keelamathur	09°57'48"	78°01'56"
83016A	Dug Well	Melur	Chittampatti	09°59'17"	78°14'35"
83075	Dug Well	Madurai East	KaruppayiUrani	09°55'50"	78°10'45"
83062A	Dug Well	Thirumangalam	Kappalur	09°50'25"	78°01'00"

Table 2. Physicochemical groundwater quality analysis results

Sl. No	Water Quality Parameter	S1	S2	S3	S4	S5	S6	S7	S8	S9	S10	S11	S12	S13	S14	S15	Permissible limit as per IS:10500:2012
1	TDS(mg/L)	347	160	180	87	614	160	580	344	200	733	450	802	2137	1866	463	500
2	NO2+NO3 (mg/L)	4	6	2	1	4	1	4	1	5	6	1	80	4	19	8	45
3	Ca(mg/L)	20	19	28	14	18	16	16	26	17	14	52	84	164	72	36	75
4	Mg(mg/L)	51.03	45	15.795	4.86	55.89	7.29	64.395	37.665	56	63.18	38.88	58.32	65.61	170.1	40.095	35
5	Na	35	0	14	7	92	37	92	58	90	161	67	106	552	322	81	Not Specified
6	K	22	40	4	3	20	4	24	13	40	15	5	19	351	86	9	Not Specified
7	Cl(mg/L)	71	71	32	32	124	21	170	78	117	216	67	145	681	737	82	250
8	SO4(mg/L)	34	60	16	5	223	14	48	13	40	120	43	127	298	288	53	200
9	CO3(mg/L)	7.19	1	3.6	0.37	0	2.07	24	8.54	1	0.7	6.46	0	0	0	2.4	200
10	HCO3(mg/L)	152.56	80	96.2	24.55	128.1	87.8	250.1	181.21	79	231.8	273.42	366	616.1	213.5	202.54	200
11	F(mg/L)	0.61	1	0.65	0.16	0.45	1.48	0.78	0.14	1	0.45	0.21	1	0.9	0.56	0.67	1
12	pH_GEN	8.7	8.5	8.6	8.2	8.1	8.4	8.4	8.7	7	8.1	8.4	7.4	8.2	7.6	8.1	6.5 -8.5
13	EC_GEN	710	720	330	130	1010	280	1010	690	340	1350	820	1340	3500	3280	780	NIL
14	HARDNESS(mg/L)l	260	150	135	55	275	70	305	220	180	295	290	450	680	880	255	200
15	SAR	0.94	0	0.52	0.41	2.41	1.92	2.29	1.7	1	4.08	1.71	2.17	9.21	4.72	0	Not Specified

Table 3. Results of basic statistic of water quality parameter from SPSS

				Missing		No. of Extremes[a]	
Univariate Statistics							
	N	Mean	Std. Deviation	Count	Percent	Low	High
TDS	13	674.08	629.607	2	13.3	0	2
NO2NO3	15	9.73	19.952	0	.0	0	2
Ca	13	43.08	42.916	2	13.3	0	1
Mg	13	51.77769	41.350297	2	13.3	0	1
Na	13	124.92	151.891	2	13.3	0	2
K	13	44.23	94.660	2	13.3	0	2
Cl	15	176.27	223.030	0	.0	0	2
SO4	13	98.62	105.808	2	13.3	0	1
CO3	13	4.2023	6.68287	2	13.3	0	1
HCO3	13	217.2215	149.40418	2	13.3	0	1
F	12	0.5883	0.37008	3	20.0	0	0
pH_GEN	15	8.160	0.4867	0	.0	2	0
EC_GEN	15	1086.00	1002.674	0	.0	0	2
HAR_Total	13	320.77	232.494	2	13.3	2	3
SAR	12	2.6733	2.42285	3	20.0	0	1
RSC	12	36.222	12.1799	3	20.0	0	0
Na_A	12	36.2217	12.17994	3	20.0	0	0
a. Number of cases outside the range (Q1 - 1.5*IQR, Q3 + 1.5*IQR).							

The researchers used standardized physicochemical data for correlation analysis using SPSS. Correlation analysis strongly emphasizes the degree of relation between the parameters. If r = +1, then the correlation between the two variables is said to be perfect and positive; if r = -1, then the correlation between the two variables is said to be perfect and negative; if r = 0, then there exists no correlation between the variables. Correlation coefficient more than 0.5 denotes the strong relationship between the parameters (Abdullah *et al.*, 2017). Tables 3 and 4 clearly show that TDS and EC have high positive correlation.

Factor Analysis

In this study, the authors used SPSS software for factor analysis, in order to reduce the huge variables numbers into few. The extraction of the factors has been done with a minimum acceptable Eigen value greater than 1.

Piper Diagram

Piper plot is a very useful diagram for visualization of common ions in water samples. It is the trilinear diagram used in hydrogeology and groundwater quality analysis. Piper plots comprises three components.

Table 4. Correlation matrix of groundwater quality parameters

	TDS	NO2+NO3	Ca	Mg	Na	K	Cl	SO4	CO3	HCO3	F	pH_GEN	EC_GEN	HARDNESS_Total	SAR	RSC	Na%
TDS	1																
NO2+NO3	0.22	1															
Ca	0.86	0.34	1														
Mg	0.8	0.26	0.47	1													
Na	0.97	0.09	0.88	0.64	1												
K	0.83	-0.02	0.88	0.36	0.93	1											
Cl	0.95	0.12	0.74	0.79	0.92	0.78	1										
SO4	0.92	0.23	0.72	0.78	0.87	0.71	0.87	1									
CO3	-0.13	-0.2	-0.19	0.05	-0.17	-0.15	-0.18	-0.29	1								
HCO3	0.8	0.32	0.9	0.48	0.83	0.79	0.62	0.66	0.08	1							
F	0.26	-0.31	0.19	0.14	0.31	0.31	0.19	0.2	0.19	0.23	1						
pH_GEN	-0.22	-0.52	-0.15	-0.22	-0.15	-0.04	-0.31	-0.3	0.41	-0.03	0.33	1					
EC_GEN	0.98	0.21	0.81	0.8	0.95	0.81	0.97	0.91	-0.17	0.74	0.18	-0.24	1				
HAR_Total	0.95	0.34	0.77	0.92	0.84	0.64	0.89	0.87	-0.05	0.74	0.19	-0.23	0.93	1			
SAR	0.91	0.06	0.82	0.58	0.96	0.89	0.84	0.84	-0.11	0.83	0.36	-0.07	0.89	0.78	1		

Table 5. Results of factor analysis

Communalities		
	Initial	**Extraction**
TDS	1.000	.997
NO2+NO3	1.000	.955
Ca	1.000	.974
Mg	1.000	.975
Na	1.000	.994
K	1.000	.978
Cl	1.000	.958
SO4	1.000	.907
CO3	1.000	.972
HCO3	1.000	.894
F	1.000	.738
pH_GEN	1.000	.929
EC_GEN	1.000	.996
HAR_Total	1.000	.982
SAR	1.000	.987
RSC	1.000	.968
Na%	1.000	.968
Extraction Method: Principal Component Analysis.		

Figure 3. Piper diagram

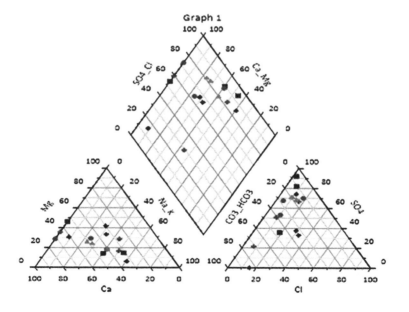

The left ternary shows the amount of cations in the given sample (e.g., Mg+, Na+, and Ca+). In the right side, the ternary shows the amount of anions present in the sample (i.e., so-, Cl-, Co3, and HCO3). The middle diamond component denotes the combination of cation and anion were represented in the picture. The above plot shows that Ca and Mg were present in higher amount, which leads to increase TDS and hardness.

CONCLUSION

In this study, the authors used multivariate statistical analysis for the statistical analysis of 15 groundwater samples. Piper diagram is used to classify the characteristics of groundwater. The results of correlation analysis show the relationships between parameters. Also, these results highlight that TDS and Conductivity (0.98), TDS and hardness (0.95), sodium and TDS (0.97), Sample 13 and 14 were identified as highly polluted samples. Thus, some treatment is necessary at sample locations 13 and 14 before supply.

ACKNOWLEDGMENT

This research received no specific grant from any funding agency in the public, commercial, or not-for-profit sectors.

REFERENCES

Anwar, K. M. (2014). Analysis of groundwater quality using statistical techniques : a case study of Aligarh city (India), *2*(5), 100–106.

Elango, L., & Subramani, T. (2005). Groundwater quality and its suitability for drinking and agricultural use in Chithar River Basin, Tamilnadu, India. *Environ Geol, 47*, 1099–1110. doi:10.1007/S00254-005-1243-0(2005)

Jamuna, M. (2018). Statistical analysis of ground water quality parameters in Erode District, Taminadu, India, (4), 84–89.

Kim, J., Kim, R., Lee, J., Cheong, T., Yum, B., & Chang, H. (2005). Multivariate statistical analysis to identify the major factors governing groundwater quality in the coastal area of Kimje, South Korea Multivariate statistical analysis to identify the major factors governing groundwater quality in the coastal area of Kimje, South Korea, (July 2018). doi:10.1002/hyp.5565

Sadat-Noori, S. M., Ebrahimi, K., & Liaghat, A. M. (2013). Groundwater quality assessment using the water quality index and GIS in Saveh-nodaran Aquifer, Iran, Enviro Earth Science. doi:10.1007/S12665-013-2770-8

Shahid, M., & Amba, G. (2018). Groundwater quality assessment of urban Bengaluru using multivariate statistical techniques. *Applied Water Science, 8*(1), 1–15. doi:10.100713201-018-0684-z

Abdullah, A. S., Selvakumar, S., & Abirami, A. M. (2017). An Introduction to Data Analytics: Its Types and Its Applications. In Handbook of Research on Advanced Data Mining Techniques and Applications for Business Intelligence. Hershey, PA: IGI Global. doi:10.4018/978-1-5225-2031-3.ch001

Thivya, C. & Chidambaram, S. (2013). A study on the significance of lithology in groundwater quality of Madurai district, Tamilnadu (India). *Environment, Development and Sustainability*. doi:10.1007/S10668-013-9439-Z

Urxqgzdwhu, Q. R. I. (n.d.)., qwhusuhwdwlrq ri *urxqgzdwhu 4xdolw\ 8vlqj 6wdwlvwlfdo $qdo\vlv iurp .rsdujdrq 0dkdudvkwud, qgld, 7. Indian water quality standards code 10500-2012.

Yidana, S. M., Yiran, G. B., Sakyi, P. A., Nude, P. M., & Banoeng-Yakubo, B. (2011). Groundwater evolution in the Voltaian basin, Ghana—an application of multivariate statistical analyses to hydrochemical data. *Nature and Science*, *3*(10), 837.

Zhang, Y., Guo, F., Meng, W., & Wang, X. Q. (2009). Water quality assessment and source identification of Daliao river basin using multivariate statistical methods. *Environmental Monitoring and Assessment*, *152*, 105–121.

Chapter 10
Probe People and Vehicle-Based Data Sources Application in Smart Transportation

Sina Dabiri
Virginia Tech, USA

Kaveh Bakhsh Kelarestaghi
ICF Incorporated LLC, USA

Kevin Heaslip
(iD) https://orcid.org/0000-0002-3393-2627
Virginia Tech, USA

ABSTRACT

Smart transportation is a framework that leverages the power of Information and Communication Technology for acquisition, management, and mining of traffic-related data sources. This chapter categorizes them into probe people and vehicles based on Global Positioning Systems, mobile phone cellular networks, and Bluetooth, location-based social networks, and transit data with the focus on smart cards. For each data source, the operational mechanism of the technology for capturing the data is succinctly demonstrated. Secondly, as the most salient feature of this study, the transport-domain applications of each data source that have been conducted by the previous studies are reviewed and classified into the main groups. Possible research directions are provided for all types of data sources. Finally, authors briefly mention challenges and their corresponding solutions in smart transportation.

INTRODUCTION

The rapid growth of various components of a city, including e-government and IT projects, technology, governance, policy, people and community, economy, built infrastructure, and the natural environment, has created a vast complex system (Chourabi et al., 2012; Saadeh et al.; 2018). Such a complicated

DOI: 10.4018/978-1-7998-0301-0.ch010

system brings about a variety of challenges and risks, ranging from air pollution and traffic congestion to an increase in the unemployment rate and adverse social effects. Making the cities "smart," using Information and Communication Technology, is one solution to manage the urban troubles and enhance cities' livability, workability, and sustainability (Council, 2013). Although the concept of the smart city is not novel, academics from different fields have defined this term in ways that are not consistent. For example, smartness in the marketing language focuses on users' perspectives, whereas the smart concept in the urban planning field is defined as new strategies for improving the quality of life and having the sustainable environment (Nam and Pardo, 2011).

A general framework of the smart city contains three layers. These layers include (1) data collection and management, (2) data analytics, and (3) service providing (Zheng et al., 2014). As shown in Figure 1, the smart city includes multiple functions such as the smart economy, smart governance, smart people, smart transportation and mobility, smart environment, and smart living (Batty et al., 2012). Notwithstanding a strong interconnection between these functions, this survey considers only "smart transportation." It should be noted that the three steps in the smart city architecture are applied to smart transportation as well.

Modern transportation systems are comprised of streets, railways, subways, traffic signals, vehicles, bicycles, and buses. The system moves people around the city for the commute to work/school, shopping, traveling, and leisure activities. The evolution of transportation networks in cities around the world gives rise to significant challenges. The principal challenge is traffic congestion due to a dramatic increase in transportation modes and population. Besides the direct adverse effects of traffic congestion on users, such as longer trip time and road rage behaviors, it also has long-term negative impacts on energy consumption and air quality (Smit et al., 2008), economic growth (Sweet, 2011), public health (Levy et al., 2010), and traffic crashes (Wang et al., 2009).

In this chapter, we investigate the currently available data sources that studies have used to develop with novel solutions to address challenges in transportation. A data source, in this study, is referred to as the technologies and systems that have the potential to collect traffic data. Data measured in the laboratories or collected by hiring subjects for a specific experiment are not within the scope of this survey. Accordingly, we are seeking to answer the following questions: What are the potential probe people and vehicle data sources in smart transportation? What type of technologies and methodologies are utilized to extract and collect transport-domain data from each traffic data source? What are the transportation-related applications associated with each traffic data source?

In pursuit of answering the first question, we categorize traffic data sources into three groups: 1) probe vehicles and people data, 2) location-based social networks data, and 3) transit data with the focus on smart cards data. The probe people and vehicle group, which refers to those moving sensors that receive spatiotemporal traffic data, are subdivided based on the technology used to collect the data: Global Positioning Systems (GPS), mobile phone cellular networks, and Bluetooth. Figure 1 illustrates the structure of the traffic data source categorization. Accordingly, the specific objectives of this study are to, first, briefly describe the operational mechanism of the technologies as mentioned above for generating traffic-related data. Secondly, for each data source, we systematically review the transport-domain applications, aiming to classify the past studies into the main groups. For the sake of feasibility, we concentrate on only representative examples in each of those main groups rather than attempt to cover all existing models exhaustively. Thirdly, about each data source, some possible areas for future research are provided.

PROBE PEOPLE AND VEHICLE DATA COLLECTION

Fixed-point sensors enable the measurement of temporal information in a particular location which results in lacking spatial traffic information which may not be representative of the network as a whole. One way to address this issue is to use vehicles and individuals that are equipped with location and communication providers as "probe people and vehicle" to sense their spatiotemporal information when they are performing their regular trips. In this study, we examine three well-established positioning tools that record spatiotemporal information of floating sensors while moving in the network. These technologies are GPS, mobile phone cellular networks, and Bluetooth. Devices equipped with these technologies can track vehicles or people's locations indexed in time to create their trajectory data. Such spatiotemporal data are useful inputs for a variety of transport-domain applications that are displayed in Figure 2 as an overall view and will be set out in the following sections. Also, a summary of the studies that used probe people and vehicle data for transport-domain applications is provided in Table 1. Before elaborating the related applications, we briefly characterize the mechanism of each technology.

Global Positioning System

Operational Mechanism of Global Positioning System

GPS is a satellite-based navigation system built up by 24 satellites that are orbiting the earth. Satellites are continually moving and powered by solar energy while backup batteries supply their energy when there is no solar power. A GPS operational mechanism incorporates three components: the satellites in the earth's orbit, the ground control stations, and the GPS-receiver devices such as mobile phones or handled GPS units (Mintsis et al., 2004). At each location on earth, at least four satellites are visible. A GPS device receives the radio signals broadcasted by the satellites. Radio signals transmit information about the current time and position of the satellite. Knowing the fact that the signal's speed is the same as the light's speed, the GPS receiver compares the difference between the transmitted time and receiving time to calculate how far a satellite is from the user's location.

Figure 1. Smart city and transportation architecture with traffic-related data sources

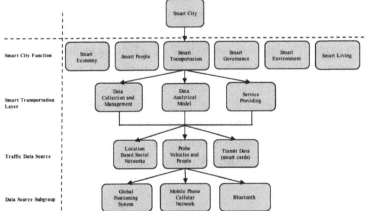

Calculating the latitude and longitude position requires obtaining the signals from at least three satellites. A user is located somewhere on the sphere surface with a radius equal to the distance from the first satellite. If the user knows the distance from the second satellite, the overlap between the first and second sphere determines the possible locations of the user. They are finding the distance from the third satellite results in two locations at intersections and the third satellite. Since one of the two points is located outside of the earth, another point is labeled as the user's location. Approximating the location using only three satellites' information may not be exact enough. The more accurate and precise location and also the latitude, longitude, and altitude requires four or more satellite signals. Thus, the object's position coordinates and its corresponding timestamp over a period can be collected using a GPS-equipped device.

Despite emerging advances in GPS technology, various factors affect its accuracy in identifying objects' position. The inability of signals to pass through solid objects such as tall buildings is one of the primary error sources. In a crowded area with skyscrapers, the signal is distorted from its main direction before reaching the receiver, which in turn results in increasing the travel time and generating errors. Orbital error, relative geometry of satellites concerning each other, and the number of visible satellites are some of the other sources for errors.

Transportation-Related Applications of GPS

A GPS trajectory also called the movement of an object, is constructed by connecting the GPS points of its GPS-enable device such as a smartphone. A GPS point, here, is denoted as (x, y, t), where x, y, and t are latitude, longitude, and timestamp, respectively. The original object trajectory is continuous while its corresponding data are discrete, which necessities linear or non-linear interpolation for finding positions between two consecutive discrete data points. Such a series of objects' chronologically ordered points can also be viewed as a set of spatial events. Spatial events are defined as consecutive occurrences of certain movements' features that are localized in the space (Andrienko et al., 2011); hence, its real definition may vary according to the specific interest of the researcher.

Vehicle fleet management and monitoring in urban and suburban areas can be considered as initial practical usage of the GPS technology (Mintsis et al., 2004). The installed GPS-enabled device in emergency, police, and transit vehicles monitors and transfers the vehicles' movement information, also well-known as Floating Car Data (FCD), to a control center via a wireless communication device. Such an operation can provide several advantages to the vehicle fleets such as detecting any traffic violation and improving the disposition of clients' orders in taxi companies. Several researchers have collated FCD to recognize current traffic states and predict its near-future conditions as the main components in designing and implementing advanced traveler traffic information systems and advanced traffic management systems. The traffic information calculated by processing and analyzing FCD consists of one or multiple layers including short-term link speed and travel time prediction (DeFabritiis et al., 2008), route travel time distribution (Rahmani, Jenelius, et al. 2015), traffic incident detection (Kerner, et al., 2005), congestion detection by estimating traffic density (Tabibiazar and Basir, 2011), dynamic routing and navigation (Gühnemann, et al., 2004), route choice behavior (Sun, et al., 2014), map-matching and path inference (Rahmani and Koutsopoulos 2013), and dynamic emission models (Gühnemann, et al. 2004).

Human mobility behaviors and their relation to traffic conditions can be characterized by constraining their movements into the underlying street network using the GPS data (Jiang et al., 2009). Analyzing the people's mobility under a vast street network leads researchers to understand and predict the traffic distribution using some GPS recorded points in each street link. For example, a power-law behavior can be fitted to the observed traffic using the speed effect of vehicles extracted from their GPS data (Jiang et al., 2009). The power-law distribution is one of the famous models that has been observed for the human mobility movement from real-world human mobility datasets in a broad spectrum of studies (Liang et al., 2012). Studying urban mobility benefits to characterize the attractiveness of various locations in the city, which is a significant factor in developing trip attraction models in urban planning (Veloso et al., 2011). Taxi GPS-based trajectories have utilized to infer driver's actions as a function of observed behaviors such as predicting actions at the next intersection, providing the fastest route to a given destination at a given departure time, and predicting the destination given a partial route (Liu et al., 2014). Moreover, vehicles and human's mobility studies based on positioning data have injected a variety of advantages in terms of inferring significant locations (Ashbrook and Starner, 2003; Agamennoni et al., 2009; Zignani and Gaito 2010; Andrienko et al., 2011; Xu and Cho 2014; Cho 2016), modes of transport, trajectory mining (Fang et al., 2009; Zheng et al., 2009; Zheng et al., 2010; Do and Gatica-Perez 2012), and location-based activities (Lin and Hsu 2014), which are discussed as follows.

Locations with frequent visits or high individual's dwell time are labeled as significant locations such as office buildings or intersections. As the object's movement is a spatial-temporal phenomenon with a sequence of events, essential locations are extracted from the occurrence of events. First, the events that happen repeatedly are detected and extracted, then important places to each group of events are determined using spatial-temporal clustering algorithms. Analysis of the aggregated events and places leads to determining the crucial locations about the subject application (Andrienko et al., 2011). Because GPS receivers are almost unable to get signals in the closed environment, office buildings as an example of significant locations can be detected by clustering the GPS points (Ashbrook and Starner 2003; Agamennoni et al., 2009; Xu and Cho 2014).

Detecting the mode of transport depends on the movement features that distinguish between modes. After dividing the GPS track into uniform segments, essential features of each segment including length, mean velocity, and maximum accelerations are employed to recognizing the transport mode(s) (Zheng et al., 2008). Nevertheless, speed and acceleration are erratic and unpredictable due to changes in traffic conditions, weather, and types of roadway. The vulnerability of essential motion characteristics forced researchers to introduce additional features such as changing rate in direction, stop rate, and velocity change rate (Zheng et al., 2008). The hand-crafted features may still cause drawbacks including vulnerability to traffic and environmental conditions as well as possessing human's bias in creating efficient features. One way to overcome these issues is by utilizing the Convolutional Neural Network (CNN) schemes that are capable of automatically driving high-level features from raw GPS trajectories (Dabiri and Heaslip, 2017).

Vehicle and human trajectory mining can be achieved by converting their movement into a sequence of spatial locations. Trajectory clustering, trajectory patterns extraction, and trajectory classification comprise the three most important groups in the trajectory mining (Lin and Hsu, 2014). In clustering, typical movement styles that are relatively close to each other are discovered and grouped into one cluster (Wu et al., 2013). The measure of closeness is considerably contingent on the research application; yet computing the distance between trajectories by temporal and spatial information is a standard technique to measure the closeness (Lee et al., 2008, Wu et al., 2013). In the trajectory pattern extraction problems, frequent (sequential) patterns, which are the routes frequently followed by an object, are discovered for predicting the future movement using association rules while the trajectory is viewed as spatiotemporal items (Cao et al., 2005). Vehicles' paths can be classified using associative classifications

Figure 2. Main applications of probe people and vehicle data in transportation systems

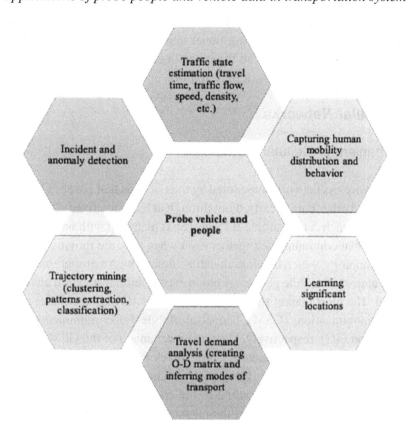

considering the trajectory patterns and users' context as features (Lee et al., 2011). The trajectories' labels are composed of every type of mobility behavior in the road network including dwell time in the current place, destination (next place), trip purpose, or activity such as shopping, parking a car, dining. Useful transportation and traffic planning applications based on GPS trajectory mining are exemplified as a carpooling system, prediction of the next destination or the current place stay duration (Fang et al., 2009; Do and Gatica-Perez, 2012), social networking service for sharing users' life experience, and travel recommendation systems (Zheng et al., 2009; Zheng et al., 2010).

Traffic anomaly detection can be categorized as a study either in traffic incident management or a type of trajectory mining (Pan et al., 2013; Liu et al., 2014). Outlier/anomaly regarding any behavior that does not conform to the expected behavior. GPS-based traffic outliers are grouped into two groups. The first group is referred to detect a driver's trajectory or a small percentage of drives' trajectories that their behavior is considerably different from others such as a fraudulent taxi driver or a traffic offender. An outstanding example of this category is the work of (Liu et al., 2014), where a fraudulent taxi behavior, who overcharges passengers for more than the actual distanced travel by altering the taximeter into a lower scale, is discovered by comparing its real speed based on GPS logs and the speed obtained from a malfunctioned taximeter. The second group investigates an occurrence of traffic anomaly by the distinction between a vast majority of trajectories. For instance, traffic anomalies caused by accidents, disasters, and unusual events are captured according to irregular drivers' routing behaviors on a sub-graph of a road network. Comparison between current suspicious drivers' actions is carried out with remaining trajectories at the same time and location or with the historical record of drivers (Pan et al., 2013). The detected outlier in the system is reported using travel information systems to the traffic management system for performing appropriate actions; in the meantime, drivers are being informed on the occurrence of ahead traffic jams, which may result in re-routing.

Mobile Phone Cellular Networks

Operational Mechanism of Cellular Networks

A cellular network is a wireless network constituted by a set of cells that covers a land area, where transmission of voice, text, and other data is carried out through at least one fixed-location Base Transceiver Station (BTS) allocated to each cell. Multiple BTS nearby is managed with one Base Station Controller. Thus, some cell phones can communicate together even when they are moving through more than one cell. Today, cellular technology, which is communicating directly with a ground-based cellular tower, has been implemented in almost all mobile phones. A broad digital cellular standard has been introduced and used across the world. The most market shares for 2G, 3G, and 4G digital cellular networks are Global System for Mobile Communications (GSM), Universal Mobile Telecommunications System (UMTS), and Long-Term Evolution (4G), respectively. Here, the mechanism of the GSM cellular network, as an example and the highest market share standard, is briefly reviewed.

When a mobile phone is connected to the BTS by exchanging signals, the network operator approximates the mobile phone location according to the coordinates of BTS, which results in having accuracy at the Cell-ID level. Although this is the most used technique for locating the mobile phone, other techniques such as Received Signal Strength method and triangulation have been proposed in the literature to increase the accuracy; however, these methods acquire extra network elements that are not essential for cellular networks to work.

In addition to the location information, parameters regarding the rate of using cellular networks help determine traffic parameters (Steenbruggen et al., 2013). Handover is the mechanism that provides a permanent connection by switching an on-going call between different BTS while a mobile phone is moving through multiple cells in the network. The Cell Dwell Time shows the duration that the phone remains with a BTS before handing off to another base station, which is an index to compare the level of traffic congestion between cell areas. Another parameter for understanding crowdedness in a zone is the Erlang that indicates how many hours a user has utilized the network- one Erlang equals to a one-person hour of phone use. It should be pointed out that all network data can be collected even when the phone is only switched on.

Transportation-Related Applications of Cellular Networks

Cell phones as a pervasive sensing device and enthusiastic adoption across most socioeconomic strata is a reliable source to record users' spatiotemporal information. Such capabilities on mobile phone data have persuaded urban researchers and authorities to employ this real-time and extensive location data on a variety of behavioral applications, ranging from inferring friendship network structure and real estate market to deriving geography of human urban activity and emergency management (Steenbruggen et al.,

2013). Besides, analyzing and modeling commuters' movement turns these large-scale spatiotemporal data into unprecedented and actionable insights in the field of urban planning and traffic management.

Mobile phone data, utilized in transport domains, can be divided into two parts: 1) spatiotemporal information which identifies the location of a phone over a period, 2) data related to the usage rate of the cellular network such as handover, Erlang, and the number of calls. One should pay heed to the fact that as long as location-based applications are the concern, it makes no difference what technology has been deployed to obtain vehicles and commuters' flow and trajectory data. As mentioned, a mobile phone is located using various techniques. The huge growth in the market share of phones with built-in GPS receivers gives room to detect the location of a phone using GPS, which is a more accurate technology with higher resolution. The cellular networks technology is still an effective way to capture real-time, cellular-signal data points; particularly, where the GPS technology is not capable of collecting movement data such as inside buildings or once the mobile phone's location service has been turned off by the user.

Moreover, on the premise that at least one cell phone is carried inside a moving vehicle, individuals' and vehicles' movements data are collected cost-effectively. *Consequently, the GPS-based travel patterns and transportation trends mentioned in section 3.1.2 have also the potential to be inferred from the movement data provided by the cellular network even though the differences in the spatial coverage and resolution of the two systems should be considered.* Thus, we focus more on potential transport-related applications of cellular networks that collate network-usage rate parameters rather than location-based information.

Traffic parameters such as origin-estimation (OD) trip matrix, travel time, speed, traffic flow, traffic density, and congestion are obtained by applying a proper procedure on the mobile phone data (Caceres et al., 2008). The ultimate goal in the OD research area is to estimate or update the average number of trips going between each pair of traffic zones, called origin and destination, during a period. The location of onboard cell phones can be determined in terms of the cells they travel through it. The centroids of cells or the location of BTS in each cell is considered as the location of the mobile phone in each cell (Caceres et al., 2007). Traffic volume is a useful parameter for understanding the traffic patterns and demand in roadways. Cellular systems can provide the traffic volume information by counting the number of vehicles moving in each cell area. Also, a transition between every two cell borders can be detected using cellular network parameters such handovers, since the handover works the same as a virtual traffic counter between borders of two cell areas (Thiessenhusen et al., 2003). The most important advantage of cellular systems compared to other methods is no need for installing additional infrastructure such as loop detectors. In cellular networks, the speed can be computed from the distance and travel time between two consecutive locations (Thiessenhusen et al., 2003). Double handovers is another way for measuring the speed. If a call is sufficiently long to produce two handovers, which means passing through two cell areas, speed is estimated using time and location information that a vehicle enters and exits the boundaries of a specific cell. The similar concept has been deployed for measuring the travel time. The time information on two adjacent handover occurrences identifies the travel time of the cell area that a phone crosses. However, as the cell area covers multiple road sections, a constant mapping process requires matching the captured travel time to the real road section that the phone has passed.

Traffic incident management is another area that can be addressed using cellular systems (Steenbruggen et al., 2013). First of all, any sudden change in traffic mentioned above parameters such as speed and travel time is a clue for an incident occurring. Furthermore, comparing the current rate of mobile phones' usage in a specific location with the historical average rate usage in the same location is a sign of a congestion situation owing to the facts that not only the number of phone users is more in heavy

Table 1. Summary of probe people and vehicle data collection transport-related applications

Application	Example	Data Source	Methodology
Learning significant locations	Exploring people movement behaviors by examining places where people make long stops or where they meet. Exploring maritime traffic, to identify anchoring areas and areas of major turns and crossings. For instance: analyzing traffic congestion; predict people movements and schedule; crowd mobility during special events.	GPS (Ashbrook and Starner, 2003, Agamennoni et al., 2009, Zignani and Gaito, 2010, Andrienko et al. 2011, Xu and Cho, 2014, Cho, 2016), Cellular (Calabrese et al., 2010, Becker et al., 2011)	Density-based clustering analysis (Andrienko et al., 2011); Post hoc clustering algorithm, Markov model (Ashbrook and Starner, 2003); Kernel density model (Agamennoni et al., 2009); Hidden Markov model, k-Nearest neighbor, Decision trees (Cho, 2016); neural network (Calabrese et al., 2010); Density-based spatial clustering (Zignani and Gaito, 2010);K-nearest neighbor and decision tree algorithms (Xu and Cho, 2014) K-means clustering algorithm (Becker et al., 2011)
Trajectory mining	Carpooling system, prediction of the next destination or the current place stay duration, social networking service for sharing users' life experience, and travel recommendation systems urban planning applications.	GPS (Fang et al., 2009, Zheng et al., 2009, Zheng et al., 2010, Do and Gatica-Perez, 2012), Cellular networks (Dash et al., 2016)	Contextual Conditional Model (Do and Gatica-Perez, 2012); particle filtering algorithm (Fang et al., 2009); change point-based segmentation method, Decision Tree-based inference model (Zheng et al., 2010); tree-based hierarchical graph, Hypertext Induced Topic Search inference model (Zheng et al., 2009); Dynamic Bayesian Network, tree-based query processing method (Dash et al., 2016)
Traffic incident and anomaly detection	Traffic incident response management purposes, identification of fraudulent taxi driver or traffic offenders.	GPS (Pan et al., 2013, Liu, et al., 2014), Cellular (Ygnace et al., 2001, Pattara-Atikom and Peachavanish, 2007)	Map-matching method, clustering analysis (Liu, Ni et al. 2014); Map-matching method, Anomaly detection (Pan et al., 2013); neural network (Ygnace et al., 2001, Pattara-Atikom and Peachavanish, 2007);
Capturing human mobility distribution & behavior	Inferring friendship network structure and real estate market, to deriving the geography of human urban activity and emergency management.	Cellular (Reades et al., 2007, Girardin et al., 2008, Rojas et al., 2008, Calabrese et al., 2011, Torkestani and Applications, 2012)	Data Fit Location Algorithm (Calabrese, Colonna et al., 2011); K-means clustering algorithm (Reades et al., 2007); Geo-visualization, Spatial data clustering (Girardin et al., 2008)
Traffic state estimation	Urban planning applications, drivers route choice behavior under road closure, traffic monitoring.	Cellular (Thiessenhusen et al., 2003, Ratti et al., 2006, Bar-Gera, 2007), Bluetooth (Hainen et al., 2011, Tsubota, et al., 2011, Wang et al., 2011, Nantes et al., 2013)	Geographical mapping of cell phone usage (Ratti, Frenchman et al. 2006); Graphical representation (Bar-Gera 2007); Descriptive statistical analysis, regression analysis (Wang, Malinovskiy et al. 2011); Descriptive statistical analysis, visualization (Hainen, Wasson et al. 2011, Tsubota, Bhaskar et al. 2011); Bayesian network (Nantes et al., 2013);
Travel demand analysis	Urban planning applications, mobility analysis	Cellular networks (Caceres et al., 2007), Bluetooth (Barceló et al., 2010)	Storing location update and transforming a cellular phone data into traffic counts (Caceres et al., 2007); Kalman Filter Approach (Barceló et al., 2010)

traffic, but users intend to communicate more with people at their trip destination. Such analysis can be fulfilled using the Erlang index. CDT, which shows the duration of being registered to a specific BTS before a handover occurs, is another measurement for identifying the level of traffic congestion. High CDT of a phone inside a probe vehicle indicates a long travel time and in turn a high level of congestion in the phone's cell area (Pattara-Atikom and Peachavanish, 2007). As Erlang is a measurement of the mobile phone usage in a sector, it can be used to estimate the number of users in a cell area, which is highly correlated with traffic density (Caceres et al., 2008). Comprehensive reviews on studies that have developed frameworks and methodologies to characterize transportation system parameters through cellular network data are available in the references (Caceres et al., 2008; Steenbruggen et al., 2013).

Bluetooth

Operational Mechanism of Bluetooth

Bluetooth is a short-range, wireless technology that uses a short-wavelength radio communication system for exchanging data between Bluetooth-equipped devices such as mobile phones and car radios. Such a standard wire-replacement protocol simplifies connection settings between devices in terms of security, network address, and permission configuration. It follows a master-slave architecture, in which a master device is communicating with one or more devices (slaves) for transferring data over a short-range and ad-hoc network in the range of 1 to 100 meters, dependent on the device's Bluetooth power class.

The Bluetooth traffic data collection system leverages Bluetooth probe devices, also called Bluetooth detectors, located adjacent to roadways for scanning Bluetooth-equipped devices inside vehicles such as driver's smartphones or built-in Bluetooth car audio. The probe device detects and records the unique Machine Access Control (MAC) address of Bluetooth-enabled vehicle that has entered the probe device's radio proximity. By knowing the location of the installed probe device, the present location and time information of those Bluetooth-enabled vehicles that have just crossed the probe device is also identified. Vehicles' trajectories can be built by recording the MAC address of their onboard Bluetooth-enabled devices at multiple stations and extracting the unique MAC address in chronological order using a sensing system with multiple probes in sequence. The stored information is then transmitted to a control server for further analysis.

Transportation-Related Applications of Bluetooth

Potential applications of the Bluetooth technology in traffic monitoring and management as a noninvasive wireless data collection method that has no impact on the existing traffic conditions have been investigated for the last decade. A vehicle's travel time between two successive Bluetooth detectors is calculated by measuring the difference in time between two consecutive sensor stations that the vehicle has crossed. Since Bluetooth detectors are capable of reading vehicles' MAC addresses within a specific range, the measured travel time is between two zones rather than two points. Hence, among multiple times that a vehicle's MAC address is scanned within each Bluetooth's detection zone, only the first-to-first or last-to-last detection needs to be matched in the travel time measurement for keeping consistency and reducing error (Wang et al., 2011). Because the locations of detectors are already known, a vehicle's travel distance can be easily computed by finding the distance between those two successive stations that the travel time measurement was calculated. The space speed is then obtained by having the travel time and the travel distance. Figure 3 displays the concept of estimating travel times and speeds using a Bluetooth sensor system. Identifying travel times and speeds based on vehicles' trajectories gives room to establish other traffic management and traveler information applications such as creating time-dependent O-D matrixes over a time interval (Barceló et al., 2010), estimating traffic distributions on alternative routes (Hainen et al., 2011), and predicting dynamics of traffic volumes (Nantes et al., 2013). Furthermore, the time duration that a Bluetooth-enabled vehicle spends to pass through the coverage area of a Bluetooth detector can be used as a quality-performance index of transportation systems. For example, in case of installing Bluetooth scanners at intersections, investigating the relationship between the link travel time and duration time leads to perceiving the level of service and delay at intersections (Tsubota et al., 2011).

Transportation Research Record: Journal of the
Transportation Research Board, 2010)

Bluetooth data are subjected to sources of errors and noise that can be categorized into three groups: spatial error, temporal error, and sampling error (Wang et al., 2011). The spatial error refers to the non-similarity of Bluetooth readers' detection zones that is dependent on the amplitude of signals emitted by home appliances and cell phones. The second source of errors stems from variations in the detection time after a vehicle enters the detection zone of a Bluetooth reader since detectable characteristics of readers such as their signal strength and sensitivity are disparate. The third group points out to the possibility of having multiple Bluetooth devices in one vehicle, which results in duplicating one instance more than once in the database. Also, cyclists and pedestrians that move fast may also be counted as vehicles since Bluetooth readers are not able to distinguish whether the scanned devices are located inside the vehicles or other transportation modes. Thus, robust and effective filtering methods have been proposed to eliminate outliers and time intervals with either low number of observations or significant fluctuation in individual observations (Haghani et al., 2010).

LOCATION-BASED SOCIAL NETWORK

Today, social networking services have attracted millions of people around the world to share their thoughts, photos, locations, and videos. According to the Business Insider Intelligence report, approximately 20% of the total time spent on the internet is on social media networks such as Facebook, Tweeter, Instagram, and Foursquare. In this context, an interesting question then arises: what type of hidden knowledge can be extracted from social network services for improving transportation systems?

Figure 3. Estimation of travel time and speed in a section using Bluetooth sensors
(Source: Data collection of freeway travel time ground truth with Bluetooth sensors.

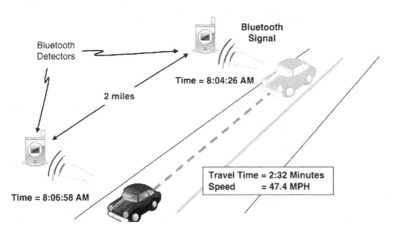

Before proceeding to describe applications of social network services in transportation systems, the most related sort of social networks alongside its distinct attributes need to be introduced. Among a variety of social networking services, the Location-Based Social Network (LBSN) that generates users' spatiotemporal information is the most useful type of social media in transportation-related applications. The most distinct feature of an LBSN is enabling users to share location-embedded social contents that result in understanding users in terms of their location (Business Insider, July 2016). A user can find and insert the location associated with the social media posts using location-acquisition technologies such as GPS that have been equipped in electronic devices. If users have permitted access location services, the geographic location of the device that a user uses to post is automatically recorded. Foursquare, Twitter, and Flicker are the most widely used examples of LBSNs.

The service for sharing the location in LBSNs is categorized into three groups: Geo-tagged-media-based, point-location-based, and trajectory-based. In the first group, the location of shared media contents such as texts and photos is automatically labeled by devices or manually by users. Flicker, Panoramio, and Twitter are the representatives for such a service. Foursquare and Yelp are the social networks where people share their current location using the check-in feature as the point-location-based group. Evolving social networks are adding an option for recording users' GPS trajectories, which results in providing new information such as travel time, speed, and duration of stay in a location (Zheng, 2011).

Users and locations are the major research topics in LBSNs although a strong correlation exists between the two (Zheng, 2011; Zheng and Xie, 2011). Location-tagged contents indicate visiting locations of users. Connecting these locations sequentially with their timestamps leads to building a trajectory for each user. By tracing the location history of users, similarities in users' behaviors and interests can be extracted. Understanding users' activities and interests bring about several applications including friend recommendations upon mutual interests, finding local experts in a region, and community discovery (Zheng, 2011). With mining users' travel sequences and processing the corresponding geo-tagged social media contents, a broad spectrum of transportation-related applications has been developed including travel recommendation systems, travel demand analysis, travel patterns and human mobility, urban planning, incident detection, emergency systems, and public transit, as depicted in Figure 4.

Transportation-Related Applications of LBSNs

Travel recommendation systems: Using the LBSNs raw information and according to individual's preferences and constraints such as time and cost, a location recommender system provides the user either stand-alone locations such as a restaurant or a series of locations in the form of a travel route. Users' Geo-tagged social media and GPS trajectories are the typical data sources for providing the sequential location recommendation (Lu et al., 2010; Yoon et al., 2012; Sun et al., 2015). For instance, geo-tagged photos generated by users in an LBSN such as Panorama can be mined to provide a customized and automatic trip plan that contains three modules including destination discovery, discovering internal paths within a destination, and travel route suggestion plans. The modules optimize the best plan according to users' travel location, intended visiting time, and potential travel duration (Lu et al., 2010). Generating incomplete paths is the main weakness of geo-tagged photos since tourists and visitors are not necessarily taking and uploading photos in all locations they are visiting. Aggregating all photos taken in the same location and designing an algorithm to produce all possible travel routes by merging incomplete travel paths is a solution to alleviate the incomplete path issue. In (Sun et al., 2015), using Flicker geotagged photos, top-ranking travel destinations and the best travel route between those destinations are detected and recommended while the main criteria for assessing the best route are the tourism popularity and the minimum distance to the destination.

Travel demand analysis is one of the active transportation research fields that models the mobility of people and vehicles to estimate travel behavior and travel demand for a specific future time window. Traditional four-step modeling, as the most widely used model, forecasts the number of trips with a specific transportation mode made on each route between every pair of traffic zones. Foursquare check-in and geo-tagged Twitter data have been utilized to generate an origin-destination model, which is the second step in the four-step modeling (Lee et al., 2015; Yang et al., 2015). The results of studies verify the reliability of O-D matrixes created upon social media data as a cost-effective and non-time-consuming approach by comparing the proposed model with expensive and large-scale approaches such as household surveys (Cebelak, 2013; Lee et al., 2015). In (Lee et al., 2015), for instance, millions of geo-tagged tweets in the Greater Los Angeles Area have been collected and mined to extract the O-D matrix. The proposed methodology consists of two steps: individual-based trajectory detection and place-based trip aggregation. Considering traffic analysis zones (TAZs) in the network, an individual OD-trip is defined when a user generates two consecutive tweets in different TAZs within a threshold time, which was set to four hours in this example. In the second step, the extracted trips for each pair of O-D are aggregated at different time windows such as an hourly or daily interval. Content analysis techniques can be applied to geo-tagged tweets to extract the distribution of travel mode choices in a specific area (Maghrebi et al., 2016).

Travel patterns and human mobility: Recently, a variety of research endeavor has sought to uncover underlying patterns of trips and human movement behavior from the social media data (Noulas, Scellato et al. 2011; Noulas, Scellato et al. 2012; Hasan and Ukkusuri 2014; Liu, Sui et al. 2014; Wu, Zhi et al. 2014; Jurdak, Zhao et al. 2015). After identifying the sequence of geo-location data for individuals, the essential characteristic of human mobility patterns, which is the displacement distribution such as power-law or exponential or a hybrid function, should be fitted according to the trajectory data collected through social media data. In the inter-urban movement analysis, the spatial interaction between cities can be determined using gravity models, in which the power-law distribution is a widely-used distance function (Liu et al., 2014). Spatial interactions between regions are rich information on understanding the spatial structure and community boundaries of a network. In intra-urban human mobility, the urban area is divided into some cells, and a trip length is considered as the distance between centers of two consecutive cells that an individual has traveled. Trip purposes can be recognized from the type of check-in locations such as home, work, entertainment, and shopping. Accordingly, an activity is defined as a triple of the particular cell area, check-in time and venue. By finding a transition probability between two activities, agent-based modeling can be deployed to reproduce the observed human mobility (Wu et al., 2014).

Furthermore, multi-day activity patterns of an individual can be inferred through Twitter messages contained Foursquare check-in data that show the type of activity (e.g., shopping and education) with the corresponding time and location (Noulas et al., 2011; Noulas et al., 2012; Liu et al., 2014; Wu et al., 2014). In spite of significant advantageous of social media data such as being large-scale, broad coverage and low cost, it accompanies with constraints including the lack of detailed description of activities, the absence of individuals' socio-demographic information, and inability to cover all activities since users are not sharing all sorts of activities (Hasan and Ukkusuri, 2014). Table 2 provides a summary of the studies that used LBSN data for transport-related applications.

Figure 4. Main applications of LBSNs in transportation systems

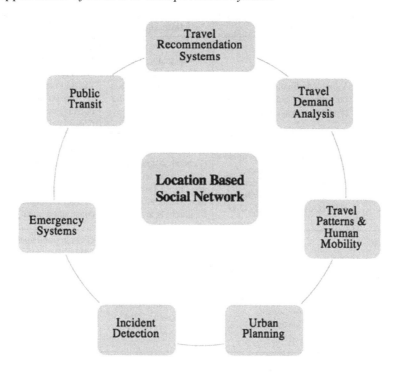

Urban planning: Geolocation social media data support the urban planning field and understanding of cities' dynamics, structures, and characters on a large scale (Cranshaw et al., 2012). City planners can use social media data for introducing metrics that show interactions between people, understanding people's problem, and identifying potential solutions for improving cities (Tasse and Hong, 2014). Mapping mobility, analyzing the design of physical spaces, and sociodemographic mapping status are examples of indices that are potential to be captured by the social media and deployed in improving the quality of life in cities. The structure of local urban areas and neighborhoods can be inferred by clustering nearby locations from check-in Foursquare data based on two criteria: spatial proximity and social proximity. While the above criterion measures the closeness of venues to each other, the latter one is the social similarity between each pair of venues based on the number of times a venue is visited (Cranshaw et al., 2012).

Monitoring traffic condition and traffic incidents: A powerful traffic monitoring and incident detection system needs access to complementary sources with a high coverage area such as social media data. On the contrary to existing data sources such as inductive-loop detectors and traffic cameras, social media data are not limited by sparse coverage, which in turn can be considered as a ubiquitous and complementary data source in traffic information systems. Among various popular social networks, Twitter has been received more attention for extracting traffic conditions due to using short messages (Carvalho, 2010; D'Andrea et al., 2015; Gu et al., 2016; Dabiri and Heaslip, 2019). The limitation on the tweet length forces people to communicate in a timely and effective fashion. Several studies have exploited text mining and machine learning techniques to automatically extract useful and meaningful information from the unstructured and irregular text of tweets to discriminate traffic-related tweets from non-traffic-related tweets. Bag-of-words representation is a typical approach for mapping tweet texts into

Table 2. Summary of location-based social network transport-related applications

Application	Examples	Data Source	Methodology
Travel recommendation systems	Destination discovery, discovering internal paths within a destination, find and rank itinerary candidates, and travel route suggestion plans.	Geo-tagged social media (Lu et al., 2010), GPS trajectories (Yoon et al., 2012, Sun et al., 2015)	The average timespan of individual fragments (AVE) and Gaussian model-based method (Lu et al., 2010); Entropy filtering method, Binary Logistic Regression (Sun et al., 2015); social itinerary recommendation framework (Yoon et al., 2012)
Travel demand analysis	O-D matrix extraction, extract the distribution of travel mode choices in a specific area, individual level travel demand models such as activity-based models, dynamic hourly O-D patterns.	Foursquare check-in (Cebelak 2013, Yang et al., 2015), and geo-tagged Twitter data (Lee et al., 2015, Maghrebi et al., 2016)	spatial lag Tobit Model and latent class regression model (Lee et al., 2015); Text mining (Maghrebi et al., 2016); Clustering analysis, regression model, and gravity model (Yang et al., 2015); doubly-constrained gravity model (Cebelak, 2013)
Travel patterns & human mobility	The consensus of people activity at a specific location and time, Determination of spatial interaction between regions, urban human movements across cities	Foursquare check-in (Noulas et al., 2011, Noulas et al., 2012, Liu et al., 2014, Wu et al., 2014), Twitter (Hasan and Ukkusuri, 2014, Jurdak et al., 2015)	Geo-temporal analysis (Noulas et al., 2011); Hierarchical mixture model (Hasan and Ukkusuri, 2014); Activity-based modeling (Wu et al., 2014); Density-based spatial clustering of applications with noise clustering (Jurdak et al., 2015); Rank-based movement model (Noulas et al., 2012); Gravity model (Liu et al., 2014)
Urban planning	Introducing metrics that show interactions between people, understanding people's problem, and identifying potential solutions for improving cities	Foursquare (Cranshaw et al., 2012), Twitter (Stefanidis et al., 2013)	Spectral clustering (Cranshaw et al., 2012); Text mining (Stefanidis et al., 2013)
Monitoring traffic condition and traffic incidents	Transportation safety, Traffic Incident Management	Twitter (Carvalho, 2010, D'Andrea et al., 2015, Gu et al., 2016, Dabiri and Heaslip, 2019)	Word-embedding tools, convolutional neural network, and recurrent neural network (Dabiri and Heaslip, 2019); Support vector machine, k-nearest neighbor (D'Andrea et al., 2015); Semi-Naive-Bayes model (Gu et al., 2016); Support vector machine (Carvalho, 2010);
Emergency systems	Choice assessment during evacuation, emergency response management	Twitter (Sakaki et al., 2010, Chaniotakis et al., 2017)	Support vector machine, Kalman and particle filtering (Sakaki et al., 2010); Density-Based Spatial Clustering (Chaniotakis et al., 2017)
Public transit	Public transit unplanned disruption, infer public perceptions of transit service	Twitter (Collins et al., 2013, Gal-Tzur et al., 2014, Casas and Delmelle, 2017)	Text mining (Gal-Tzur et al., 2014, Casas and Delmelle, 2017); Sentiment Strength Detection Algorithm (Collins et al., 2013)

numerical feature vectors before feeding them into a supervised learning algorithm for the classification task (Carvalho, 2010; D'Andrea et al., 2015). However, it creates a high-dimensional sparse matrix, which calls for compressing the numerical vector space into only a set of traffic-related keywords to limit the number of features (Gu et al., 2016). Although such a strategy engenders a much-lower-dimensional matrix, the immediate critique of using a pre-defined set of keywords as features is that the vocabulary may not include all essential traffic-related keywords.

Furthermore, bag-of-words representation completely ignores the temporal order of words, which leads to missing the semantic and syntactic relationship between words in a tweet. One way to address these shortcomings is to utilize the semantic similarity between the words in a tweet using word embeddings tools (Pereira et al., 2017). A word embedding model maps millions of words to numerical feature vectors in such a way that similar words tend to be closer to each other in vector space. However, the most efficient way of detecting traffic-related tweets without needing a pre-defined set of keywords is to deploy deep learning architectures such as CNN (Dabiri and Heaslip, 2018). These models, which are very popular in semantic analysis and text categorization, have high capability to capture both local features of words as well as global and temporal sentence semantics (Zhou et al., 2015).

Emergency systems: A significant and growing body of literature has investigated the role of social media in emergency and disaster management systems since the past decade (Sakaki et al., 2010, Chaniotakis et al., 2017). Imran et al. (2015) reviewed the cutting-edge computational methods for processing and converting the raw information of social media into meaningful and actionable insights for managing emergency events such as natural disasters. The study presented methods on social media data acquisition and preprocessing, detection and tracking emergency events, clustering and classification of messages for extracting and aggregating useful information, as well as applications of semantic technologies in a crisis. Twitter is the most widespread social media platform used in the literature of emergency management due to its distinct features such as the ability of users to post tweets on a real-time basis during emergency cases and providing a higher number of queries and data. In (Sakaki et al., 2010), as a well-structured example for the application of Twitter in the real-time emergency event detection. The target event is first detected by classifying tweets into positive/negative labels using the support vector machine algorithm when features include keywords in a tweet, the number of words in a tweet, and the word before and after the query word. Afterward, the location and trajectory of the target event are estimated using probabilistic spatiotemporal models such as Kalman and particle filtering.

Public Transit: In the public transportation research area, social media data have been utilized to assess transit systems operation according to public opinion and attitudes that can be inferred by a sentiment analysis (Collins et al., 2013; Gal-Tzur et al., 2014; Casas and Delmelle, 2017; Rashidi et al., 2017). The strategies that public transportation agencies have deployed to arrange their social media programs, associated goals, and measurements for assessing their programs have been studied in the reference (Liu et al., 2016). Results of statistical analysis, that was applied to information collected through online surveys filled out by top transit agencies, indicate that social media can be used to proliferate transit environmental benefits, safety, and livability improvements. A comprehensive literature review on the role of social media in managing public transit has been conducted in the reference (Pender et al., 2014).

SMART CARD AND AUTOMATED PASSENGER COUNTER

Smart card Automated Fare Collection (AFC) and Automated Passenger Counter (APC) are technologies for collecting public transit data in order to both describe the spatial-temporal patterns of passengers' behavioral and evaluate transit facilities. These two sources of transit data can be supported by openly General Transit Feed Specification (GTFS) file, which contains publicly-accessible public transportation scheduled operations (e.g., daily service patterns) and network geometry information (e.g., stop locations) that have been published by transit agencies (Lee and Hickman, 2013). Although APC has been mainly designed for counting passengers in and out of public transport modes (e.g., bus and subway), the primary purpose of a smart card is to collect revenue. In the following sections, we briefly present how AFC and APC work and what kinds of raw data they can collect. Then, we review the studies that have disclosed applications of these emerging technologies on transit system planning and operations. Compared to relatively few historical studies in the area of APC, a great deal of previous research into public transit data has focused more on smart card systems on account of its ubiquity and ability to collect a large-scale spatiotemporal data. Thus, we mainly elaborate on the applications of smart card systems even although the data provided by both systems are interchangeable and complementary in several applications.

Operational Mechanisms of Smart Card AFC and APC

The smart card is a device designed for storing data and equipped with an embedded chip on which information is stored. Smart card technologies can be divided into two groups: contact and contactless card. In the contact type, the embedded chip is not covered with plastic, which the card needs to have a direct physical contact to be connected with a reader. However, in the contactless smart card, the card contains not only the chip, which is completely embedded into the card but also an antenna. The latter component enables the card to have remote communication with radio frequency identification (RFID) devices through high-frequency waves (Pelletier et al., 2011). By bringing the card close to a reader, the chip is powered through the electromagnetic field of the reader. Then, wireless communication, based on high-frequency waves, is established to transfer data between the card and the reader. As the contactless cards have become more secure, more transit agencies in the US have used contactless smart cards for AFC systems since the 1990s (Alliance, 2011). By moving a smart card close to the reader of a public transport station, a set of typical data are being collected: validation status of the smart card (expiration date, compatibility with type of service), status of transaction (boarding acceptance/refusal, transfer), card ID, route ID/direction, stop/bus ID, date and time of operation (Pelletier et al., 2011). All data stored in a reader are transferred to the central server regularly.

An APC is a device installed on transit vehicles such as a bus and a light train to count the passengers who get off and get in the transit vehicle at each station. The goal is to improve the accuracy of counting ridership at the disaggregate level, which is appropriate for service planning and schedule routes. APCs work based on two types of electronic units: infrared beams and video cameras. The beams of infrared lights shine down to the person crossing the stairwells, where the laser is located mounted on the ceiling (Boyle, 1998). The way that a person breaks the beam determines boarding or alighting activities. Video counting system is the second type of units used in APCs, which increases the accuracy of counting by 98% in comparison to the manual counting collected by a checker or onboard surveys (TransitWiki, April 2016). A video-based-people counter sums up how many passengers get on/off. Definite merit of video camera systems is the ability to verify the counting process by watching the video back. In addition to counting people on and off the vehicle, the dwell time at each station and the passengers flow around stations are other important features recorded by APCs. Integrating APC systems with onboard GPS receivers leads to detecting the location of counted passengers. Similarly to smart card systems, the stored data are wirelessly transferred to a central server.

Transportation-Related Applications of Smart Card Systems

In addition to the aforementioned smart card data, boarding date/time/location are other principal data provided by smart cards for further analysis in understanding public transit operations and users' motilities. GTFS and APC data, as complementary sources, have also been combined with the smart card data source in order to enhance the ability to realize various aspects of transit networks. The location of bus/ subway stations can be obtained directly from the database of operating transit agencies (i.e., GTFS). For those transactions made inside the bus, the boarding location is derived by matching database from other positioning systems such as GPS receivers equipped in the bus fleet. Unfortunately, alighting information for tap-in-only stations cannot be accessed through smart card systems, yet a group of research scientists has developed models for forecasting alighting locations (Trépanier et al., 2007). The route profile load can be computed using individual boarding and alighting location data. Also, the absence of socio-demographic information of cardholders in the smart card data calls for other complementary sources such as household surveys to enrich smart card data before conducting strategic planning (Trepanier et al., 2009). Pelletier et al. (2011) reviewed the literature on the use of smart card data in public transit systems in their solid survey paper. They classified studies in this context into three folds: 1) long-term planning including transit demand modeling and behavior analysis, 2) service and schedule adjustment, 3) operational metrics. Accordingly, we have mainly reviewed studies that have been conducted after 2011, and refer the remaining literature to the work of Pelletier et al. Figure 5 displays three main groups of smart card applications in transportation fields.

Transit operational-level analysis: Summarizing and visualizing the main characteristics of public transit data (Utsunomiya et al., 2006; Trépanier et al., 2009; Lin et al., 2013; Ma and Wang, 2014) are first steps to take benefits of such large-scale data. Basic statistic tools such as Exploratory Data Analysis (EDA) and probability distribution models empower practitioners to transfer a large volume of data into useful information on transit services and operations (Utsunomiya et al., 2006). Distribution of minimum-distance access to transit stations, daily smart card usage frequency and bus/subway stations usage frequency are examples of hidden knowledge in public transport services that can be quickly revealed through EDA as well as descriptive and numerical statistics (Utsunomiya et al., 2006). Furthermore, in order to enhance users' satisfaction and transit management systems, performance indicators on both transit operation (e.g., schedule adherence) as well as transit usage (e.g., passenger kilometers/hours) need to be calculated for every individual run segment, bus stop, route, and day (Trépanier et al., 2009).

Travel pattern and behavior analysis: Understanding passengers' travel behavior leads to a variety of applications (Hofmann et al., 2009; Ma et al., 2013; Mohamed et al., 2014; Goulet-Langlois et al., 2016; Ma et al., 2017). First of all, authorities can provide oriented services and information for identifiable classes of similar behaviors (Kieu et al., 2014). Assessing the performance of a transit network, detecting irregularities, forecasting demand, service adjustments, analysis of minimum-access distribution to transit stations, are other application examples of passengers' behavioral analysis (Mohamed et al., 2014). Data mining techniques can also be applied to individual level trip chains to extract passengers' travel pattern and travel regularity (Ma et al., 2013). Both high-likelihood travel patterns of an individual rider and similar patterns among all transit riders (aggregated level) have been recognized using data mining algorithms. After identifying passengers' travel pattern, the relationship between passengers' socioeconomic characteristics and travel patterns can be discovered by assigning passengers to their residential area retrieved from boarding information. Classifying users' travel behavior at the trip level is another implication of mining smart card transactions. Transfer journeys can be distinguished from

Figure 5. Main applications of smart card data in transportation systems

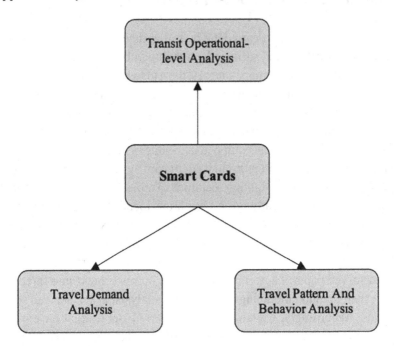

single journeys according to the boarding information (Hofmann et al., 2009). Summary of the smart card and automated passenger counter transport-related applications are provided in Table 3.

Travel demand analysis: Transit O-D matrix is valuable information for not only making a balance between supply and demand in transit networks but a better understanding of individuals' travel patterns. For transit stations without a tap-out facility, a user's trip, that contains segment(s) between two consecutive boarding and alighting locations, is constructed by estimating the alighting points while the boarding location is already obtained through smart card transactions (Munizaga and Palma, 2012). Several assumptions have made in literature for predicting the alighting station of a segment and in turn the trip destination. Two primary assumptions include: the origin of the next trip is the destination of the previous trip and the origin of the first daily trip is the destination of the last one (Barry et al., 2002). The nearest station to the next boarding bus stop within a walking distance threshold or an allowable transfer time (e.g., 400 meters or 5 minutes) is chosen as the alighting station of the previous trip. Discovering the alighting positions by following the location and time of the next boarding point is the main idea to infer the final destination of the trip and as its consequence to create the transit O-D matrix (Munizaga and Palma 2012). Recently, the transit systems equipped with a tap-out device payment have provided alighting points data, which give room to assessing the previous methods and assumptions in the literature (Alsger et al., 2015). Trip purpose or type of activity is another salient feature in the travel demand modeling that can be derived from users' information, temporal information, and spatial information (Lee and Hickman, 2014). Applying a rule-based classification on the mentioned features leads to building a trip-purpose inference model.

Table 3. Summary of smart card and automated passenger counter transport-related applications

Application	Example	Data Source	Methodology
Transit operational-level analysis	Distribution of minimum-distance access to transit stations, daily smart card usage frequency and bus/subway stations usage frequency, transit operation and usage performance indicators	Chicago Transit Authority smartcard (Utsunomiya et al., 2006); Société de transport de l'Outaouais smartcard fare collection (Trépanier et al., 2009); Beijing automated fare collection and automated vehicle location systems (Ma and Wang, 2014); City of Jinan, China GPS and automatic fare collection system data (Lin et al., 2013)	Exploratory Data Analysis and probability distribution models (Utsunomiya et al., 2006); Database queries (Trépanier et al., 2009); Bayesian decision tree algorithm, geospatial data model (Ma and Wang, 2014); Artificial neural network (Lin et al., 2013)
Travel pattern and behavior analysis	Assessing the performance of a transit network, detecting irregularities, forecasting demand, service adjustments, analysis of minimum-access distribution to transit stations	Rennes Metropolitan area smart card data (Mohamed et al., 2014); Beijing automated fare collection Data (Ma et al., 2013, Ma et al., 2017); Smart card data at entry point (Hofmann et al., 2009); London smart card data, and London Travel Demand Survey (Goulet-Langlois et al., 2016)	Expectation-Maximization algorithm (Mohamed et al., 2014); Density-based Spatial Clustering, K-Means++ clustering algorithm and rough-set theory (Ma et al., 2013); Iterative classification algorithm (Hofmann et al., 2009); Density-based Spatial Clustering (Ma et al,. 2017); Agglomerative hierarchical clustering (Goulet-Langlois et al., 2016)
Travel demand analysis	Developing Transit O-D matrix, Trip purpose or type of activity	Chile public transit automated fare collection automatic, vehicle location systems, and passenger counts Data (Munizaga and Palma 2012); South East Queensland smart card data (Alsger et al., 2015); New York City Transit's automated fare collection (Barry et al., 2002); Minneapolis/St. Paul metropolitan area automated fare collection data (Lee and Hickman, 2014); Chicago Transit automated fare collection and automated vehicle location data (Zhao et al., 2007); São Paulo, Brazil farecard and GPS data (Farzin, 2008)	Method to estimate travel sequence based on boarding information (Munizaga and Palma, 2012); Proposed O-D estimation algorithm, and sensitivity analysis (Alsger et al., 2015) (Barry et al., 2002); Rule-based classification, decision tree, chi-square automatic interaction detection (Lee and Hickman, 2014); destination inference algorithm (Zhao et al., 2007, Farzin, 2008)

POTENTIAL FUTURE RESEARCH DIRECTIONS

Probe People and Vehicle Data

A future research direction on prediction of traffic states using the FCD data is to incorporate other factors including (a) prevailing traffic conditions (e.g., percentage of heavy vehicles, local bus stations, and pedestrians flow), (b) roadway geometric conditions (e.g., grade and parking conditions), (c) weather conditions, and (d) traffic signal control settings. Moreover, the extent of uncertainty and reliability of many postulated distributions for traffic states such as travel time have not been considered. Consequently, in addition to the FCD data, other passive and active crowd sensing technologies can be employed to reduce the uncertainty in forecasting traffic states' distributions for large-scale traffic networks.

Notwithstanding the discoveries in the human mobility probability distribution, the rapid changes in the peoples' lifestyle, due to the growth of technologies and modern quality of thoughts, have been imposing changes in the human mobility patterns, which call for the continuous assessments on the human mobility. Thus, the movement data should be linked with their contextual spatial, environmental, and socio-demographic information to generate frameworks for a better understanding of human behavior. One particular characteristic of human mobility is the choice of travel modes. In this sense, the transportation mode inference from GPS trajectories is limited to a few transport modes such as walk, bus, train, bike, and car. However, according to the FHWA, vehicles are categorized into 13 classes, where the passenger car is only one of them. Integrating spatiotemporal trajectory data with other sources of information provides the opportunity for advanced learning algorithms to extract additional movement features, which ultimately leads to classifying motorized vehicles into more subgroups.

Mobility studies that are constructed based on crowdsourced data convey several inevitable challenges such as randomness and a fixed sampling rate (Lin and Hsu, 2014), which necessitate innovative strategies to relieve their effects. For instance, since users' personal information such as habits, social activities, and sickness can be inferred using only their location (Ghaffari et al., 2016), their privacy and security need to be protected when users' location-based data are collected, or implications of their mobility pattern analysis are published. Syntactic models of anonymity and differential privacy are two widely acceptable privacy models that target privacy-preserving data mining (PPDM) and privacy-preserving data publishing (PPDP). A compromise between the risk of privacy violation and utility of data should be taken since privacy models such as noise addition techniques might directly impact on original data and learning algorithms (Clifton and Tassa, 2013).

Location-Based Social Network

Spatial-temporal sparsity is the main issue in the LBSNs information as users do not necessarily share geo-tagged media contents during their travels, reflecting in incomplete trajectories. One way to deal with data sets that contain missing values is to apply advanced strategies such as collaborative filtering. Then, missing values are imputed with a set of plausible values that represent the uncertainty about the correct values, which ultimately results in transferring the sparse original dataset to a dense one. Applying the graphical inference models, such as Bayesian network and Markov random field, in the dense geo-tagged tweets or check-in records results in a more reliable framework for individuals' activities and mobility behaviors.

As mentioned, low-cost and large-scale location-embedded social media data have considerable potentials in advanced travel demand modeling and long-term urban planning goals. However, an activity-based travel demand modeler might encounter a couple of challenges for extracting the information regarding in-home activities, different types of daily activities, and future activities for all individuals from their limited shared social media contents (Rashidi et al., 2017). Other than deploying advanced learning frameworks to extract semantic and syntactic information on social media contents, enriching social media data by adding other sources (e.g., GIS-based land use information and GPS-based trajectories) clarifies unknown details in travel demand models including the number of daily trips with different purposes, visiting locations, travel routes, and transport modes. Like crowd sensing technologies, privacy issues need to be considered so that the personal information is untraceable and irretrievable from outputs of travel demand models.

Smart Card and Automated Passenger Counter

The smart card AFC technology has become acceptable as an efficient, profitable, and convenient payment system in public transport networks for both transit agencies and customers. As a result, millions of transactions are completed in each operating day, which ends up generating vast amounts of smart card records in exponential growth. Beyond the dire need for complex distributed storage and computing platforms such as Hadoop, advanced preprocessing techniques are essential to organizing such massive data. Accordingly, before using learning algorithms for extracting travel patterns, improved data validation and data cleansing methods need to be applied to smart card data to ensure that high-quality data are fed into learning algorithms. However, a majority of studies that use smart card data for developing transport-domain applications have focused on only the analytical part without paying heed to the fact that the final implications of travel patterns and demand analyses are primary contingent on to what extent the smart card data are validated, clean, accurate, complete, fitted, and uniform.

One of the main challenges in entry-only AFC systems is the scarcity of alighting information because no facility exists to record the smart card fare data when passengers alight and exit the station. The fundamental assumptions, mentioned in the previous sections, for estimating the alighting location might be invalid in many circumstances. In case that a passenger's journey contains an intermediate segment with a transport mode other than a bus or a train, the assumption that the most likely destination of a passenger trip is near to the origin of the next trip does not hold anymore. Also, passengers' mode choices towards the work might be different from returning home; where, for instance, a passenger uses buses in the early morning while they can carpool with family on the way back to the home. Thus, the assumption that the origin of the first daily trip is near to the destination of the last one might be invalid. Many research endeavors have been conducted to validate only the destination inference based on external data sources such as O-D trip surveys or integration of AFC with other facilities such as Automated Vehicle Location technologies to overcome the issue (Nunes et al., 2016). In this context, a future research direction for inferring a high percentage of the journey destination, when the assumptions are broken, and no external infrastructure is available, is to incorporate other spatiotemporal sources such as related GSM data to directly extract the alighting information from the fused data.

SMART TRANSPORTATION: CURRENT CHALLENGES, SOLUTIONS, AND CONCLUSIONS

We have elaborated on the ways that multi-modal data sources and technologies have been deployed to build advanced models in a wide range of transport domains. However, data sensing and management, and analytic models (i.e., the leading steps in the smart city and smart transportation) have been confronting some common challenges. The purpose of this section is to warn the readers that although deploying and mining the data above sources bring many benefits, it is imperative to also pay heed to the critical challenges accompanied with data sensing, acquisition, management, and analytical processes. Here, we only introduce obstacles and provide references for more details. It should be noted that many issues that are introducing in this section still needs more advanced strategies, which in turn can be thought of as future research directions in smart transportation.

Despite the fact that many types of fixed-point traffic sensors have been installed throughout transportation networks, they have not covered all segments of the network. Installing the stationary sensing infrastructures in the majority portions of the city incurs large expenditures and redundant data, which is not a practical and intelligent approach. One way to meet this challenge, as mentioned, is to leverage probe vehicles and people as movable sensors by tracking their mobile phones, commuting smart cards, and social media activities. Although moving sensors support to realize the mobility behavior and travel patterns, on the opposite side, they produce new issues such privacy protection, non-uniform distributed sensors, and generating unstructured data (Zheng et al., 2014). Regarding the invasion of privacy, collecting the individuals' GPS trajectories discloses the person's private affairs such as their visited locations and daily routines. Also, the data generated by moving sensors possess inherent issues. One is the sparsity and non-uniformity that originates from the fact that people are not necessarily turning on their device's location service to sharing information, which results in the lack of the sufficient movement data in some regions. Another type of issue, unlike the sources designed specifically for traffic management, is that the data produced by people are coming in a variety of formats such as texts and images, which are not only unstructured with a variable rate of resolution but require an initial inspection to assure they are traffic-related data.

Another existence challenge is that traffic data are received from various heterogeneous sources even for solving a specific issue. Data sources for constructing origin-destination matrix, for example, involves household interviews, mobile cellular networks, and GPS trajectories. Thus, while a majority of existing data mining algorithms analyze data from one source (e.g., computer vision with images and natural language processing with texts), advanced machine learning and data-fusion algorithms are indispensable to exploit mutual knowledge from multi-source and heterogeneous data. Furthermore, in many transportation scenarios such as adaptive traffic signal control and in-vehicle collision systems, it is imperative to respond on a real-time basis. This calls for not only cutting-edge data management systems to organize the multiple-source traffic but advanced data fusion and analytical tools to promptly make the optimal decision in responding to dynamic changes in traffic conditions.

Each of the sensors mentioned above generates extensive traffic data that can be deployed for predicting and managing traffic states, including travel time, vehicular speed, vehicle classification, traffic density, and traffic volume. However, measurement errors, uncertainty in an individual source, inability to collect all kinds of data, incapability of a sensor to detect all types of objects, sparseness, and limitation to a specific point or a small area are the major weaknesses of deploying only a single data source. Fortunately, with the advent of technologies and ability to collect data from multiple sources, data fusion techniques are applied to independent but complementary data sources in order to obtain a better understanding of mobility behavior and travel patterns that could not be attained from a single data source. The primary goal of data fusion algorithms in smart transportation is to develop an improved control system that fuses the knowledge learned from a set of heterogeneous data sources. Data fusion frameworks aim to provide sufficient information for estimating traffic parameters and making appropriate decisions that ensure users' safety and an efficient traffic stream (El Faouzi et al., 2011). Many traffic engineering problems and applications that have been explained throughout this paper can also be addressed by fusing those heterogeneous sources of information. Examples of data fusion application in transportation can be found in (Pan et al., 2013), (Kusakabe and Asakura, 2014), and (Itoh et al., 2014) studies.

The similarity between the phases in the smart transportation framework and stages in the Big Data phenomenon (i.e., data generation, data sensing, data management, and data analysis (Chen, Mao et al. 2014)) hints at considering the smart transportation as an application of Big Data. As a consequence, in addition to what was explained in the previous paragraphs, the smart transportation involves challenges in Big Data, which are enumerated as heterogeneous and inconsistent data, the large scale of data, visualization raw and analyzed data, redundancy reduction and data compression, and safety and confidentiality of data. A broad spectrum of studies in literature and other ongoing research has been seeking to address these challenges. However, as scrutinizing the methods for resolving the mentioned issues is out of this article's scope, further details on Big Data problems are referred to the references (Labrinidis and Jagadish, 2012; Chen et al., 2014; Wu et al., 2014).

CONCLUSION

A land transportation network is any platform that permits people to move from an origin to a destination with various modes such as car, bus, train, walking, bicycle. Despite the mobility and access benefits of transportation networks, cities have always been struggling with transportation-related issues due to the ever-growing increase in the population and the number of vehicles. Smart transportation, as a cost-effective and significant component of the smart city, has the promising goal to relieve issues and improve cities' livability, workability, and sustainability by developing intelligent and novel transport models. The ultimate efficacy of smart transportation architectures is highly contingent on the quality of deployed data and characteristics of the analytical techniques into which the data are fed. In order to investigate the available traffic-related data sources, we introduce the widely used technologies that are either designed specifically for traffic purposes or potential to be thought of as a traffic data source.

Smart transportation, as one of the smart city functions, is also pursuing the main goals of smart cities, which are ameliorating a city's livability, workability, and sustainability. In the livability context, which seeks for a better quality of life for citizens, the use of Information and Communication Technology (ICT) and advanced traffic data analytics leads to trip time reduction, coordinated traffic signal controls, in-vehicle collision-avoidance systems, ridesharing apps, and public safety. In the workability aspects, which means having better jobs and economic status, business appointments and collaborations between industry owners can be achieved through a reliable transportation network. Moreover, business owners intend to invest in locations with high mobility accessibility. Sustainability is accomplished by efficiently using natural, human, and financial sources while they are not entirely used up or ruined permanently. In pursuit of fulfilling sustainable sources, smart technologies and knowledge discovery can be deployed to reduce air and noise pollution by means of efficient technologies such as route navigation systems. Furthermore, properly assigning budgets for expanding transit services in areas with higher transportation demand results in not only making balance between demand and supply but minimizing energy consumption and pollutant emissions.

Researchers and authorities have always been looking for smart solutions for resolving endless challenges caused by transportation activities. Data collected using fixed-point sensors only present traffic states for the location where they were installed. Many ITS applications require the traffic information in a wider area. A cost-effective approach is to optimize the transportation network by mining the hidden knowledge in traffic-related data that are sensed through ICT infrastructures. In addition to the traffic flow sensors that have mainly been designed for recording vehicles' presence and passage, other existing technologies that have the potential to record people's mobility can be leveraged for improving the transportation network performance. People actively participate in the process of recording data about different aspects of their life using technologies such as GPS systems, cellular networks, social media services, and smart card AFCs. Such diverse and advanced sensing technologies deliver a large-scale data regarding mobility behavior of people and vehicles. Which can be fed into the advanced learning algorithms and data analytic models so as to bring a variety of advantages, including modern traffic management and control strategies, robust vehicle detection systems, mining mobility patterns, travel demand analysis, incident detection, air quality measurement, urban planning, and emergency systems. The acquisition, management, and mining of traffic-related data constitute the core components of smart transportation.

Keeping in view the significance of having effective data acquisition and management systems as well as high-quality data analytic models in the smart transportation, in this paper, we investigate the current and widely data sources that various studies have used to come up with novel ideas and address challenges in the issues pertaining to transportation systems. The survey aimed to identify the transport-domain application of widely-used-traffic data sources. For each data source, we not only explained the operational mechanism of the technology used for generating and extracting data but also divided all related applications into major groups and indicated some possible future research directions. The key strength of this study is the provision of an exhaustive guideline for readers to perceive the potential applications of multiple traffic data sources, which in turn assist them to comprehend the existing literature very promptly and brainstorm the future work. We also succinctly discussed the current issues and the corresponding solutions of smart transportation.

REFERENCES

Agamennoni, G., Nieto, J., & Nebot, E. (2009). Mining GPS data for extracting significant places. In *Proceedings IEEE International Conference on Robotics and Automation, 2009. ICRA'09*. IEEE. 10.1109/ROBOT.2009.5152475

Alliance, S. C. (2011). *Transit and contactless open payments: an emerging approach for fare collection.* White paper, Princeton: Smart Card Alliance.

Alsger, A. A., Mesbah, M., Ferreira, L., & Safi, H. (2015). Use of smart card fare data to estimate public transport origin–destination matrix. *Transportation Research Record: Journal of the Transportation Research Board, 2535*(1), 88–96. doi:10.3141/2535-10

Andrienko, G., Andrienko, N., Hurter, C., Rinzivillo, S., & Wrobel, S. (2011). From movement tracks through events to places: Extracting and characterizing significant places from mobility data. In *Proceedings 2011 IEEE Conference on Visual Analytics Science and Technology (VAST)*. IEEE. 10.1109/VAST.2011.6102454

Ashbrook, D., & Starner, T. (2003). Using GPS to learn significant locations and predict movement across multiple users. *Personal and Ubiquitous Computing, 7*(5), 275–286. doi:10.100700779-003-0240-0

Bar-Gera, H. (2007). Evaluation of a cellular phone-based system for measurements of traffic speeds and travel times: A case study from Israel. *Transportation Research Part C: Emerging Technologies, 15*(6), 380-391.

Barceló, J., Montero, L., Marqués, L., & Carmona, C. (2010). Travel time forecasting and dynamic origin-destination estimation for freeways based on Bluetooth traffic monitoring. *Transportation Research Record: Journal of the Transportation Research Board, 2175*(1), 19–27. doi:10.3141/2175-03

Barry, J., Newhouser, R., Rahbee, A., & Sayeda, S. (2002). Origin and destination estimation in New York City with automated fare system data. *Transportation Research Record: Journal of the Transportation Research Board,* 1817(1), 183-187.

Batty, M., Axhausen, K. W., Giannotti, F., Pozdnoukhov, A., Bazzani, A., Wachowicz, M., ... Portugali, Y. (2012). Smart cities of the future. *The European Physical Journal. Special Topics, 214*(1), 481–518. doi:10.1140/epjst/e2012-01703-3

Becker, R. A., Caceres, R., Hanson, K., Loh, J. M., Urbanek, S., Varshavsky, A., &Volinsky, C. (2011). A tale of one city: Using cellular network data for urban planning. *IEEE Pervasive Computing, 10*(4), 18-26.

Boyle, D. K. (1998). Passenger counting technologies and procedures.

Businessinsider. (2016, July). Social media engagement: the surprising facts about how much time people spend on the major social networks. Retrieved from http://www.businessinsider.com/social-media-engagement-statistics-2013-12

Caceres, N., Wideberg, J., & Benitez, F. (2007). Deriving origin destination data from a mobile phone network. *IET Intelligent Transport Systems, 1*(1), 15–26. doi:10.1049/iet-its:20060020

Caceres, N., Wideberg, J., & Benitez, F. (2008). Review of traffic data estimations extracted from cellular networks. *IET Intelligent Transport Systems, 2*(3), 179–192. doi:10.1049/iet-its:20080003

Calabrese, F., Colonna, M., Lovisolo, P., Parata, D., & Ratti, C. (2011). Real-time urban monitoring using cell phones: A case study in Rome. *IEEE Transactions on Intelligent Transportation Systems, 12*(1), 141-151.

Calabrese, F., Pereira, F. C., Di Lorenzo, G., Liu, L., & Ratti, C. (2010). The geography of taste: analyzing cell-phone mobility and social events. In *Proceedings International Conference on Pervasive Computing,* Springer. 10.1007/978-3-642-12654-3_2

Cao, H., Mamoulis, N., & Cheung, D. W. (2005). Mining frequent spatio-temporal sequential patterns. In *Proceedings Fifth IEEE International Conference on Data Mining (ICDM'05),* pp. 8-pp. IEEE.

Carvalho, S. (2010). Real-time sensing of traffic information in twitter messages.

Casas, I. & Delmelle, E. C. (2017). Tweeting about public transit—Gleaning public perceptions from a social media microblog. *Case Studies on Transport Policy, 5*(4), 634-642.

Cebelak, M. K. (2013). *Location-based social networking data: doubly-constrained gravity model origin-destination estimation of the urban travel demand for Austin,* TX.

Chaniotakis, E., Antoniou, C., & Pereira, F. C. (2017). Enhancing resilience to disasters using social media. In *Proceedings 2017 5th IEEE International Conference on Models and Technologies for Intelligent Transportation Systems (MT-ITS).* IEEE. 10.1109/MTITS.2017.8005602

Chen, M., Mao, S., & Liu, Y. (2014). Big data: A survey. *Mobile Networks and Applications, 19*(2), 171–209. doi:10.100711036-013-0489-0

Cho, S.-B. J. N. (2016). Exploiting machine learning techniques for location recognition and prediction with smartphone logs. *Neurocomputing, 176,* 98-106.

Chourabi, H., Nam, T., Walker, S., Gil-Garcia, J. R., Mellouli, S., Nahon, K., . . . Scholl, H. J. (2012). Understanding smart cities: An integrative framework. In *Proceedings 2012 45th Hawaii International Conference on System Science (HICSS)*. IEEE. 10.1109/HICSS.2012.615

Clifton, C. & Tassa, T. (2013). On syntactic anonymity and differential privacy. In *Proceedings 2013 IEEE 29th International Conference on Data Engineering Workshops (ICDEW)*. IEEE. 10.1109/ICDEW.2013.6547433

Collins, C., Hasan, S., & Ukkusuri, S. (2013). A novel transit rider satisfaction metric: Rider sentiments measured from online social media data. *Journal of Public Transportation, 16*(2), 2.

Council, S. C. (2013). Smart cities readiness guide. The planning manual for building tomorrow's cities today.

Cranshaw, J., Schwartz, R., Hong, J. I., & Sadeh, N. (2012). The livehoods project: Utilizing social media to understand the dynamics of a city. In *Proceedings International AAAI Conference on Weblogs and Social Media*.

D'Andrea, E., Ducange, P., Lazzerini, B., & Marcelloni, F. (2015). Real-time detection of traffic from twitter stream analysis. *IEEE Transactions on Intelligent Transportation Systems, 16*(4), 2269–2283.

Dabiri, S., & Heaslip, K. (2018). Introducing a Twitter-based traffic information system using deep learning architectures. *IEEE Transactions on Intelligent Transportation Systems*.

Dabiri, S. & Heaslip, K. (2017). Inferring transportation modes from GPS trajectories using a convolutional neural network. *Transportation Research Part C: Emerging Technologies*.

Dabiri, S. & Heaslip, K. (2019). Developing a Twitter-based traffic event detection model using deep learning architectures. *Expert Systems with Applications, 118*, 425-439.

Dash, M., Koo, K. K., Krishnaswamy, S. P., Jin, Y., & Shi-Nash, A. (2016). Visualize people's mobility-both individually and collectively-using mobile phone cellular data. In *2016 17th IEEE International Conference on Mobile Data Management (MDM)*. IEEE.

De Fabritiis, C., Ragona, R., & Valenti, G. (2008). Traffic estimation and prediction based on real time floating car data. In *Proceedings 11th International IEEE Conference on Intelligent Transportation Systems, 2008. ITSC 2008*. IEEE. 10.1109/ITSC.2008.4732534

Do, T. M. T., & Gatica-Perez, D. (2012). Contextual conditional models for smartphone-based human mobility prediction. *Proceedings of the 2012 ACM conference on ubiquitous computing*, ACM. 10.1145/2370216.2370242

El Faouzi, N.-E., Leung, H., & Kurian, A. (2011). Data fusion in intelligent transportation systems: Progress and challenges–A survey. *Information Fusion, 12*(1), 4–10. doi:10.1016/j.inffus.2010.06.001

Fang, H., Hsu, W.-J., & Rudolph, L. (2009). Cognitive personal positioning based on activity map and adaptive particle filter. *Proceedings of the 12th ACM International Conference on Modeling, Analysis and Simulation of Wireless and Mobile Systems*. ACM. 10.1145/1641804.1641873

Farzin, J. M. (2008). Constructing an automated bus origin–destination matrix using farecard and global positioning system data in Sao Paulo, Brazil. *Transportation Research Record, 2072*(1), 30-37.

Gal-Tzur, A., Grant-Muller, S. M., Kuflik, T., Minkov, E., Nocera, S., & Shoor, I. (2014). The potential of social media in delivering transport policy goals. *Transport Policy, 32,* 115-123.

Ghaffari, M., Ghadiri, N., Manshaei, M. H., & Lahijani, M. S. (2016). P4QS: a peer to peer privacy preserving query service for location-based mobile applications. *arXiv preprint arXiv:1606.02373.*

Girardin, F., Calabrese, F., Fiore, F. D., Ratti, C., & Blat, J. (2008). *Digital footprinting: Uncovering tourists with user-generated content.* Institute of Electrical and Electronics Engineers.

Goulet-Langlois, G., Koutsopoulos, H. N., & Zhao, J. (2016). Inferring patterns in the multi-week activity sequences of public transport users. *Transportation Research Part C: Emerging Technologies, 64,* 1-16.

Gu, Y., Qian, Z., & Chen, F. (2016). From Twitter to detector: Real-time traffic incident detection using social media data. *Transportation Research Part C, Emerging Technologies, 67,* 321–342. doi:10.1016/j.trc.2016.02.011

Gühnemann, A., Schäfer, R.-P., Thiessenhusen, K.-U., & Wagner, P. (2004). Monitoring traffic and emissions by floating car data.

Haghani, A., Hamedi, M., Sadabadi, K., Young, S., & Tarnoff, P. (2010). Data collection of freeway travel time ground truth with bluetooth sensors. *Transportation Research Record: Journal of the Transportation Research Board, 2160*(1), 60–68. doi:10.3141/2160-07

Hainen, A., Wasson, J., Hubbard, S., Remias, S., Farnsworth, G., & Bullock, D. (2011). Estimating route choice and travel time reliability with field observations of Bluetooth probe vehicles. *Transportation Research Record: Journal of the Transportation Research Board, 2256*(1), 43–50. doi:10.3141/2256-06

Hasan, S., & Ukkusuri, S. V. (2014). Urban activity pattern classification using topic models from online geo-location data. *Transportation Research Part C, Emerging Technologies, 44,* 363–381. doi:10.1016/j.trc.2014.04.003

Hofmann, M., Wilson, S. P., & White, P. (2009). Automated identification of linked trips at trip level using electronic fare collection data. Transportation Research Board 88th Annual Meeting.

Imran, M., Castillo, C., Diaz, F., & Vieweg, S. (2015). Processing social media messages in mass emergency: A survey. [CSUR]. *ACM Computing Surveys, 47*(4), 67. doi:10.1145/2771588

Itoh, M., Yokoyama, D., Toyoda, M., Tomita, Y., Kawamura, S., & Kitsuregawa, M. (2014). Visual fusion of mega-city big data: an application to traffic and tweets data analysis of metro passengers. In *Proceedings 2014 IEEE International Conference on Big Data.* IEEE. 10.1109/BigData.2014.7004260

Jiang, B., Yin, J., & Zhao, S. (2009). Characterizing the human mobility pattern in a large street network. *Physical Review. E, 80*(2), 021136. doi:10.1103/PhysRevE.80.021136 PMID:19792106

Jurdak, R., Zhao, K., Liu, J., AbouJaoude, M., Cameron, M., & Newth, D. (2015). Understanding human mobility from Twitter. *PLoS One, 10*(7). doi:10.1371/journal.pone.0131469 PMID:26154597

Kerner, B., Demir, C., Herrtwich, R., Klenov, S., Rehborn, H., Aleksic, M., & Haug, A. (2005). *Traffic state detection with floating car data in road networks. In Proceedings 2005 IEEE Intelligent Transportation Systems, 2005.* IEEE.

Kieu, L. M., Bhaskar, A., & Chung, E. (2014). Transit passenger segmentation using travel regularity mined from Smart Card transactions data.

Kusakabe, T., & Asakura, Y. (2014). Behavioural data mining of transit smart card data: A data fusion approach. *Transportation Research Part C, Emerging Technologies, 46,* 179–191. doi:10.1016/j.trc.2014.05.012

Labrinidis, A., & Jagadish, H. V. (2012). Challenges and opportunities with big data. In *Proceedings of the VLDB Endowment International Conference on Very Large Data Bases, 5*(12), 2032–2033. doi:10.14778/2367502.2367572

Lee, J.-G., Han, J., Li, X., & Cheng, H. (2011). Mining discriminative patterns for classifying trajectories on road networks. *IEEE Transactions on Knowledge and Data Engineering, 23*(5), 713–726. doi:10.1109/TKDE.2010.153

Lee, J.-G., Han, J., Li, X., & Gonzalez, H. (2008). TraClass: Trajectory classification using hierarchical region-based and trajectory-based clustering. In *Proceedings of the VLDB Endowment International Conference on Very Large Data Bases, 1*(1), 1081–1094. doi:10.14778/1453856.1453972

Lee, J. H., Gao, S., Janowicz, K., & Goulias, K. G. (2015). Can Twitter data be used to validate travel demand models? IATBR 2015-WINDSOR.

Lee, S., & Hickman, M. (2013). Are transit trips symmetrical in time and space? evidence from the twin cities. *Transportation Research Record: Journal of the Transportation Research Board, 2382*(1), 173–180. doi:10.3141/2382-19

Lee, S. G., & Hickman, M. (2014). Trip purpose inference using automated fare collection data. *Public Transport (Berlin), 6*(1), 1–20. doi:10.100712469-013-0077-5

Levy, J. I., Buonocore, J. J., & Von Stackelberg, K. (2010). Evaluation of the public health impacts of traffic congestion: A health risk assessment. *Environmental Health, 9*(1), 1. doi:10.1186/1476-069X-9-65 PMID:20979626

Liang, X., Zheng, X., Lv, W., Zhu, T., & Xu, K. (2012). The scaling of human mobility by taxis is exponential. *Physica A, 391*(5), 2135–2144. doi:10.1016/j.physa.2011.11.035

Lin, M., & Hsu, W.-J. (2014). Mining GPS data for mobility patterns: A survey. *Pervasive and Mobile Computing, 12,* 1–16.

Lin, Y., Yang, X., Zou, N., & Jia L. (2013). Real-time bus arrival time prediction: Case study for Jinan, China. *Journal of Transportation Engineering, 139*(11), 1133-1140.

Liu, J. H., Shi, W., Elrahman, O. A., Ban, X., & Reilly, J. M. (2016). Understanding social media program usage in public transit agencies. *International Journal of Transportation Science and Technology, 5*(2), 83–92. doi:10.1016/j.ijtst.2016.09.005

Liu, S., Ni, L. M., & Krishnan, R. (2014). Fraud detection from taxis' driving behaviors. *IEEE Transactions on Vehicular Technology, 63*(1), 464–472. doi:10.1109/TVT.2013.2272792

Liu, Y., Sui, Z., Kang, C., & Gao, Y. (2014). Uncovering patterns of inter-urban trip and spatial interaction from social media check-in data. *PLoS One, 9*(1). doi:10.1371/journal.pone.0086026 PMID:24465849

Lu, X., Wang, C., Yang, J.-M., Pang, Y., & Zhang, L. (2010). Photo2trip: generating travel routes from geo-tagged photos for trip planning. *Proceedings of the 18th ACM international conference on Multimedia*, ACM. 10.1145/1873951.1873972

Ma, X., Liu, C., Wen, H., Wang, Y., & Wu, Y.-J. (2017). Understanding commuting patterns using transit smart card data. *Journal of Transport Geography, 58*, 135-145.

Ma, X. & Wang, Y. (2014). Development of a data-driven platform for transit performance measures using smart card and GPS data. *Journal of Transport Geography, 140*(12), 04014063.

Ma, X., Wu, Y.-J., Wang, Y., Chen, F., & Liu, J. (2013). Mining smart card data for transit riders' travel patterns. *Transportation Research Part C, Emerging Technologies, 36*, 1–12. doi:10.1016/j.trc.2013.07.010

Maghrebi, M., Abbasi, A., & Waller, S. T. (2016). Transportation application of social media: Travel mode extraction. In *Proceedings 2016 IEEE 19th International Conference on Intelligent Transportation Systems (ITSC)*. IEEE. 10.1109/ITSC.2016.7795779

Mintsis, G., Basbas, S., Papaioannou, P., Taxiltaris, C., & Tziavos, I. (2004). Applications of GPS technology in the land transportation system. *European Journal of Operational Research, 152*(2), 399–409. doi:10.1016/S0377-2217(03)00032-8

Mohamed, K., Côme, E., Baro, J., & Oukhellou, L. (2014). Understanding passenger patterns in public transit through smart card and socioeconomic data.

Munizaga, M. A., & Palma, C. (2012). Estimation of a disaggregate multimodal public transport Origin–Destination matrix from passive smartcard data from Santiago, Chile. *Transportation Research Part C, Emerging Technologies, 24*, 9–18. doi:10.1016/j.trc.2012.01.007

Nam, T., & Pardo, T. A. (2011). Conceptualizing smart city with dimensions of technology, people, and institutions. *Proceedings of the 12th Annual International Digital Government Research Conference: Digital Government Innovation in Challenging Times*. ACM. 10.1145/2037556.2037602

Nantes, A., Billot, R., Miska, M., & Chung, E. (2013). Bayesian inference of traffic volumes based on Bluetooth data.

Noulas, A., Scellato, S., Lambiotte, R., Pontil, M., & Mascolo, C. (2012). A tale of many cities: Universal patterns in human urban mobility. *PLoS One, 7*(5), e37027. doi:10.1371/journal.pone.0037027 PMID:22666339

Noulas, A., Scellato, S., Mascolo, C., & Pontil, M. (2011). An empirical study of geographic user activity patterns in foursquare. *ICwSM, 11*, 70–573.

Nunes, A. A., & Dias, T. G. (2016). Passenger journey destination estimation from automated fare collection system data using spatial validation. *IEEE Transactions on Intelligent Transportation Systems*, *17*(1), 133–142. doi:10.1109/TITS.2015.2464335

Pan, B., Zheng, Y., Wilkie, D., & Shahabi, C. (2013). Crowd sensing of traffic anomalies based on human mobility and social media. In *Proceedings of the 21st ACM SIGSPATIAL International Conference on Advances in Geographic Information Systems*. ACM. 10.1145/2525314.2525343

Pattara-Atikom, W. & Peachavanish, R. (2007). Estimating road traffic congestion from cell dwell time using neural network. In *Proceedings 2007 7th International Conference on ITS Telecommunications*. IEEE. 10.1109/ITST.2007.4295824

Pelletier, M.-P., Trépanier, M., & Morency, C. (2011). Smart card data use in public transit: A literature review. *Transportation Research Part C, Emerging Technologies*, *19*(4), 557–568. doi:10.1016/j.trc.2010.12.003

Pender, B., Currie, G., Delbosc, A., & Shiwakoti, N. (2014). Social media use during unplanned transit network disruptions: A review of literature. *Transport Reviews*, *34*(4), 501–521. doi:10.1080/01441647.2014.915442

Pereira, J., Pasquali, A., Saleiro, P., & Rossetti, R. (2017). Transportation in social media: an automatic classifier for travel-related tweets. *Portuguese Conference on Artificial Intelligence*. Springer. 10.1007/978-3-319-65340-2_30

Rahmani, M., Jenelius, E., & Koutsopoulos, H. N. (2015). Non-parametric estimation of route travel time distributions from low-frequency floating car data. *Transportation Research Part C, Emerging Technologies*, *58*, 343–362. doi:10.1016/j.trc.2015.01.015

Rahmani, M., & Koutsopoulos, H. N. (2013). Path inference from sparse floating car data for urban networks. *Transportation Research Part C, Emerging Technologies*, *30*, 41–54. doi:10.1016/j.trc.2013.02.002

Rashidi, T. H., Abbasi, A., Maghrebi, M., Hasan, S., & Waller, T. S. (2017). Exploring the capacity of social media data for modelling travel behaviour: Opportunities and challenges. *Transportation Research Part C, Emerging Technologies*, *75*, 197–211. doi:10.1016/j.trc.2016.12.008

Ratti, C., Frenchman, D., Pulselli, R. M., & Williams, S. (2006). *Mobile landscapes: using location data from cell phones for urban analysis, 33*(5), 727-748.

Reades, J., Calabrese, F., Sevtsuk, A., & Ratti, C. (2007). Cellular census: Explorations in urban data collection. IEEE Pervasive computing, 6(3), 30-38.

Rojas, F., Calabrese, F., Krishnan, S., Ratti, C. J. N., & Studies, C. (2008). *Real Time Rome, 20*(3), 247–258.

Saadeh, M., Sleit, A., Sabri, K. E., & Almobaideen, W. (2018). Hierarchical architecture and protocol for mobile object authentication in the context of IoT smart cities. Journal of Network and Computer Applications, 121, 1-19.

Sakaki, T., Okazaki, M., & Matsuo, Y. (2010). Earthquake shakes Twitter users: real-time event detection by social sensors. In *Proceedings of the 19th International Conference on World Wide Web*. Raleigh, NC: ACM. 851-860. 10.1145/1772690.1772777

Smit, R., Brown, A., & Chan, Y. (2008). Do air pollution emissions and fuel consumption models for roadways include the effects of congestion in the roadway traffic flow? *Environmental Modelling & Software, 23*(10), 1262–1270. doi:10.1016/j.envsoft.2008.03.001

Steenbruggen, J., Borzacchiello, M. T., Nijkamp, P., & Scholten, H. (2013). Mobile phone data from GSM networks for traffic parameter and urban spatial pattern assessment: A review of applications and opportunities. *GeoJournal, 78*(2), 223–243. doi:10.100710708-011-9413-y

Stefanidis, A., Cotnoir, A., Croitoru, A., Crooks, A., Rice, M., & Radzikowski, J. (2013). Demarcating new boundaries: mapping virtual polycentric communities through social media content. Cartography and Geographic Information Science, 40(2), 116-129.

Sun, D., Zhang, C., Zhang, L., Chen, F., & Peng, Z.-R. (2014). Urban travel behavior analyses and route prediction based on floating car data. *Transportation Letters, 6*(3), 118–125. doi:10.1179/1942787514Y.0000000017

Sun, Y., Fan, H., Bakillah, M., & Zipf, A. (2015). Road-based travel recommendation using geo-tagged images. *Computers, Environment and Urban Systems, 53,* 110–122. doi:10.1016/j.compenvurbsys.2013.07.006

Sweet, M. (2011). Does traffic congestion slow the economy? *Journal of Planning Literature, 26*(4), 391–404. doi:10.1177/0885412211409754

Tabibiazar, A. & Basir, O. (2011). Kernel-based modeling and optimization for density estimation in transportation systems using floating car data. In *Proceedings 2011 14th International IEEE Conference on Intelligent Transportation Systems (ITSC)*. IEEE. 10.1109/ITSC.2011.6083098

Tasse, D. & Hong, J. I. (2014). Using social media data to understand cities.

Thiessenhusen, K., Schafer, R., & Lang, T. (2003). *Traffic data from cell phones: a comparison with loops and probe vehicle data*. Germany: Institute of Transport Research, German Aerospace Center.

Torkestani, J. A. (2012). Mobility prediction in mobile wireless networks. Journal of Network and Computer Applications, 35(5), 1633-1645.

TransitWiki. (2016, April). Automated Passenger Counter. Retrieved from https://www.transitwiki.org/TransitWiki/index.php/Category:Automated_Passenger_Counter

Trépanier, M., Morency, C., & Agard, B. (2009). Calculation of transit performance measures using smartcard data. *Journal of Public Transportation, 12*(1), 5. doi:10.5038/2375-0901.12.1.5

Trepanier, M., Morency, C., & Blanchette, C. (2009). Enhancing household travel surveys using smart card data. Transportation Research Board 88th Annual Meeting.

Trépanier, M., Tranchant, N., & Chapleau, R. (2007). Individual trip destination estimation in a transit smart card automated fare collection system. *Journal of Intelligent Transport Systems, 11*(1), 1–14. doi:10.1080/15472450601122256

Tsubota, T., Bhaskar, A., Chung, E., & Billot, R. (2011). Arterial traffic congestion analysis using Bluetooth Duration data.

Utsunomiya, M., Attanucci, J., & Wilson, N. (2006). Potential uses of transit smart card registration and transaction data to improve transit planning. *Transportation Research Record: Journal of the Transportation Research Board(1971),* 119-126.

Veloso, M., Phithakkitnukoon, S., Bento, C., Fonseca, N., & Olivier, P. (2011). Exploratory study of urban flow using taxi traces. *First Workshop on Pervasive Urban Applications (PURBA) in conjunction with Pervasive Computing,* San Francisco, CA.

Wang, C., Quddus, M. A., & Ison, S. G. (2009). Impact of traffic congestion on road accidents: A spatial analysis of the M25 motorway in England. *Accident; Analysis and Prevention, 41*(4), 798–808. doi:10.1016/j.aap.2009.04.002 PMID:19540969

Wang, Y., Malinovskiy, Y., Wu, Y.-J., Lee, U. K., & Neeley, M. (2011). *Error modeling and analysis for travel time data obtained from Bluetooth MAC address matching.* Department of Civil and Environmental Engineering, University of Washington.

Wu, H.-R., Yeh, M.-Y., & Chen, M.-S. (2013). Profiling moving objects by dividing and clustering trajectories spatiotemporally. *IEEE Transactions on Knowledge and Data Engineering, 25*(11), 2615–2628. doi:10.1109/TKDE.2012.249

Wu, L., Zhi, Y., Sui, Z., & Liu, Y. (2014). Intra-urban human mobility and activity transition: Evidence from social media check-in data. *PLoS One, 9*(5), e97010. doi:10.1371/journal.pone.0097010 PMID:24824892

Wu, X., Zhu, X., Wu G.-Q., & Ding, W. (2014). Data mining with big data. *IEEE transactions on knowledge and data engineering, 26(1),* 97-107.

Xu, H., & Cho, S. B. (2014, December). Recognizing semantic locations from smartphone log with combined machine learning techniques. In 2014 IEEE 11th Intl Conf on Ubiquitous Intelligence and Computing and 2014 IEEE 11th Intl Conf on Autonomic and Trusted Computing and 2014 IEEE 14th Intl Conf on Scalable Computing and Communications and Its Associated Workshops (pp. 66-71). IEEE. 10.1109/UIC-ATC-ScalCom.2014.128

Yang, F., Jin, P. J., Cheng, Y., Zhang, J., & Ran, B. (2015). Origin-destination estimation for non-commuting trips using location-based social networking data. *International Journal of Sustainable Transportation, 9*(8), 551–564. doi:10.1080/15568318.2013.826312

Ygnace, J. L. (2001). Travel time/speed estimates on the french rhone corridor network using cellular phones as probes. Final report of the SERTI V program, INRETS, Lyon, France.

Yoon, H., Zheng, Y., Xie, X., & Woo, W. (2012). Social itinerary recommendation from user-generated digital trails. Personal and Ubiquitous Computing, 16(5), 469-484.

Zhao, J., Rahbee, A., & Wilson, N. H. (2007). Estimating a rail passenger trip origin-destination matrix using automatic data collection systems. Computer-Aided Civil and Infrastructure Engineering, 22(5), 376-387.

Zheng, Y. (2011). Location-based social networks: Users. Computing with spatial trajectories, Springer: 243-276.

Zheng, Y., Capra, L., Wolfson, O., & Yang, H. (2014). Urban computing: Concepts, methodologies, and applications. [TIST]. *ACM Transactions on Intelligent Systems and Technology*, *5*(3), 38. doi:10.1145/2629592

Zheng, Y., Li, Q., Chen, Y., Xie, X., & Ma, W.-Y. (2008). Understanding mobility based on GPS data. *Proceedings of the 10th international conference on Ubiquitous computing*, ACM.

Zheng, Y., Liu, L., Wang, L., & Xie, X. (2008). Learning transportation mode from raw gps data for geographic applications on the web. *Proceedings of the 17th international conference on World Wide Web*, ACM. 10.1145/1367497.1367532

Zheng, Y., & Xie, X. (2011). Location-based social networks: Locations. In Computing with spatial trajectories. New York, NY: Springer. 277-308.

Zheng, Y., Xie, X., & Ma, W.-Y. (2010). GeoLife: A collaborative social networking service among user, location and trajectory. *IEEE Data Eng. Bull.*, *33*(2), 32–39.

Zheng, Y., Zhang, L., Xie, X., & Ma, W.-Y. (2009). Mining interesting locations and travel sequences from GPS trajectories. *Proceedings of the 18th International Conference on World Wide Web*. ACM. 10.1145/1526709.1526816

Zhou, C., Sun, C., Liu, Z., & Lau, F. (2015). A C-LSTM neural network for text classification. *arXiv preprint arXiv:1511.08630.*

Zignani, M., & Gaito, S. (2010). *Extracting human mobility patterns from gps-based traces. Wireless Days (WD), 2010 IFIP*. IEEE.

Chapter 11
Application of Machine Learning Methods for Passenger Demand Prediction in Transfer Stations of Istanbul's Public Transportation System

Hacer Yumurtaci Aydogmus
Alanya Alaaddin Keykubat University, Turkey

Yusuf Sait Turkan
Istanbul University-Cerrahpasa, Turkey

ABSTRACT

The rapid growth in the number of drivers and vehicles in the population and the need for easy transportation of people increases the importance of public transportation. Traffic becomes a growing problem in Istanbul which is Turkey's greatest urban settlement area. Decisions on investments and projections for the public transportation should be well planned by considering the total number of passengers and the variations in the demand on the different regions. The success of this planning is directly related to the accurate passenger demand forecasting. In this study, machine learning algorithms are tested in a real world demand forecasting problem where hourly passenger demands collected from two transfer stations of a public transportation system. The machine learning techniques are run in the WEKA software and the performance of methods are compared by MAE and RMSE statistical measures. The results show that the bagging based decision tree methods and rules methods have the best performance.

DOI: 10.4018/978-1-7998-0301-0.ch011

INTRODUCTION

Predicting what will happen in the future using the available data has always been of interest. The ability to predict the future course of a time series and the values is a continuing issue, the importance of which is still going up in various fields such as biology, physics, mathematics, engineering, economics and statistics. Studies related to public transportation hold a crucial position among studies involving estimation. Public transportation is of great importance for people's quality of social life and economic and social development of cities. In order to ensure that people reach their jobs as quickly as possible, as well as to improve their access to social services such as health and education, municipalities should provide cheap and safe public transportation within their borders. In order to establish public transportation plans and to manage public transport effectively, passenger statistics should be monitored for different regions and the amount of passengers in future periods should be estimated.

Istanbul is the greatest urban settlement area in Turkey. Traffic becomes a growing problem in Istanbul where approximately more than 800 new vehicles hit the roads and nearly 13 million passengers are transported per day. Therefore, besides the efficient management of public transportation, new public transportation investments are also of great importance. In order to make investment decisions on public transport, it is vital to accurately estimate the public transportation demands of different regions or stations. Decisions on investments and projections for the public transportation should be well planned by considering the total number of passengers and the variations in the demand on the different regions. The success of such planning is directly related to the accurate passenger demand forecasting.

Statistical methods and intelligent techniques can be used in the prediction of the public transport demand. Four-stage model, land use models and time series methods are influential classical methods. In recent years, for the estimation of passenger demands hybrid methods are used together with classical methods. In many complex transportation forecasting problems, it is hard to understand the relationships between different variables. Therefore, it is seen that artificial techniques are more preferred in recent years.

The aim of this study is to determine the effectiveness of different machine learning algorithms for prediction of passenger demand in transfer stations of a public transportation system. For this purpose, various machine learning algorithms have been tried in a real-world demand forecasting problem. To examine the forecast performance of machine learning algorithms, five-year daily passenger traffic data of two selected transfer stations in Istanbul are used in the experiment to see the prediction accuracy measured by Mean Absolute Error (MAE), Root Mean Square Error(RMSE) and Correlation Coefficient (R). The results show that some bagging (decision trees and rules) algorithms are very successful and they can even be used to predict passenger demand in transfer stations.

BACKGROUND

Passenger demand estimation problem in public transport can be categorized into long term and short term demand forecasting problems. Long-term public transport passenger forecasting is used for long-term planning, strategic decisions and investments in public transport, while short-term forecasting is more effective in operational decisions. Conventional demand forecasting methods are generally classified as univariate time series approaches and multivariate demand modeling approaches. Multivariate demand modelling approaches can be undertaken using a conventional four-step travel planning model or direct demand models. Travel planning model including the steps of trip generation, trip distribution, mode choice, and assignment has been used in many demand forecasting applications (Bar-Gera and Boyce, 2003; Blainey and Preston, 2010; Dargay et al., 2010; Jovicic and Hansen, 2003; Owen and Philips, 1987; Preston, 1991; Wardman and Tyler, 2000; Wardman, 2006).

The data driven model includes linear and nonlinear estimation methods. Time series models (Preez and Witt, 2003; Williams et al.1998) and Kalman filtering model (Ye et al., 2006) are the most common linear forecasting methods. SARIMA, which is the most known model among all time series models, can take into consideration the periodicity of the time series when estimating, which can increase the estimation success. However, the success of SARIMA is significantly reduced if the time series is non-linear.

When the studies in recent years are examined, it is seen that the most widely used methods are nonparametric regression (Clark, 2003; Smith et al., 2003), gaussian maximum likelihood model (Lin, 2001), neural network algorithm (Huang Y., and Pan, 2011; Tsung-Hsien et al., 2005) and support vector machine (Chen et al., 2011; Zhang and Xie, 2007).

In the recent studies on passenger estimation, studies on airline passenger estimation and bus passenger estimation stand out. For example, Cyril et al. (2018) used a univariate time series ARIMA model to forecast the inter-district public transport travel demand and in 2019 they studied the application of time-series method for forecasting public bus passenger demand. They used Electronic Ticketing Machines, which generated the big data, based time-series for issuing tickets and collecting fares. Suh and Ryerson (2019) built a methodology that is grounded in established airport passenger demand forecasting practices and is able to significantly improve the accuracy of aviation demand forecasting models. Efendigil and Eminler (2017) compared artificial intelligence methods and regression analysis technique. They also examined 114 academic publications in airport passenger demand and observed that artificial intelligence techniques are becoming more preferable over econometric models.

The human brain is capable of intelligent data processing, and artificial neural networks are precisely developed to mimic this data processing. In artificial neural networks, synaptic weights can be adjusted to automatically match the input-output relationship of the analyzed system through a learning process (Hagan et al. 1996; Wei and Wu 1997). It is seen that artificial intelligence techniques, especially artificial neural networks, are used in many public transportation forecasting studies (Ding and Chien 2000; Chien et al. 2002; Chen et al. 2004).

In this study, many machine learning methods have been applied to a real-life problem in one of the vital bus transfer stations of Istanbul. In this way, the most successful algorithms in these types of problems were investigated and the parameters that might lead to success and failures were investigated.

Machine Learning Techniques of Estimation

Fischer and Lehner (2013) have classified forecasting methods into three categories: time series, casual and machine learning. According to them ARIMA, multiple linear regression and exponential smoothing are time series techniques; ARMAX and multivariate linear regression are causal techniques; support vector machines, bayesian networks, neural networks and decision trees are machine learning techniques.

Machine learning is a frequently used concept and research area in recent years. This approach is applied in academic research or industrial fields. Before presenting the definition of machine learning, it would be appropriate to consider briefly the concept of learning. We can say that human life consists of what they learn. As a multi-faceted event learning processes include the acquisition of new declarative knowledge, the development of motor and cognitive skills through instruction or practice, the organization of new knowledge into general, effective representations, and the discovery of new facts and theories through observation and experimentation (Carbonell et al.,1983). Also, Laplante (2000) gave another definition for the learning as "generally, any scheme whereby experience or past actions and reactions are automatically used to change parameters in an algorithm".

The history of machine learning extends to the Neural Networks (1943) and Turing Test (1950). Basic definition of machine learning in the literature is "the field of scientific study that concentrates on induction algorithms and on other algorithms that can be said to 'learn'" (Provost and Kohavi, 1998). Beam and Kohane (2018) also presented a broad definition as "machine learning was originally described as a program that learns to perform a task or make a decision automatically from data, rather than having the behavior explicitly programmed".

Machine Learning is considered a natural outgrowth of the intersection of Statistics and Computer Science (Mitchell, 2006). Data mining and machine learning are intersect concepts.

Nowadays, machine learning methods are applied in many different fields. For example, there are many researches in the field of civil engineering (Kewalramani and Gupta, 2006; Heshmati et al., 2008; Chou and Tsai, 2012; Chou and Pham, 2013; Yumurtacı Aydogmus et al., 2015a), stock market (Choudhry and Garg, 2008; Patel et al., 2015; Shen et al., 2012), oil price forecasting (Yumurtacı Aydogmus et al., 2015b; Abdullah and Zeng, 2010), network security (Haq et al., 2015), spam detection (Crawford et al., 2015; Kolari et al., 2006), medicine (Deo, 2015; Obermeyer and Emanuel, 2016), healthcare (Chen et al., 2017); wind speed (Türkan et al., 2016; Salcedo-Sanz et al., 2011) etc.

There are two main types of machine learning: unsupervised learning and supervised learning. Unsupervised machine learning technique is "clustering" and supervised machine learning techniques are "classification" and "regression". Supervised learning trains a model on known input and output data and so that it can predict future outputs. Unsupervised learning finds hidden patterns or intrinsic structures in input data. While the classification techniques predict discrete responses (for example whether an email is genuine or spam), regression techniques predict continuous responses. Clustering is used for exploratory data analysis (gene sequence analysis, and object recognition etc.) (MathWorks, 2016).

Some of the well-known machine learning methods and the methods used in the study are as follows:

Support Vector Machines

Support Vector Machines was firstly used to solve a classification problem by Vapnik in 1995. SVM is a method, which considers the input data in an n dimension space as two sets of vectors. This method is a recently popularized method and uses structural risk minimization in problem solving along with empirical risk minimization, which is also used by ANN and other learning machines. While empirical risk minimization tries to reduce the error rate between examples, structural risk minimization lowers the error rate upper limit that could be on all examples. Thus, SVM reduce the risk of getting trapped in local extreme points in problem solving (Vapnik, 1999, as cited in Türkan et al., 2011). In classification, SVM method uses VC (Vapnik-Chervonenkis dimension) based structural risk minimization. SVM is an inductive method used in the machine learning and support vectors are defined as the vectors passing from the closest samples of the two data sets, which maximizes the distance between two samples (Türkan and Yumurtacı Aydoğmuş, 2016). SVM is a type of core-based artificial intelligence technique used to solve the regression and classification problems. It was firstly used to solve a classification problem by Vapnik in 1995. Successful forecasting studies were performed with support vector regression on the issue of time series forecasting in different fields such as production forecasting, speed of traffic flow forecasting and financial time series forecasting. (Castro-Neto et al, 2009; Hsu, et al, 2009; Huang &

Tsai, 2009; Kim, 2003; Lu, et al, 2009; Mohandes, et al, 2004; Pai& Lin, 2005; Pai, et al, 2009; Tay & Cao, 2003; Thissen et al, 2003).

Artificial Neural Network

Artificial Neural Network (ANN) has emerged as a result of the studies trying to imitate the working principle of human brain in artificial systems. Generally, ANN is a complex system consisting of the connection of various neurons in human brain, or the simple processors that artificially connect with each other at different levels of interaction (Gülez, 2008). Artificial Neural Networks (ANN) is one of artificial intelligence techniques and has a widespread use in the different areas (Kecman, 2001). ANN, unlike traditional statistical methods, does not need model assumptions and can work on any kind of non-linear function without making any prior assumptions (McNelis, 2004, as cited in Türkan et al., 2011). Many researchers concluded that ANN, which applies the Empirical Risk Minimization Principle, was more useful than traditional statistical methods (Hansen & Nelson, 1997, as cited in Türkan et al., 2011).

Multilayer Perceptron (MLP), or namely back propagation network, developed in 1986 by Rumelhart et al. The MLP works with supervisory learning principle and in the learning phase, both inputs and the corresponding outputs are shown. The task of the network is to produce output for each input. The learning rule of the MLP is the extended version of the Delta Rule, which is based on the least square method and called as Extended Delta Rule (Güllü et al., 2007).

M5P

Decision trees are the most preferred predictive method because of their high performance and easy to understand graphical interpretation (Bozkır et al., 2019).

The M5P method is strong because it implements a decision tree as much as linear regression to predict a continuous variable. Moreover, M5P implements regression trees and model trees. The M5P algorithm consists of three stages (Bragaet al., 2007):

1. Creating trees,
2. Prune the tree and
3. Smoothing the tree.

This tree model is used for numerical estimation. M5P keeps a linear regression model that predicts the class value of samples that reach each leaf. In determining which feature is best for separating the T part of the training data that reaches a specific node using the split criterion. The standard deviation of the class in T is considered a measure of the error at that node, and each feature at that node is tested by calculating the expected reduction in error. The feature selected to split maximises the expected error reduction at that node. The standard deviation reduction (SDR) calculated by the following formula is the reduction of expected errors.

$$SDR = sd\left(T\right) - \sum \frac{\left|T_i\right|}{\left|T\right|} * sd\left(T_i\right) \qquad (1)$$

Linear regression models in leaves aim to predict continuous numerical qualities. They resemble piecemeal linear functions. They eventually come together and a nonlinear function is formed. It is aimed to build a model that correlates a target value of the training cases with the values of their input qualities. The quality of the model is measured by accurately estimating the target values of situations that are not usually seen. The division process ends only when the standard deviation is a fraction smaller than the standard deviation of the original sample set, or when the number of samples remains several (Onyari and Ilunga, 2013).

Weka program provides M5P algorithm that is used to create classification and regression tree with a multivariate linear regression model where *p* stands for prime. This algorithm provides linear model as classes with some percent of approximated errors (Bhargava et al., 2013).

M5'Rules

That is the method of producing rules from trees and quite simple. It works the following way: a learning tree (in this instance the model trees) is applied to the full training data set and a pruned tree is learned. Then, the best leaf (according to some intuitions) is made into the rule, and finally the tree is discarded.

All samples covered by the rule are removed from the data set. The process is applied repeatedly to the remaining samples and operations end when all samples are covered by one or more rules. This is the basic separation and conquest strategy for the rules of learning; however, instead of creating a single rule, as usually, a full model tree is built at each stage, and its "best" leaf is made into a rule. This avoids the over-pruning potential called hasty-generalization (FW98). Unlike PART, which uses the similar strategy for categorical prediction, M5'Rules creates complete trees instead of partially discovered trees. Creating partial trees leads to more computational efficiency and does not affect the size and accuracy of the resulting rules (Holmes et al., 1999).

REP Tree Algorithm

The Reduced Error Pruning Tree applies additional reduced-error pruning as a post-processing step of the C4.5 algorithm (Gulenko et al., 2016). Fundamentally REP Tree is a quick decision tree learning and it builds a decision tree based on information benefit or reducing the variance. The basis of pruning of this algorithm is the use of "reduced error pruning" along with back over fitting. It gently sorts values for numerical attribute once and it handles the missing values with embedded method by C4.5 in fractional instances (Mohamed et al., 2012).

This algorithm was first proposed by Quinlan. In REP Tree algorithm, more than one tree is created in different iterations with the logic of regression tree. Then, the best decision trees are chosen. The algorithm is based on the following principles (Aksu ve Karaman, 2017):

1. Minimize error caused by variance,
2. The principle of knowledge acquisition with entropy.

Random Forrest Algorithm

Random Forest (RF) is the popular and very effective algorithm put forward by Breiman in 2001. This algorithm is based on model collection ideas for both regression and classification problems. The method combines the idea of bagger with random feature selection. RF's principle is to assemble many binary decision trees that are created using several bootstrap samples from the sample L (Learning) and selecting a random subset of explaining factors X at each node. Defining parameters of RF models;

1. The number of trees,
2. Number of predictors.

RF can handle large data sets, such as other tree-based methods, with numerous predictors with automatic factor selection and without factor deletion. Random forests increase the diversity among classification trees by resampling the data with modification and randomly changing the predictive variable sets over the different tree induction processes.

The RF method has advantages over other methods. These are (Naghibi et al., 2016):

1. It does not require assumptions about the distribution of explanatory factors,
2. Allows the use of categorical and numerical factors without repeating the use of indicative (or dummy) factors,
3. It can take into account interactions and nonlinear relationships between factors.

KStar

KStar is an instance-based learning algorithm that uses entropy distance measure and based on distance between instances (Malhotra and Singh, 2011). Entropy is used as distance assessment, and this property provides numerous benefits. A consistency of approach in real, symbolic, missing value attributes makes it important. A sample-based algorithm prepared for symbolic values fails on real value properties. Therefore, there is no complete theoretical basis. Approaches that succeed in the real values property are used as ad hoc approaches made to address symbolic attributes. Similar problems arise when missing values are addressed by classifiers. Generally, the missing values are considered a separate value and used instead of the average value, otherwise it is simply ignored (Madhusudana et al., 2016).

Radial Basis Function Neural Networks (RBFNetwork)

An RBFNetwork is a type of a feed-forward neural network comprised of three layers: input, hidden and output layers (Türkan et al., 2016). RBF networks consist of a three-tier architecture:

1. Input layer,
2. A single hidden layer using radial functions that give the name of the network as the event function,
3. Output layer (Çuhadar, 2013)

The operating principle of an "RBF Network" can be described as the process of determining the relationship between input and output by determining the radial-based functions with appropriate width and center values in the hidden layer based on the input data, and creating linear combinations with the appropriate weight values of the outputs produced by these functions in the output layer. RBF networks have been used as an alternative neural network to MLP, in applications involving the solution of problems such as prediction and function approximation, due to their shorter training time compared to MLP and their convergence to the best solution without local minimums (Demirkoparan et al., 2010).

Random Tree

RandomTree is also a regression-based decision tree algorithm. Trees built by RandomTree consider randomly selected attributes at each node. It performs no pruning. Also has an option to allow prediction of class probabilities based on a hold-out set (backfitting) (Witten et al., 2011). The classification works as follows (Thornton et al., 2013):

1. Random trees classifier takes vector with input property,
2. It classifies the vector with every tree in the forest and subtracts the class label, which takes the majority of the "votes".
3. In the event of a setback, the classifier's response is the average of the responses on all trees in the forest.

Bagging

Bagging or Bootstrap Aggregating improves the performance of classification models by creating various sets of the training sets and this method was proposed by Leo Breiman in 1996 (Malhotra and Singh, 2011). The bagging method is based on the procedure of mixing the outcomes of the models that result from the learning of weak training data, obtained by creating different combinations of training data, by the basic-learners. In this sense, bagging is a voting method. In Bagging, the process of creating different combinations of training data is based on the bootstrap method. This method is similar to cross-validation and is an alternative. The aim is to create multiple samples from a single sample. In the meantime, new samples are created by displacement from the original sample (Timur et al., 2011).

Locally Weighted Learning

Instance-based algorithm is used by LWL and this method assigns instance weights. LWL can be applied both classification and regression (Jiawei and Kamber, 2001, as cited in Türkan et al., 2016)

LWL depends on the distance function. There are many different approaches to define a distance function. In LWL learning, distance functions do not need to provide formal mathematical models for distance measurement. The term of scaling is used to have been reserved the weight term for the contribution of individual points (not dimensions) in the regression. Distance functions may be asymmetric and nonlinear. Thus, a distance along a given dimension may depend on whether the value of query point for the dimension is larger or smaller than the value of stored point for that dimension. The distance along a dimension may also depend on the values being compared (Atkeson et al., 1997).

Lazy IBK

Lazy IBK is also known by the K-NN (K nearest neighbor) algorithm. Lazy IBK is is an instance-based learning method for classifying objects based on the closest training examples in the feature space (Adebowale et al., 2013). Comparative study of selected data mining algorithms used for intrusion detection.

International Journal of Soft Computing and Engineering (IJSCE), 3(3), 237-241.). The amount of data/ nearest neighbors is determined by the user stated by k (Fitri, 2014). Lazy IBK has successful applications in classification problems. Solomatine et al. (2006) explored the applicability of lazy (instance-based) learning methods and results showed that lazy IBk is accurate predictor. Peng et al. (2009) developed a performance metric to evaluate the quality of classifiers for software defect prediction and according to the results Lazy IBk (K-nearest-neighbor) algorithm is among the first three methods of 13 classifier algorithms. Fitri (2014) studied to find out the best performance of Naïve Bayesian, Lazy-IBK, Zero-R and Decision Tree-J48. The test results show that the Bayesian naïve algorithm has the best accuracy in the cross-validation test mode, and Bayesian naïve algorithm and lazy IBk have the best accuracy in the percentage-validation test mode.

Problem Definition and Data Set

Today, crowded cities are overwhelming and people need transportation service during the day. For private organizations or public corporations providing transportation services, meeting customer demand at the right place at the right time is of great importance. Predicting the number of passengers (demand) is one of the most important elements for public transport management, in order to make the vehicle and personnel schedules and to determine the necessary investments. With advances in technology, faster and more efficient solutions are focused. Machine learning methods are one of the current methods for transportation management problems.

In this study, metrobus line which is the busiest passenger transportation line in Istanbul is investigated. Metrobus line, given in the Figure1, the longest public transport line in Istanbul, has a total of 45 stations, eight of which are transfer stations. This line, which connects Asia and Europe, runs 535 buses on separated roads and transports an average of 800,000 passengers per day.

In the problem, daily and hourly the number of passenger data of five years (from 2015 to 2019 June) for two main stations were used. These stations are transfer stations and there have been no major changes in their environment in the last five years, therefore no parameter other than time data is use. The inputs for the problem are time values including year, month, day and hour. The output is the number of passengers. The stations named M1 and M2 given in the Figure2 are among the most active transfer stations with many universities and dormitories around them. There are more than 1 million records in the entire data set and in order to reduce the computing time, we limit our study with two stations. The data belonging to the first four years were used for learning stage and those belonging to the last year were used for test stage.

Figure 1. The Metrobus Line
Source: metrobusharitasi.com, 2018

In terms of transportation, the availability of more vehicles during peak hours is one of the important factors that will ensure customer satisfaction and reduce waiting time. In order to make this planning, supervised learning was carried out with data sets and the results were presented to the corporation. Correlation coefficient, root mean square error and mean absolute error are used as statistical techniques to compare machine learning methods success.

Statistical Measures

Successes of the methods employed were compared using different statistical measures. Mean Absolute Error (MAE), Root Mean Square Error (RMSE) and Correlation Coefficient (R) are among the widely used measures that are based on the notion of "mean error". Correlation analysis is performed to determine the level and direction of relationship between two variables. Correlation coefficient takes a value between -1 and +1. Here, the absolute magnitude of numbers refers to the level of correlation between variables, and the numbers' signs (positive or negative) express the direction of correlation. The positive correlation coefficient means the rise (or decline) in the values of a variable increases (or decreases) those of the other.

$$R = \frac{n\sum yy' - \left(\sum y\right)\left(\sum y'\right)}{\sqrt{\left(\sum y^2\right) - \left(\sum y\right)^2}\sqrt{\left(\sum y'^2\right) - \left(\sum y'\right)^2}} \tag{2}$$

The relationship between the level of correlation between variables and the value the correlation coefficient may be given as follows: If it is in the range 0-0.25 the level of correlation is very weak, 0.26-0.49 weak, 0.50-0.69 medium, 0.70-0.89 strong, and 0.90-1.00 very strong. Root Mean Square Error (RMSE) is used to determine the rate of error between calculation values and model forecasts. As

Figure 2. Istanbul railway network map and M1, M2 Metrobus transfer stations
Source: metrobusharitasi.com, 2018

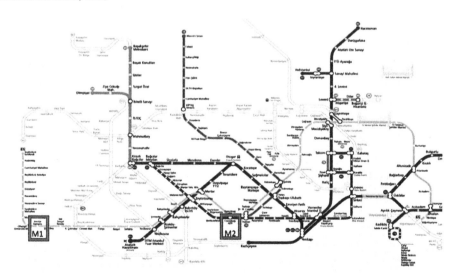

RMSE approaches zero, the forecasting capacity of the model increases. For the problem addressed in this paper; and, expressed in the following RMSE equation, refer to the model's forecasting results and realized values, respectively. N, on the other hand, shows the number of forecasts.

$$RMSE = \sqrt{\frac{\sum_{0-1}^{N} \left(o_i - t_i\right)^2}{N}} \tag{3}$$

Mean Absolute Error (MAE) questions the absolute error between calculation values and model forecasts. As MAE approaches zero, the forecasting capacity of the model increases. In the following equation, and refer to the model's forecasting results and realized values, respectively (Türkan et al., 2011).

$$MAE = \frac{1}{N} \sum_{Y'=1}^{N} \left|o_i - t_i\right| \tag{4}$$

RESULTS

In the study, by using machine learning methods in WEKA software platform the number of passengers in the M1 and M2 stations are predicted. The results of the prediction were compared by statistical measures R, MAE and RMSE. Learning stage was performed by using data of first 48 months, and by using data of last 5 months the test was performed. Table 1 and 2 show the results of some machine learning methods for M1 and M2 stations respectively. The methods are ranked from good to bad according to the values in the test phase.

According to the results obtained for M1 station, four of the five most successful methods are in the bagging structure (Table 1). According to values for test phase, the correlation coefficient for the eight methods was higher than 0,9 and the level of correlation is very strong. MAE and RMSE values of the first four methods are very close and there is only one passenger difference between them for MAE test values.

According to the results obtained for M2 station, three of the five most successful methods are in the bagging structure (Table 2). According to values for test phase, the correlation coefficient for the eight methods was higher than 0,9 and the level of correlation is very strong. MAE and RMSE values of the first three methods are very close and there is only three passenger difference between them for MAE test values.

The MAE and RMSE values in Table 1 and 2 are shown in Figure 3 and 4 for stations M1 and M2, respectively.

The R values obtained for M1 and M2 stations are shown in Figure 5.

Table 1. The abilities of the methods for the prediction of the number of the passengers for M1 station

Methods	Type	Train / Test	R	MAE	RMSE	Number of Instances
Bagging M5 Rules	Meta /Rules	Train	0,9049	38,4712	454,0787	34599
		Test	0,9114	239,1067	458,5407	3601
Bagging REP Tree	Meta /Trees	Train	0,9366	180,2343	374,1572	34599
		Test	0,9095	239,1950	462,8889	3601
M5 P	Trees	Train	0,9078	231,6832	447,3046	34599
		Test	0,9091	240,0613	464,6997	3601
Bagging M5 P	Meta /Trees	Train	0,9084	229,9234	446,0061	34599
		Test	0,9106	240,1593	459,8334	3601
Bagging-Lazy IBK	Meta /Lazy	Train	0,9877	75,1689	169,0644	34599
		Test	0,9011	244,1227	487,5597	3601
M5 Rules	Rules	Train	0,9044	239,5068	454,9735	34599
		Test	0,9077	245,8961	468,9580	3601
REP Tree	Trees	Train	0,9196	203,9997	419,0296	34599
		Test	0,9036	247,2436	477,4424	3601
Bagging Randomizable Filtered Classifier	Meta	Train	0,9803	103,9136	215,5580	34599
		Test	0,9070	256,7968	469,2043	3601
Random Committee	Meta	Train	0,9999	8,9139	14,6959	34599
		Test	0,8687	269,5896	571,7861	3601
Lazy IBK	Lazy	Train	1	0	0	34599
		Test	0,8686	269,895	571,9134	3601
Randomizable Filtered Classifier	Meta	Train	1	0	0	34599
		Test	0,8686	269,8950	571,9134	3601
Random Tree	Tree	Train	0,9999	10,0129	17,2456	34599
		Test	0,8687	270,0506	571,8926	3601
K Star	Lazy	Train	0,8764	384,8665	577,8713	34599
		Test	0,8781	405,6397	598,4694	3601
LWL	Lazy	Train	0,6840	510,0190	778,3020	34599
		Test	0,6987	529,1702	802,1674	3601

CONCLUSION

Transport is of great importance for countries for their economic and social development. Passenger transport has a critical role in enabling people to reach basic social services.

In this study, it can be concluded that machine learning methods are highly successful in the passenger flow estimation and results can be used for decisions on investments and projections for the public transportation. Furthermore, the success of the estimation was measured with the use of different inputs

Table 2. The abilities of the methods for the prediction of the number of the passengers for M2 station

Methods	Type	Train / Test	R	MAE	RMSE	Number of Instances
Bagging M5 P	Meta /Trees	Train	0,9150	332,8420	585,0478	34585
		Test	0,9179	339,3311	605,8028	3601
Bagging M5 Rules	Meta /Rules	Train	0,9136	338,1297	589,6476	34585
		Test	0,9174	342,2148	607,0451	3601
M5 P	Trees	Train	0,9125	338,9278	593,1908	34585
		Test	0,9160	342,5776	613,3095	3601
M5 Rules	Rules	Train	0,9113	342,7935	597,0392	34585
		Test	0,9156	345,2501	614,1827	3601
Bagging-Lazy IBK	Meta /Lazy	Train	0,9884	108,3782	222,7395	34585
		Test	0,9003	355,7866	671,7106	3601
Bagging REP Tree	Meta /Trees	Train	0,9370	276,0961	507,1352	34585
		Test	0,9149	356,4012	614,7465	3601
REP Tree	Trees	Train	0,9201	310,5068	568,1633	34585
		Test	0,9144	363,3687	617,0568	3601
Bagging Randomizable Filtered Classifier	Meta	Train	0,9894	114,6319	215,3494	34585
		Test	0,9035	366,2529	653,6433	3601
Random Committee	Meta	Train	0,9999	14,2861	23,3542	34585
		Test	0,8691	391,0943	782,0272	3601
Random Tree	Trees	Train	0,9998	15,2580	25,6270	34585
		Test	0,8691	391,3976	782,1465	3601
Lazy IBK	Lazy	Train	1	0	0	34585
		Test	0,8690	391,6870	781,9324	3601
Randomizable Filtered Classifier	Meta	Train	1	0	0	34585
		Test	0,8252	492,6240	902,4183	3601
K Star	Lazy	Train	0,8791	483,8534	721,0367	34585
		Test	0,8756	510,1458	761,9977	3601
LWL	Lazy	Train	0,6324	851,3814	1123,4483	34585
		Test	0,6504	876,9572	1162,6019	3601

as well as by estimation on time basis. In this study, 5-year of data of passenger numbers is used and MAE, RMSE and R statistics are used to evaluate the machine learning methods.

In this study, it can be concluded that the first five methods are respectively; bagging with M5Rules and M5P, M5 P, bagging with Rep Tree and Lazy IBK as seen in Table 3.

It is seen from the results of the study that bagging methods as ensemble learning methods have a clear superiority. In this aspect, this study also investigates the potential usage of the bagging methods. Bagging M5 Rules is superior than M5 P and RepTree. Table 3 shows that bagging ensemble learning methods and the methods-based rules and trees are quite successful in the prediction of the passenger

Figure 3. MAE and RMSE values of test phase for M1 Metrobus transfer stations

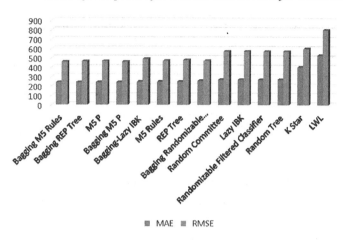

Figure 4. MAE and RMSE values of test phase for M2 Metrobus transfer stations

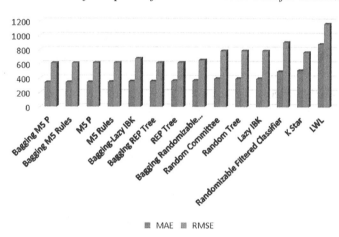

Figure 5. R values of test phase for M1 and M2 Metrobus transfer stations

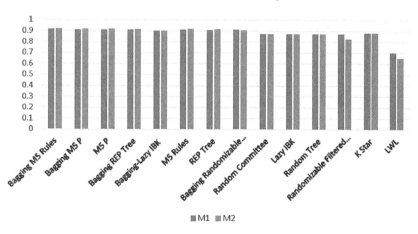

Table 3. Ranking of the first five methods for M1 and M2 stations

Methods	Type	Rank of M1 Station	Rank of M2 Station
Bagging M5 Rules	Meta /Rules	1	2
Bagging M5 P	Meta /Trees	4	1
M5 P	Trees	3	3
Bagging REP Tree	Meta /Trees	2	6
Bagging-Lazy IBK	Meta /Lazy	5	5

numbers. For this reason, it can be stated that passenger number forecasting made by these methods may help in decision making for investments and projections for the public transportation.

Further studies will apply for estimating the number of passengers for future with the methods presented in this study. The number of vehicles will be scheduled by these methods. With the schedule thus obtained, labor costs, fuel costs, vehicle maintenance costs, and even the density of metrobus traffic will be reduced.

ACKNOWLEDGMENT

The authors would like to acknowledge the support of the Istanbul Metropolitan Municipality and Istanbul Electricity, Tramway and Tunnel General Management for providing the data that were used in this study.

REFERENCES

Abdullah, S. N. & Zeng, X. (2010, July). Machine learning approach for crude oil price prediction with Artificial Neural Networks-Quantitative (ANN-Q) model. In *The 2010 International Joint Conference on Neural Networks* (IJCNN) (pp. 1-8). IEEE.

Adebowale, A., Idowu, S. A., & Amarachi, A. (2013). Comparative study of selected data mining algorithms used for intrusion detection. [IJSCE]. *International Journal of Soft Computing and Engineering*, *3*(3), 237–241.

Aksu, M. Ç. & Karaman, E. (n.d.). Link analysis and detection in a web site with decision trees. *Acta INFOLOGICA*, 1(2), 84-91.

Atkeson, C. G., Moore, A. W., & Schaal, S. (1997). Locally weighted learning. In *Lazy learning* (pp. 11–73). Dordrecht, The Netherlands: Springer. doi:10.1007/978-94-017-2053-3_2

Bar-Gera, H., & Boyce, D. (2003). Origin-based algorithms for combined travel forecasting models. *Transportation Research Part B: Methodological*, *37*(5), 405–422. doi:10.1016/S0191-2615(02)00020-6

Beam, A. L., & Kohane, I. S. (2018). Big data and machine learning in health care. *Journal of the American Medical Association, 319*(13), 1317–1318. doi:10.1001/jama.2017.18391 PMID:29532063

Bhargava, N., Sharma, G., Bhargava, R., & Mathuria, M. (2013). Decision tree analysis on J48 algorithm for data mining. In Proceedings of International Journal of Advanced Research in Computer Science and Software Engineering, 3(6).

Blainey, S. P., & Preston, J. M. (2010). Modelling local rail demand in South Wales. *Transportation Planning and Technology, 33*(1), 55–73. doi:10.1080/03081060903429363

Bozkır, A. S., Sezer, E., & Bilge, G. (2019). *Determination of the factors influencing student's success in student selection examination (OSS) via data mining techniques. IATS'09.* Karabük Turkiye.

Braga, P. L., Oliveira, A. L., & Meira, S. R. (2007, September). Software effort estimation using machine learning techniques with robust confidence intervals. In *7th International Conference on Hybrid Intelligent Systems (HIS 2007)* (pp. 352-357). IEEE. 10.1109/HIS.2007.56

Breiman, L. (2001). Random forests. *Machine Learning, 45*(1), 5–32. doi:10.1023/A:1010933404324

Carbonell, J. G., Michalski, R. S., & Mitchell, T. M. (1983). An overview of machine learning. In *Machine learning* (pp. 3–23). Morgan Kaufmann.

Castro-Neto, M., Jeong, Y. S., Jeong, M. K., & Han, L. D. (2009). Online-SVR for shortterm traffic flow prediction under typical and atypical traffic conditions. *Expert Systems with Applications, 36*(3), 6164–6173. doi:10.1016/j.eswa.2008.07.069

Chen, M., Hao, Y., Hwang, K., Wang, L., & Wang, L. (2017). Disease prediction by machine learning over big data from healthcare communities. *IEEE Access: Practical Innovations, Open Solutions, 5*, 8869–8879. doi:10.1109/ACCESS.2017.2694446

Chen, M., Liu, X. B., Xia, J. X., & Chien, S. I. (2004). A dynamic bus-arrival time prediction model based on APC data. *Computer-Aided Civil and Infrastructure Engineering, 19*, 364–376.

Chen, Q., Li, W., & Zhao, J. (2011). The use of LS-SVM for short-term passenger flow prediction. *Transport, 26*(1), 5–10. doi:10.3846/16484142.2011.555472

Chien, I.-Jy., Ding, Y., & Wei, C. (2002). Dynamic bus arrival time prediction with artificial neural networks. *Journal of Transportation Engineering, 128*(5), 429–438.

Choudhry, R., & Garg, K. (2008). A hybrid machine learning system for stock market forecasting. *World Academy of Science, Engineering and Technology, 39*(3), 315–318.

Clark, S. (2003). Traffic prediction using multivariate nonparametric regression. *Journal of Transportation Engineering, 129*(2), 161–168.

Crawford, M., Khoshgoftaar, T. M., Prusa, J. D., Richter, A. N., & Al Najada, H. (2015). Survey of review spam detection using machine learning techniques. *Journal of Big Data, 2*(1), 23. doi:10.118640537-015-0029-9

Çuhadar, M. (2013). Modeling and forecasting inbound tourism demand to Turkey by MLP, RBF and TDNN artificial neural networks: A comparative analysis. *Journal of Yaşar University, 8*(31).

Cyril, A., Mulangi, R. H., & George, V. (2018, August). Modelling and forecasting bus passenger demand using time series method. In *2018 7th International Conference on Reliability, Infocom Technologies and Optimization (Trends and Future Directions)(ICRITO)* (pp. 460-466). IEEE.

Cyril, A., Mulangi, R. H., & George, V. (2019). Bus passenger demand modelling using time-series techniques-big data analytics. *The Open Transportation Journal, 13*(1).

Dargay, J., Clark, S., Johnson, D., Toner, J., & Wardman, M. (2010). A forecasting model for long distance travel in Great Britain. *12th World Conference on Transport Research*, Lisbon, Portugal.

Demirkoparan, F., Taştan, S., & Kaynar, O. (2010). Crude oil price forecasting with artificial neural networks. *Ege Academic Review, 10*(2), 559–573.

Deo, R. C. (2015). Machine learning in medicine. *Circulation, 132*(20), 1920–1930. doi:10.1161/CIRCULATIONAHA.115.001593 PMID:26572668

Ding, Y. & Chien, S. (2000). The prediction of bus arrival times with link-based artificial neural networks. In *Proceedings of the International Conference on Computational Intelligence & Neurosciences (CI&N)—Intelligent Transportation Systems* (pp. 730–733), Atlantic City, NJ.

Du Preez, J., & Witt, S. F. (2003). Univariate versus multivariate time series forecasting: An application to international tourism demand. *International Journal of Forecasting, 19*(3), 435–451. doi:10.1016/S0169-2070(02)00057-2

Efendigil, T., & Eminler, Ö. E. (2017). The importance of demand estimation in the aviation sector: A model to estimate airline passenger demand. *Journal of Yaşar University, 12*, 14–30.

Fischer, U., & Lehner, W. (2013). Transparent forecasting strategies in database management systems. In E. Zimányi (Ed.), *European Business Intelligence Summer School* (pp. 150–181). Cham, Switzerland: Springer. Retrieved from https://books.google.com.tr/

Fitri, S. (2014). Perbandingan Kinerja Algoritma Klasifikasi Naïve Bayesian, Lazy-Ibk, Zero-R, Dan Decision Tree-J48. *Data Manajemen dan Teknologi Informasi (DASI), 15*(1), 33.

Gulenko, A., Wallschläger, M., Schmidt, F., Kao, O., & Liu, F. (2016, December). Evaluating machine learning algorithms for anomaly detection in clouds. *In 2016 IEEE International Conference on Big Data (Big Data)* (pp. 2716-2721). IEEE.

Gulez, K. (2008). Yapay sinir ağlarının kontrol mühendisliğindeki uygulamaları (Applications of artificial neural networks in control engineering). (Lecture notes, 2008) Yıldız Teknik University. Retrieved from http://www.yildiz.edu.tr/~gulez/3k1n.pdf

Güllü, H., Pala, M., & Iyisan, R. (2007). Yapay sinir ağları ile en büyük yer ivmesinin tahmin edilmesi (Estimating the largest ground acceleration with artificial neural networks). *Sixth National Conference on Earthquake Engineering*, Istanbul, Turkey.

Hagan, M. T., Demuth, H. B., & Beale, M. (1996). *Neural network design*. Boston, MA: PWS.

Haq, N. F., Onik, A. R., Hridoy, M. A. K., Rafni, M., Shah, F. M., & Farid, D. M. (2015). Application of machine learning approaches in intrusion detection system: A survey. *IJARAI-International Journal of Advanced Research in Artificial Intelligence*, *4*(3), 9–18.

Holmes, G., Hall, M., & Prank, E. (1999, December). Generating rule sets from model trees. *In Australasian Joint Conference on Artificial Intelligence* (pp. 1-12). Berlin, Germany: Springer.

Hsu, S. H., Hsieh, J. J. P.-A., Chih, T. C., & Hsu, K. C. (2009). A two-stage architecture for stock price forecasting by integrating self-organizing map and support vector regression. *Expert Systems with Applications*, *36*(4), 7947–7951. doi:10.1016/j.eswa.2008.10.065

Huang, C. L., & Tsai, C. Y. (2009). A hybrid SOFM–SVR with a filter-based feature selection for stock market forecasting. *Expert Systems with Applications*, *36*(2), 1529–1539. doi:10.1016/j.eswa.2007.11.062

Huang, Y., & Pan, H. (2011, April). Short-term prediction of railway passenger flow based on RBF neural network. In *Proceedings of the 4th International Joint Conference on Computational Sciences and Optimization, CSO 2011* (pp. 594–597). 10.1109/CSO.2011.240

Istanbul Railway Network and Metrobus Line. Retrieved from http://www.metrobusharitasi.com/metrobus-haritasi

Jiawei, H. & Kamber, M. (2001). Data mining: concepts and techniques. San Francisco, CA: Morgan Kaufmann, 5.

Jovicic, G., & Hansen, C. O. (2003). A passenger travel demand model for Copenhagen. *Transportation Research Part A, Policy and Practice*, *37*(4), 333–349. doi:10.1016/S0965-8564(02)00019-8

Kim, K. J. (2003). Financial time series forecasting using support vector machines. *Neurocomputing*, *55*(1-2), 307–319. doi:10.1016/S0925-2312(03)00372-2

Kolari, P., Java, A., Finin, T., Oates, T., & Joshi, A. (2006, July). Detecting spam blogs: A machine learning approach. In *Proceedings of the national conference on artificial intelligence* (Vol. 21, No. 2, p. 1351). Menlo Park, CA: AAAI Press.

Laplante, P. A. (Ed.). (2000). Dictionary of computer science, engineering and technology. Boca Raton, FL: CRC Press.

Lin, W. H. (2001). A Gaussian maximum likelihood formulation for short-term forecasting of traffic flow. In *Proceedings of the 2001 IEEE Intelligent Transportation Systems* (pp. 150–155), Oakland, CA.

Lu, C. J., Lee, T. S., & Chiu, C. C. (2009). Financial time series forecasting using independent component analysis and support vector regression. *Decision Support Systems*, *47*(2), 115–125. doi:10.1016/j.dss.2009.02.001

Madhusudana, C. K., Kumar, H., & Narendranath, S. (2016). Condition monitoring of face milling tool using K-star algorithm and histogram features of vibration signal. *Engineering Science and Technology*. *International Journal (Toronto, Ont.)*, *19*(3), 1543–1551.

MathWorks. (2016). Introducing Machine Learning. Retrieved from https://www.academia.edu/31538731/ Introducing_Machine_Learning

Mitchell, T. M. (2006). *The discipline of machine learning* (Vol. 9). Pittsburgh, PA: Carnegie Mellon University, School of Computer Science, Machine Learning Department.

Mohamed, W. N. H. W., Salleh, M. N. M., & Omar, A. H. (2012, November). A comparative study of reduced error pruning method in decision tree algorithms. In *2012 IEEE International Conference on Control System, Computing and Engineering* (pp. 392-397). IEEE. 10.1109/ICCSCE.2012.6487177

Mohandes, M. A., Halawani, T. O., Rehman, S., & Hussain, A. A. (2004). Support vector machines for wind speed prediction. *Renewable Energy, 29*(6), 939–947. doi:10.1016/j.renene.2003.11.009

Naghibi, S. A., Pourghasemi, H. R., & Dixon, B. (2016). GIS-based groundwater potential mapping using boosted regression tree, classification and regression tree, and random forest machine learning models in Iran. *Environmental Monitoring and Assessment, 188*(1), 44. doi:10.100710661-015-5049-6 PMID:26687087

Obermeyer, Z., & Emanuel, E. J. (2016). Predicting the future—Big data, machine learning, and clinical medicine. *The New England Journal of Medicine, 375*(13), 1216–1219. doi:10.1056/NEJMp1606181 PMID:27682033

Onyari, E. K., & Ilunga, F. M. (2013). Application of MLP neural network and M5P model tree in predicting streamflow: A case study of Luvuvhu catchment, South Africa. *International Journal of Innovation, Management and Technology, 4*(1), 11.

Owen, A. D., & Philips, G. D. A. (1987). An econometric investigation into the characteristics of railway passenger demand. *Journal of Transport Economics and Policy, 21*, 231–253.

Pai, P. F., & Lin, C. S. (2005). A hybrid ARIMA and support vector machines model in stock price forecasting. *Omega, 33*(6), 497–505. doi:10.1016/j.omega.2004.07.024

Pai, P. F., Yang, S. L., & Chang, P. T. (2009). Forecasting output of integrated circuit industry by support vector regression models with marriage honey-bees optimization algorithms. *Expert Systems with Applications, 36*(7), 10746–10751. doi:10.1016/j.eswa.2009.02.035

Patel, J., Shah, S., Thakkar, P., & Kotecha, K. (2015). Predicting stock market index using fusion of machine learning techniques. *Expert Systems with Applications, 42*(4), 2162–2172. doi:10.1016/j.eswa.2014.10.031

Peng, Y., Kou, G., Wang, G., Wang, H., & Ko, F. I. (2009). Empirical evaluation of classifiers for software risk management. *International Journal of Information Technology & Decision Making, 8*(04), 749–767. doi:10.1142/S0219622009003715

Preston, J. (1991). Demand forecasting for new local rail stations and services. *Journal of Transport Economics and Policy, 25*, 183–202.

Provost, F., & Kohavi, R. (1998). Glossary of terms. *Journal of Machine Learning, 30*(2-3), 271–274.

Salcedo-Sanz, S., Ortiz-Garcı, E. G., Pérez-Bellido, Á. M., Portilla-Figueras, A., & Prieto, L. (2011). Short term wind speed prediction based on evolutionary support vector regression algorithms. *Expert Systems with Applications*, *38*(4), 4052–4057. doi:10.1016/j.eswa.2010.09.067

Shen, S., Jiang, H., & Zhang, T. (2012). *Stock market forecasting using machine learning algorithms* (pp. 1–5). Stanford, CA: Department of Electrical Engineering, Stanford University.

Smith, B. L., Williams, B. M., & Oswald, R. K. (2002). Comparison of parametric and nonparametric models for traffic flow forecasting. *Transportation Research Part C, Emerging Technologies*, *10*(4), 303–321. doi:10.1016/S0968-090X(02)00009-8

Solomatine, D. P., Maskey, M., & Shrestha, D. L. (2006, July). Eager and lazy learning methods in the context of hydrologic forecasting. In *The 2006 IEEE International Joint Conference on Neural Network Proceedings* (pp. 4847-4853). IEEE.

Suh, D. Y., & Ryerson, M. S. (2019). Forecast to grow: Aviation demand forecasting in an era of demand uncertainty and optimism bias. *Transportation Research Part E, Logistics and Transportation Review*, *128*, 400–416. doi:10.1016/j.tre.2019.06.016

Tay, F. E. H., & Cao, L. J. (2003). Support vector machine with adaptive parameters in financial time series forecasting. *IEEE Transactions on Neural Networks*, *14*(6), 1506–1518. doi:10.1109/TNN.2003.820556 PMID:18244595

Thissen, U., van Brakel, R., de Weijer, A. P., Melssen, W. J., & Buydens, L. M. C. (2003). Using support vector machines for time series prediction. *Chemometrics and Intelligent Laboratory Systems*, *69*(1-2), 35–49. doi:10.1016/S0169-7439(03)00111-4

Thornton, C., Hutter, F., Hoos, H. H., & Leyton-Brown, K. (2013, August). Auto-WEKA: Combined selection and hyperparameter optimization of classification algorithms. In *Proceedings of the 19th ACM SIGKDD international conference on Knowledge discovery and data mining* (pp. 847-855). ACM. 10.1145/2487575.2487629

Timur, M., Aydın, F., & Akıncı, T. Ç. (2011). The prediction of wind speed of Göztepe district of Istanbul via machine learning method. *Electronic Journal of Machine Technologies*, *8*(4), 75–80.

Tsung-Hsien, T., Chi-Kang, L., & Chien-Hung, W. (2005). Design of dynamic neural networks to forecast short-term railway passenger demand. *Journal of the Eastern Asia Society for Transportation Studies*, *6*, 1651–1666.

Türkan, Y. S., Erdal, H., Ekinci, A. (2011). *Predictive stock exchange modeling for the developing Balkan countries: An application on Istanbul stock exchange*. Vol. 4, 34-51.

Türkan, Y. S., & Yumurtacı Aydogmuş, H. (2016). Passenger demand prediction for fast ferries based on neural networks and support vector machines. *Journal of Alanya Faculty of Business*, *8*(1).

Türkan, Y. S., Yumurtacı Aydogmus, H., & Erdal, H. (2016). The prediction of the wind speed at different heights by machine learning methods. [IJOCTA]. *An International Journal of Optimization and Control: Theories & Applications*, *6*(2), 179–187.

Wardman, M. (2006). Demand for rail travel and the effects of external factors. *Transportation Research Part E, Logistics and Transportation Review, 42*(3), 129–148. doi:10.1016/j.tre.2004.07.003

Wardman, M., & Tyler, J. (2000). Rail network accessibility and the demand for inter- urban rail travel. *Transport Reviews, 20*(1), 3–24. doi:10.1080/014416400295310

Wei, C., & Wu, K. (1997). Developing intelligent freeway ramp metering control systems. In *Conf. Proc., National Science Council in Taiwan, 7C*(3), 371–389.

Weka. Retrieved from http://www.cs.waikato.ac.nz/~ml/weka/

Williams, B. M., Durvasula, P. K., & Brown, D. E. (1998). Urban free-way traffic flow prediction: Application of seasonal autoregressive integrated moving average and exponential smoothing models. *Transportation Research Record: Journal of the Transportation Research Board, 1644*(1), 132–141. doi:10.3141/1644-14

Ye, Z. R., Zhang, Y. L., & Middleton, D. R. (2006). Unscented Kalman filter method for speed estimation using single loop detector data. *Transportation Research Record: Journal of the Transportation Research Board, 1968*(1), 117–125. doi:10.1177/0361198106196800114

Yumurtacı Aydogmus, H., Ekinci, A., Erdal, H. İ., & Erdal, H. (2015b). Optimizing the monthly crude oil price forecasting accuracy via bagging ensemble models. *Journal of Economics and International Finance, 7*(5), 127–136. doi:10.5897/JEIF2014.0629

Yumurtacı Aydogmus, H., Erdal, H. İ., Karakurt, O., Namli, E., Turkan, Y. S., & Erdal, H. (2015a). A comparative assessment of bagging ensemble models for modeling concrete slump flow. *Computers and Concrete, 16*(5), 741–757. doi:10.12989/cac.2015.16.5.741

Zhang, Y., & Xie, Y. (2007). Forecasting of short-term freeway volume with v-support vector machines. *Transportation Research Record: Journal of the Transportation Research Board, 2024*(1), 92–99. doi:10.3141/2024-11

Chapter 12
Metaheuristics Approaches to Solve the Employee Bus Routing Problem With Clustering-Based Bus Stop Selection

Sinem Büyüksaatçı Kiriş
iD https://orcid.org/0000-0001-7697-3018
Istanbul University-Cerrahpasa, Turkey

Tuncay Özcan
iD https://orcid.org/0000-0002-9520-2494
Istanbul University-Cerrahpasa, Turkey

ABSTRACT

Vehicle routing problem (VRP) is a complex problem in the Operations Research topic. School bus routing (SBR) is one of the application areas of VRP. It is also possible to examine the employee bus routing problem in the direction of SBR problem. This chapter presents a case study for capacitated employee bus routing problem with data taken from a retail company in Turkey. A mathematical model was developed based on minimizing the total bus route distance. The number and location of bus stops were determined using k-means and fuzzy c-means clustering algorithms. LINGO optimization software was utilized to solve the mathematical model. Then, due to NP-Hard nature of the bus routing problem, simulated annealing (SA) and genetic algorithm (GA)-based approaches were proposed to solve the real-world problem. Finally, the performances of the proposed approaches were evaluated by comparing with classical heuristics such as saving algorithm and nearest neighbor algorithm. The numerical results showed that the proposed GA-based approach with k-means performed better than other approaches.

DOI: 10.4018/978-1-7998-0301-0.ch012

INTRODUCTION

Vehicle routing problem (VRP) has become one of the most widely studied topics in the field of Operations Research since it was introduced by Dantzig and Ramser (1959) in the study of "Truck Dispatching Problem". In 1965, Clarke and Wright (1964) further developed the model and presented the most known heuristic solution for VRP. The first paper used the term "vehicle routing" in the title can be shown as Golden, Magnanti, and Nguyan (1972). After the computers became more effective in the 1990s, the researches for VRP has increased considerably and ever-larger problems have begun to be solved.

The classical VRP designs optimal delivery routes where each vehicle only travels one route, each vehicle has the same characteristics and there is only one central depot. The goal of the VRP is to find a set of least-cost vehicle routes such that each customer is visited exactly once by one vehicle, each vehicle starts and ends its route at the depot, and the total demand of customers does not exceed the capacity of the vehicles (Braekers et al., 2016).

As can be understood from the above definition of the problem, the VRP is a rather complicated problem because it contains the dynamically changing information over time, such as time-dependent travel times, time windows for pickup and delivery, capacity constraints and input information like demand. For this reason, it is included in the combinatorial optimization problems in literature.

There are many variants of the VRP developed for different situations in literature. According to one of the classifications presented by Eksioglu et al. (2009), the VRP can be categorized generally by applied methods, characteristics of the problem (e.g. number of stops on route, customer service demand quantity, request times of new customers, service/waiting times, time window structure), physical characteristics (geographical location of customers, number of depot, type, number or capacity of vehicles, time window type, travel time, transportation cost), information characteristics (e.g. availability, quality) and data characteristics (real data or synthetic).

In this study, we deal with employee bus routing problem that arises in transporting the employees of one of the retail companies, which is located in Istanbul, Turkey. The remainder of this paper is as follows: The next section provides a brief review of the literature on the school bus routing problem. In the third part, the background of the case and the mathematical model developed in this study are described. In the fourth part, the analytic approaches developed using metaheuristics and clustering algorithms for solving the employee bus routing problem are detailed. Also, this part shows an application of the proposed approaches for the employee bus routing problem of a retail company. Finally, the results and conclusions are presented.

LITERATURE REVIEW

Employee bus routing problems are described in the literature as "school bus routing problem (SBRP)". School bus routing is one of the application areas of VRP (Toth & Vigo, 2002). In this section, only the literature related to SBRP is presented within the scope of the study.

The SBRP is concerned with the transportation service for a group of spatially distributed students, which is provided between students' residences and schools. The general aim of the problem is to find service routes that provide this service equally to all students, in line with transportation constraints (Bowerman et al., 1995). As can be seen from this explanation, although the majority of the problem is either to minimize the total travel time (Angel et. al., 1972; Newton & Thomas, 1974; Russell & Morrel,

1986; Chen et. al., 1988, 1990; Thangiah & Nygard, 1992; Bektaş & Elmastaş, 2007) or distance (El-legood et. al., 2015), there are studies in the literature that include different objective functions such as minimizing total student walking distance (Bowerman et al., 1995; Riera-Ledesma & Salazar-González, 2013), minimizing total student travel time (Bennett & Gazis, 1972; Thangiah & Nygard, 1992; Li & Fu, 2002), minimizing costs (Schittekat et. al., 2006, 2013; Kinable et. al, 2014), minimizing routes length (Corboren et al., 2002; Riera-Ledesma & Salazar-González, 2012), minimizing number of buses needed (Bodin & Berman, 1979; Swersey & Ballard, 1984; Braca et al., 1997; Bektaş & Elmastaş, 2007), maximizing number of students in each route (Dulac et. al, 1980) and balancing bus loads (Chen et. al, 1988, 1990). Besides some of the probable limitations of this problem can be expressed as follows: not exceeding the current capacity of each vehicle, ensuring that the total time that the students spend on the road does not exceed a prescribed value, and taking each student from a certain point at certain intervals.

According to Desrosiers et al. (1981), the SBRP is in fact divided into several sub-problems. These are data preparation, bus stop selection (student assignment to stops), bus route generation, school bell time adjustment and route scheduling. In order to solve the main problem, these problems must be solved step by step.

It is also possible to classify the SBRP in terms of the differences in its characteristics. The number of schools, surrounding type (urban or rural), allowance of mixed loading, the homogeneous or hetero-geneous fleet can be given as an example for these characteristics.

In the literature, the SBRP was solved by exact methods, heuristic algorithms and metaheuristic algo-rithms depending on the size and complexity of the problem. Integer programming (Bektaş & Elmastaş, 2007; Fügenschuh, 2009) and mixed-integer programming (Chen et al., 2015) are some of the examples of exact methods. Some other used heuristics for solving the SBRP are saving algorithm (Bennett & Gazis, 1972), nearest neighbor algorithm (Faraj et al, 2014) and sweep algorithm (Park et al., 2012). The metaheuristics that are simulated annealing (Spada et al, 2005; Samadi-Dana et al., 2017) tabu search (Spada et al, 2005; Rashidi et al., 2009), genetic algorithm (Minocha & Tripathi, 2014; Kang et al., 2015; Oluwadare et al., 2018; Ünsal & Yiğit, 2018), ant colony algorithm (Arias-Rojas, 2012; Euchi & Mraihi, 2012; Yigit & Unsal, 2016; Mokhtari & Ghezavati, 2018), harmony search (Geem, 2005; Kim & Park, 2013) have also been proposed for SBRP. In addition, some studies, which combine heuristics and metaheuristics (Shafahi et al., 2018), are also available in the literature. Interested readers can glance at Park and Kim (2010) and Ellegood et al. (2019), which give an extensive and detailed literature review for the classification of SBRP.

CASE STUDY BACKGROUND

This study was conducted in one of Turkey's largest retail company that operates in Istanbul. There are 822 employees working in the company and the transportation of these employees are provided by 54 buses. The existing transportation routes are created with a completely intuitive perspective. Hence, this led to the following problems:

- The capacities of the buses are not used effectively. Thus, transportation costs per capita increase.
- The occupancy rates of buses are unbalanced.
- Some of the employees have to get on the taxi / public transport etc. to reach the existing stops. This situation adds extra costs to the company and negatively affects employees' satisfaction.

In order to solve these problems, a mathematical model has been developed, which minimizes the total distance for employee bus routing problem.

MATHEMATICAL MODEL

The mathematical model developed in this study is a capacity-constrained SBRP model, in which the start and end points are the same. There are also a large number of buses that can be used and they are heterogeneous.

The assumptions of this study are as follows:

1. Each selected bus moves in one route.
2. Each route ends at the company's head office.
3. Each bus stop is only visited by one bus.
4. The number of employees in each selected bus should not exceed the capacity of the bus.

Before formulating the model, the decision variables are defined.

$$x_{ijk} = \begin{cases} 1, & \text{if stop } j \text{ comes before stop } i \text{ for bus } k \\ 0, & \text{otherwise} \end{cases}$$

$$y_{ik} = \begin{cases} 1, & \text{if stop } i \text{ is visited by bus } k \\ 0, & \text{otherwise} \end{cases}$$

where K is the set of buses available, S is set of main points (represents the company's head office), B is the set of potential stops, I is the set of all arcs between stops ($I = S \bigcup B$). The mathematical model of the problem is the following:

$$\min \sum_{i \in B} \sum_{j \in B} c_{ij} \sum_{k \in K} x_{ijk} \tag{1}$$

s.t.

$$\sum_{k \in K} y_{ik} = |K| \, \forall i \in S \tag{2}$$

$$\sum_{k \in K} y_{ik} = 1 \, \forall i \in B \tag{3}$$

$$\sum_{i=R} x_{ijk} = y_{jk} \ \forall j \in B, k = 1, \dots, n \tag{4}$$

$$\sum_{j \in B} x_{ijk} = y_{ik} \ \forall i \in B, k = 1, \dots, n \tag{5}$$

$$\sum_{i,j \in I} x_{ijk} * D_j \leq W_k \ k = 1, \dots, n \tag{6}$$

$$f_{ik} - f_{jk} + n * x_{ijk} \leq n - 1 \ \forall i, j \in B, i \neq j, k = 1, \dots, n \tag{7}$$

$$f_{ik} \leq n * x_{ijk} \ \forall i, j \in B, i \neq j, k = 1, \dots, n \tag{8}$$

$$f_{ik} \geq 0 \ \forall i \in B, k = 1, \dots, n \tag{9}$$

$$x_{ijk} \in \{0,1\} \ \forall i, j \in B, i \neq j, k = 1, \dots, n \tag{10}$$

$$y_{ik} \in \{0,1\} \ \forall i \in B, k = 1, \dots, n \tag{11}$$

where c_{ij} is the distance between stop i and stop j. D_j is the number of total employees who are picked up from stop i. W_k shows the capacity of bus k.

The objective function (1) minimizes the total bus route length. Constraint (2) provides that the company's head office is on all routes. Constraint (3) ensures that each stop to which eligible employee are assigned is only included on one route. Constraints (4) and (5) specify that a bus that visits a stop also leaves it and that the bus only visits the stop on its own route. Constraint (6) is the bus capacity constraint. Constraints (7), (8) and (9) prevent the formation of sub-tours using flow variables (f_{ik}). Constraints (10) and (11) guarantees that all decision variables are binary.

APPLICATION

Figure 1 presents the schematic diagram of the proposed framework to solve the employee bus routing problem discussed in this study. According to the proposed framework, the bus routing problem consists of three stages: data collection and preparation, bus stop selection and bus route generation. These stages can be detailed as follows.

Data Collection and Preparation

In this stage, firstly, the addresses of 822 employees were obtained from the company's database. These addresses were converted into GPS coordinates using Google Maps. Also, the coordinates of the company's head office were added as the center. The latitude-longitude coordinates of the employees' home locations are shown in Figure 2.

Figure 1. The proposed framework in this study

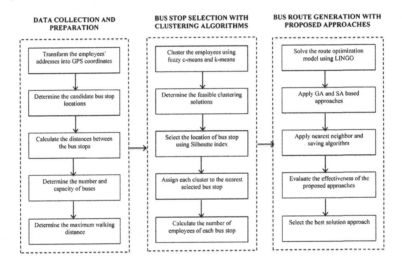

Figure 2. The coordinates of the employee addresses

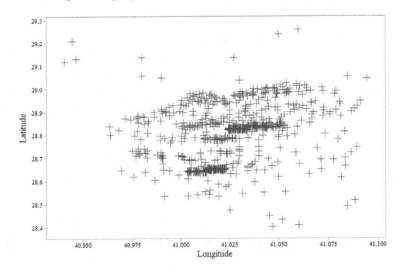

Then, the coordinates of all bus stops in Istanbul were obtained. The distances of these bus stops were calculated using Google Maps API.

In this study, the company had 54 buses for employees transportation service. According to the actual data taken from the company, the numbers and capacities of these buses are as shown in Table 1.

In the employee bus routing problem, employees usually walk to the nearest potential bus stop. This walking distance is critical for employee satisfaction. At this point, maximum walking distance (w_{max}) constraint is widely used in the literature. This constraint ensures that a bus stop is located within maximum walking distance for each employee. In this study, the maximum walking distance was taken as 800 meters.

In addition to the real-world bus routing problem, ten test problems involving different numbers of employees and buses were generated to evaluate the effectiveness of the proposed approaches. These test instances are given in Table 2.

Bus Stop Selection with Clustering Algorithms

The selection of bus stop locations has critical importance in bus routing problems. In this study, the employees were clustered by coordinates to determine the optimal bus stop. At this point, k-means and fuzzy c-means clustering algorithms were used. These algorithms can be summarized as follows:

Table 1. The number and capacities of buses

Type	Number of Buses	Capacity
Type 1	43	16
Type 2	2	23
Type 3	6	27
Type 4	3	45

Table 2. The test problems and real-world problem in this study

ID	Employee	Bus	Capacity	w_{max}
1	50	4	16	500
2	80	6	16	500
3	100	8	16	600
4	120	10	16	600
5	150	12	16	700
6	200	15	16	700
7	250	20	16	800
8	250	20	16	900
9	300	25	16	900
10	400	30	16	800
Real	822	54	Heter.	800

K-Means Clustering Algorithm

The k-means algorithm was introduced by MacQueen in 1967. It is a well-known clustering algorithm that has been used extensively in many fields of application for many years. The assignment mechanism of the k-means algorithm allows each data to belong to only one cluster. Therefore, it is a sharp clustering algorithm and tends to find equal-sized spherical clusters. It is also classified as a partitional or nonhierarchical clustering method (Huand, 1998).

The general logic of the k-means algorithm is to divide a data set consisting of n data objects into k sets, which is given as an input parameter. The aim is to ensure that the clusters obtained at the end of the partitioning process have the maximum intra-cluster similarities and the minimum inter-cluster similarities. Cluster similarity is measured by the mean value of the distances between the center of gravity of the cluster and other objects in the cluster. Although different approaches are used as distance value, Euclidean distance is the most common unit of measurement. The cluster similarity is defined as Equation 12.

$$J\left(c_k\right) = \sum_{x_i \in c_k} x_i - \mu_k^2 \tag{12}$$

where μ_k is the center of gravity of the kth cluster, x_i is the data object. The objective function of the K-Means clustering algorithm is as follows:

$$J\left(C\right) = \sum_{k=1}^{K} \sum_{x_i \in c_k} x_i - \mu_k^2 \tag{13}$$

The higher the value of the objective function indicates that the objects in the cluster are far from the cluster center. Likewise the lower value is the indicator that the objects are closer to the cluster center (Kiriş and Tüysüz, 2017).

The general steps of the k-means algorithm are as follows:

1. Choose k- the number of clusters.
2. Choose k points as a center of clusters.
3. Calculate the distance and assign each point to the nearest cluster.
4. If stopping criterion is met then stop.
5. Recalculate center of clusters.
6. Go to step 3.

Although the selection of the number of clusters in the first step depends on the data set, various systematic approaches have been developed to determine this number (Ray & Turi, 1999; Tibshirani et al., 2001; Sugar & James, 2003; Chiang & Mirkin, 2007). In the second step, first selection of center points is usually random. The stopping criterion in the fourth step may be the maximum number of iterations, the smallest determined change value etc.

Fuzzy C-Means Clustering Algorithm

According to the fuzzy approach, each object has a membership degree within the interval of [0,1]. Fuzzy c-means clustering algorithm allows the objects to be the members more than one clusters at the same time. However, the membership degree of any object should be 1 in total. Closeness of an object to any cluster, makes the membership degree of that object to that cluster greater. Fuzzy c-means clustering algorithm has a minimization objective as J_m shown as below (Hendalianpour et al., 2017).

$$J_{m,1\leq m<\infty} = \sum_{i=1}^{N}\sum_{j=1}^{C}\mu_{ij}^{m} * d_{ij}^{2}$$

(14)

In the formula, μ_{ij}^{m} shows the membership degrees. d_{ij} is the distance between ith data and centre of related cluster. As Euclidean distance, d_{ij} is calculated as Equation 15. x is the object under observation, while m shows the centroids of clusters.

$$d_{ij} = \sum_{j=1}^{C}\left(x_i - m_j\right)$$

(15)

Algorithm starts with a random selection of membership degrees. Secondly, centroid vectors are calculated as follows.

$$m_j = \frac{\sum_{i=1}^{N}\mu_{ij}^{m} * x_i}{\sum_{i=1}^{N}\mu_{ij}^{m}}$$

(16)

According to m_j centroid values, the membership degrees are recalculated as Equation 17.

$$\mu_{ij} = \frac{1}{\sum_{i=1}^{C}\frac{\left(x_i - m_j\right)}{\left(x_i - m_C\right)}^{\frac{2}{m-1}}}$$

(17)

The general steps of the fuzzy c-means algorithm is as follows (EtehadTavakol et al., 2010):

1. Initialize the number of clusters C and the membership μ_{ij}^{m}.
2. Calculate the cluster centers m_i.
3. Update the membership value μ_{ij}.
4. If distortion is less than a specified value, then stop. Else go to step 2.

In this study, k-means and fuzzy c-means algorithms were used to find the optimal number and location of the bus stops. These methods were run from 10 to 300 bus stop locations, in increments of 10.

For a feasible clustering, one or more bus stop should be selected at a distance of not more than 800 meters from the home of each employee. In this direction, maximum distance values for each clustering algorithm are shown in Figure 3. Here, maximum distance refers to the largest distance between employees' home locations and their assigned cluster centers.

As can be seen in Figure 3, the number of clusters (bus stops) should be greater than 160 for k-means clustering algorithm and greater than 180 for fuzzy c-means clustering algorithm to ensure the maximum walking distance constraint.

There are many indexes in the literature to evaluate the validity of clustering. In this study, the optimum number of bus stops was determined using Silhouette index. Equation 18 shows the calculation of the local Silhouette index (Liu et al., 2005):

$$S\left(i\right) = \frac{b\left(i\right) - a\left(i\right)}{\max\left(a\left(i\right), b\left(i\right)\right)} \tag{18}$$

where $a(i)$ is the average distance of i-point to the other points in the same cluster and $b(i)$ is the smallest of the average distance of i-point to all points in the other clusters.

The global Silhouette index is as follows:

$$S\left(C\right) = \frac{1}{n}\sum_{i=1}^{n}S\left(i\right) \tag{19}$$

where n is the number of data points. The Silhouette index takes a value between -1 and 1. The larger Silhouette index shows the better clustering (Liu et al., 2017).

Figure 3. The maximum distance values of the clustering algorithms

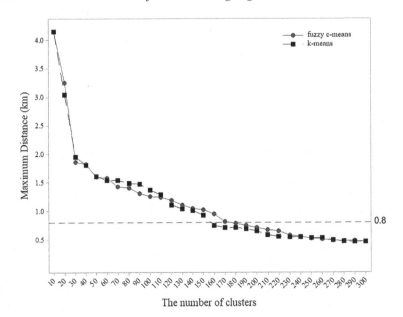

The Silhouette values for clustering solutions with different cluster numbers were calculated. Figure 4 presents that k-means algorithm with 220 clusters has the highest Silhouette index in the feasible clustering solutions.

According to the above data obtained, the centroids of 220 clusters were calculated and assigned to the nearest potential bus stop. Thus, the employees were assigned to the selected stops. Finally, the numbers of employees were calculated for each selected stop point.

The representation of the stop points and the coordinates of addresses of employees on the two-dimensional plane are shown in Figure 5.

Table 3 shows the assigned stop points and walking distances of each employee and Table 4 shows the number of employees for each stop point.

Similarly, the above steps were applied to the bus stop selection for the test problems.

Bus Route Generation with Proposed Approaches

After the bus stop locations were determined, the bus route optimization was performed. The bus stop locations obtained from the second stage, the location of the company's head office, the distances of these locations, the numbers of employees at each selected bus stop, the numbers and capacities of the available buses were used as the input data for this stage.

In this stage, firstly, LINGO 16.0 optimization software was used to solve the mathematical model of the bus routing problem. For the first test problem (50, 4), LINGO 16.0 was run for about 12 hours and found local optimal result instead of global optimum as a result of about 1.5 billion iterations. For the other test problems and the real-world problem, the LINGO 16.0 software was run for 24 hours and during this time 1 trillion iterations were done for each problem, but this software was not able to find a local or global optimal solution. This is because the employee bus routing problem is a combinato-

Figure 4. The silhouette index values of the clustering solutions

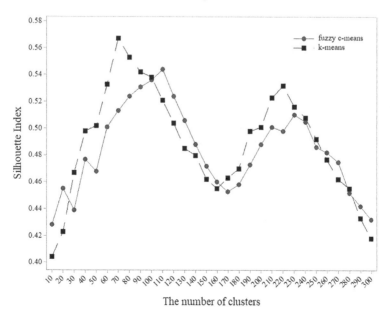

Figure 5. The coordinates of the selected stop points and the home locations of the employees

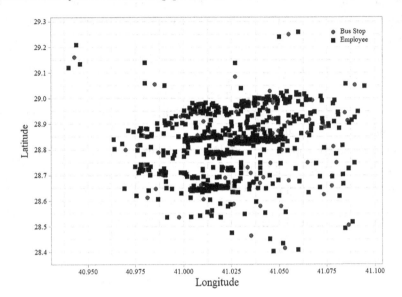

rial optimization problem. The bus routing problem has been classified as NP-Hard and no polynomial algorithms exist to find the optimal solution for the large-scale bus routing problems. In this case, the heuristic and metaheuristic algorithms are used to find good solutions with a reasonable computational time

In this study, simulated annealing and genetic algorithm based approaches were proposed to generate the optimal route among all selected bus stop locations. These approaches are summarized below.

Simulated Annealing Based Approach to Bus Route Optimization

The initial ideas underlying the simulated annealing (SA) are based on the work presented by Metropolis et al. (1953), which simulates the development of a layer in the heat bath up to the "thermal equilibrium" with Monte Carlo technique. In 1983, Kirkpatrick et al. reported that such a simulation could be used to solve difficult combinatorial optimization problems in order to converge to an optimal result and presented the simulated annealing algorithm.

The SA algorithm starts with a high temperature (T) value and an initial solution to avoid local minimums. In each calculation step, the algorithm generates a large number of solutions from the neighborhoods of the previous solution and reduces the temperature according to a predetermined rule. A random solution is chosen among these solutions and it is either accepted as a new solution or rejected. The iterations continue with the new solution (if accepted) or with the previous solution (if not accepted). Many different criteria (number of iterations, minimum temperature, etc.) can be used to stop the algorithm.

The choice of the right initial temperature is crucially important for SA. If (T) value is too high, the minimum value cannot be easily reached. If (T) value is too low, the system may be trapped in a local minimum.

Another important issue is how to reduce the temperature. The cooling process should be slow enough to allow the system to stabilize easily. In literature, there are two cooling schedules, which are linear and

Table 3. Assignment of each employee to the stop points and their walking distance

Employee #	Assigned Cluster #	Assigned Stop Point #	Walking Distance (km)
1	Cluster33	14	0.638
2	Cluster12	91	0.538
3	Cluster18	23	0.376
4	Cluster188	109	0.108
5	Cluster172	152	0.405
6	Cluster61	11	0.434
7	Cluster36	206	0.314
8	Cluster24	191	0.355
9	Cluster36	206	0.187
10	Cluster36	206	0.569
11	Cluster10	59	0.615
12	Cluster36	206	0.388
		⋮	
821	Cluster134	150	0.541
822	Cluster14	45	0.375

Table 4. The number of employees for each stop point

Stop Point #	D_j
1	2
2	3
3	1
4	6
5	2
6	4
7	8
8	3
9	5
10	6
11	7
12	2
	⋮
219	3
220	3

Table 5. The parameter values of the Simulated Annealing Algorithm

Parameter	Value
Error tolerance (ε)	0.5
Maximum number of iteration	10000
Maximum number of sub-iteration	80
Initial temperature (T)	100
Temperature cooling factor (α)	0.98

geometric. A geometric cooling schedule decreases the temperature by a cooling factor α $(0<\alpha<1)$, so that T is replaced by T. In practice, α=0.7~0.99 is commonly used (Yang, 2010).

The parameter values of the SA algorithm used in this study are given in Table 5.

Genetic Algorithm Based Approach to Bus Route Optimization

The genetic algorithm (GA), which is inspired by Darwin and Mendel's theories of evolution and genetics, is an evolutionary algorithm and provides solutions optimization and machine learning problems.

In his book named "Adaptation in Natural and Artificial Systems", John Holland (1975) developed the GA technique by combining information particles from parents in different combinations to form new individuals. Innovations of this technique were genetic operators such as crossing, mutation and selection (Glover & Kochenberger, 2006).

The genetic algorithm has two main advantages over traditional algorithms. One of them is the ability to deal with complex problems and the other is parallelism. The objective function can be easily handled by genetic algorithm, whether it is continuous or discontinuous, stationary or transient, linear or nonlinear. Multiple gene constructs may also be suitable for parallel implementation (Koziel & Yang, 2011).

The genetic algorithm starts with the creation of the initial population. Each individual in the initial population is a candidate solution and is called as "chromosome". Each chromosome is encoded in a string (binary or decimal). The chromosomes with the best fitness values according to the objective function are transferred to the next population (generation) while the others with low fitness values are not allowed to survive. Instead, new individuals are produced from good chromosomes using natural selection mechanism such as crossing and mutation. The crossover of two parent strings produces offsprings (new solutions) by swapping part or genes of the chromosomes. On the other hand, mutation is carried out by flipping some digits of a string, which generates new solutions. These steps are repeated until the best solution is achieved according to the determined fitness function.

Table 6. The parameter values of the genetic algorithm

Parameter	Value
Population size	20
Maximum number of iteration	10000
Elitism rate (P_e)	0.20
Crossover rate (P_c)	0.80
Mutation rate (P_m)	0.05

The parameter values of the GA used in this study are given in Table 6.

In this study, the performances of the proposed approaches were evaluated by comparing with classical heuristics such as saving algorithm and nearest neighbor algorithm. The details of these algorithms are explained below.

Saving Algorithm

The saving algorithm is a heuristic method developed by Clarke and Wright in 1964 to solve the vehicle routing problem. The algorithm does not guarantee the optimum solution, but generally provides a fairly good solution. Basically, the saving algorithm is based on combining two routes in one route and makes a saving from the cost.

The Saving Algorithm starts with the creation of the saving matrix. The formula $S_{ij}=c_{i0}+c_{0j}-c_{ij}$ is taken into account when constructing the matrix. S_{ij} is the saving as a result of combining the stop point i and stop point j, c is the distance between the points. Then the values in the matrix are sorted in descending order and became ready to be solved. The saving algorithm can be executed in $(n^2\log n)$ time (Laporte, 1992).

The algorithm proceeds as following:

1. Make n routes: $v_0 \rightarrow v_i \rightarrow v_0$, for each $i \geq 1$.
2. Compute the savings for merging delivery locations i and j, which is given by $S_{ij}=c_{i0}+c_{0j}-c_{ij}$, for all $i,j \geq 1$ and $i \neq j$.
3. Sort the savings in descending order and create "list of savings".
4. Starting with the topmost entry in the list of savings, merge the two routes (i,j) associated with the largest savings. The following three situations need to be considered here.
    a. If neither i nor j has already been assigned to a route, then a new route is initiated including both i and j.
    b. If exactly one of the two points (i or j) has already been included in an existing route and that point is not interior to that route then the link (i,j) is added to that same route. If the point is interior and not violating the capacity then add (i,j) to the same route. If it's violating the capacity, make a new route with the point i.
    c. If both i and j have already been included in two different existing routes and neither point is interior to its route, then the two routes are merged by connecting i and j. If they are interior then the merge cannot be done.
5. Return to step 3 and repeat until no additional savings can be achieved (Laporte, 1992; Sule, 2001; Caccetta et al., 2013).

Nearest Neighbor Algorithm

The nearest neighbor (NN) algorithm, which is one of the simplest and most understandable algorithms for the travelling salesman problem, was presented by Bellmore and Nemhauser (1968). It is an intuitive method that uses positional data to determine vehicle routes. In this method, the vehicle's overall route is determined by adding the closest node to a particular node on the route. This is repeated until all nodes are contained in the route (Golden et al., 1980).

The NN algorithm has advantages such as ease of implementation and giving quite good results. However, the algorithm is also rather slow if the set has many data. The running time for the NN algorithm is n^2 computations (Johnson & McGeoch, 1997).

This algorithm has been used in this study as part of the conformity function calculation because it produces fast and effective results such as saving algorithm. The stops were combined with the nearest stop respectively, starting from the initial point. This process continued until the capacity limitation was reached and then the other routing was started.

RESULTS

In this study, the proposed bus stop selection and bus route generation algorithms were coded in MATLAB 2017a. Additionally, the MATLAB optimization toolbox was used for the GA and SA based approaches. All calculations were performed on Intel i7-4870HQ 2.50 GHz CPU with 16 GB RAM. As previously mentioned, in addition to the real-world bus routing problem, 10 test problems were created to compare the performance of the proposed approaches. For each test problem, 10 independent replications were carried out in the GA and SA based approaches and the best solutions were considered. The minimum total distance was used as the objective function. The total distance and the number of routes obtained by the proposed approaches are given in Table 7.

The numerical results indicate that the GA and SA based approaches result in better total distance than the saving algorithm and nearest neighbor algorithm.

Table 7. Computational results of the proposed approaches

ID	Total Distance (km)					Number of Routes				
	LINGO	SA	GA	Saving	NN	LINGO	SA	GA	Saving	NN
1	108.56	108.13	107.12	115.88	121.78	4	4	4	4	4
2	--	182.65	180.32	201.34	215.12	--	6	6	5	5
3	--	203.58	208.19	231.13	272.65	--	8	8	7	7
4	--	215.45	213.84	238.41	281.66	--	9	9	8	8
5	--	269.89	264.55	312.21	341.34	--	12	12	10	10
6	--	334.43	338.21	392.11	435.48	--	15	15	13	13
7	--	396.08	398.23	462.23	495.54	--	18	19	16	16
8	--	385.12	379.68	455.24	488.28	--	19	19	16	16
9	--	455.28	441.33	504.49	541.24	--	23	24	19	19
10	--	558.37	541.57	609.11	678.32	--	28	29	25	25
Real	--	915.14	908.18	1075.65	1208.86	--	46	46	49	49

CONCLUSION

School bus routing problem (SBRP) is a special case of vehicle routing problem (Mohktari and Ghezavati, 2018). The employee bus routing problem is very similar to the SBRP. This problem has critical importance for companies in terms of transportation cost and employee satisfaction. In this direction, this chapter aims to develop an analytical framework for the solution of employee bus routing problem by using metaheuristic approaches and clustering algorithms. This framework consists of three stages: data collection and preparation, bus stop selection and bus route generation. In the data collection and preparation stage, data such as GPS coordinates of employees' home locations, locations of candidate bus stop points, distances between the stop points, number and capacity of available buses were collected. In the bus stop selection stage, k-means and fuzzy c-means algorithms were applied to cluster the employees by their coordinates. The optimum number and location of bus stops were determined with Silhouette index by considering the maximum walking distance constraint. In the bus route generation stage, firstly, LINGO optimization software was utilized to solve the employee bus routing model. Then, because the problem is NP-Hard, simulated annealing and genetic algorithm-based approaches were proposed. The effectiveness of the proposed approaches were evaluated by comparing them with the nearest neighbor algorithm and saving algorithm. At this point, a real-world problem and test problems were presented with data taken from a retail company to illustrate the application of the proposed approaches.

The numerical results showed that the proposed SA and GA based approaches are efficient for providing good results for large-sized employee bus routing problems. While the SA based approach gives the best results for test problems 3, 6 and 7, the GA based approach gives the best results for other test problems and real-world problem. The solutions in Table 7 show that the GA based approach with k-means has yielded the best performance on average.

The optimization bus route generation by using different metaheuristic and hyper-heuristic can be considered for further research. In addition, different clustering algorithms and validity indices can be used to solve the bus stop selection problem.

ACKNOWLEDGMENT

This research received no specific grant from any funding agency in the public, commercial, or not-for-profit sectors.

REFERENCES

Angel, R. D., Caudle, W. L., Noonan, R., & Whinston, A. N. D. A. (1972). Computer-assisted school bus scheduling. *Management Science*, *18*(6), B-279–B-288. doi:10.1287/mnsc.18.6.B279

Arias-Rojas, J. S., Jiménez, J. F., & Montoya-Torres, J. R. (2012). Solving of school bus routing problem by ant colony optimization. *Revista EIA*, (17), 193-208.

Bektaş, T., & Elmastaş, S. (2007). Solving school bus routing problems through integer programming. *The Journal of the Operational Research Society*, *58*(12), 1599–1604. doi:10.1057/palgrave.jors.2602305

Bellmore, M., & Nemhauser, G. L. (1968). The traveling salesman problem: A survey. *Operations Research*, *16*(3), 538–582. doi:10.1287/opre.16.3.538

Bennett, B. T. & Gazis, D. C. (1972). School bus routing by computer. *Transportation Research/UK/, 6*(4).

Bodin, L. D., & Berman, L. (1979). Routing and scheduling of school buses by computer. *Transportation Science*, *13*(2), 113–129. doi:10.1287/trsc.13.2.113

Bowerman, R., Hall, B., & Calamai, P. (1995). A multi-objective optimization approach to urban school bus routing: Formulation and solution method. *Transportation Research Part A, Policy and Practice*, *29*(2), 107–123. doi:10.1016/0965-8564(94)E0006-U

Braca, J., Bramel, J., Posner, B., & Simchi-Levi, D. (1997). A computerized approach to the New York City school bus routing problem. *IIE Transactions*, *29*(8), 693–702. doi:10.1080/07408179708966379

Braekers, K., Ramaekers, K., & Van Nieuwenhuyse, I. (2016). The vehicle routing problem: State of the art classification and review. *Computers & Industrial Engineering*, *99*, 300–313. doi:10.1016/j. cie.2015.12.007

Caccetta, L., Alameen, M., & Abdul-Niby, M. (2013). An improved Clarke and Wright algorithm to solve the capacitated vehicle routing problem. *Engineering, Technology & Applied Scientific Research*, *3*(2), 413–415.

Chen, D. S., Kallsen, H. A., Chen, H. C., & Tseng, V. C. (1990). A bus routing system for rural school districts. *Computers & Industrial Engineering*, *19*(1-4), 322–325. doi:10.1016/0360-8352(90)90131-5

Chen, D. S., Kallsen, H. A., Chen, H. C., & Tseng, V. C. (1990). A bus routing system for rural school districts. *Computers & Industrial Engineering*, *19*(1-4), 322–325. doi:10.1016/0360-8352(90)90131-5

Chen, X., Kong, Y., Dang, L., Hou, Y., & Ye, X. (2015). Exact and metaheuristic approaches for a bi-objective school bus scheduling problem. *PLoS One*, *10*(7), e0132600. doi:10.1371/journal.pone.0132600 PMID:26176764

Chiang, M. M. T. & Mirkin, B. (2007). Experiments for the number of clusters in k-means. In *Proceedings Portuguese Conference on Artificial Intelligence* (pp. 395-405). Berlin, Germany: Springer. 10.1007/978-3-540-77002-2_33

Chopde, N. R., & Nichat, M. (2013). Landmark based shortest path detection by using A* and Haversine formula. *International Journal of Innovative Research in Computer and Communication Engineering*, *1*(2), 298–302.

Clarke, G., & Wright, J. W. (1964). Scheduling of vehicles from a central depot to a number of delivery points. *Operations Research*, *12*(4), 568–581. doi:10.1287/opre.12.4.568

Corberán, A., Fernández, E., Laguna, M., & Marti, R. (2002). Heuristic solutions to the problem of routing school buses with multiple objectives. *The Journal of the Operational Research Society*, *53*(4), 427–435. doi:10.1057/palgrave.jors.2601324

Dantzig, G. B., & Ramser, J. H. (1959). The truck dispatching problem. *Management Science*, *6*(1), 80–92. doi:10.1287/mnsc.6.1.80

Desrosiers, J., Ferland, J. A., Rousseau, J.-M., Lapalme, G., & Chapleau, L. (1981). An overview of a school busing system. In N. K. Jaiswal (Ed.), *Scientific Management of Transport Systems* (pp. 235–243). Amsterdam, The Netherlands: North-Holland.

Dulac, G., Ferland, J. A., & Forgues, P. A. (1980). School bus routes generator in urban surroundings. *Computers & Operations Research, 7*(3), 199–213. doi:10.1016/0305-0548(80)90006-4

Eksioglu, B., Vural, A. V., & Reisman, A. (2009). The vehicle routing problem: A taxonomic review. *Computers & Industrial Engineering, 57*(4), 1472–1483. doi:10.1016/j.cie.2009.05.009

Ellegood, W. A., Campbell, J. F., & North, J. (2015). Continuous approximation models for mixed load school bus routing. *Transportation Research Part B: Methodological, 77*, 182–198. doi:10.1016/j.trb.2015.03.018

Ellegood, W. A., Solomon, S., North, J., & Campbell, J. F. (2019). School bus routing problem: contemporary trends and research directions. *Omega*, 102056. doi:10.1016/j.omega.2019.03.014

Etehad Tavakol, M., Sadri, S., & Ng, E. Y. K. (2010). Application of K-and fuzzy c-means for color segmentation of thermal infrared breast images. *Journal of Medical Systems, 34*(1), 35–42. PMID:20192053

Euchi, J., & Mraihi, R. (2012). The urban bus routing problem in the Tunisian case by the hybrid artificial ant colony algorithm. *Swarm and Evolutionary Computation, 2*, 15–24. doi:10.1016/j.swevo.2011.10.002

Faraj, M. F., Sarubbi, J. F. M., Silva, C. M., Porto, M. F., & Nunes, N. T. R. (2014, October). A real geographical application for the school bus routing problem. In *17th International IEEE Conference on Intelligent Transportation Systems (ITSC)* (pp. 2762-2767). IEEE.

Fügenschuh, A. (2009). Solving a school bus scheduling problem with integer programming. *European Journal of Operational Research, 193*(3), 867–884.

Geem, Z. W. (2005). School bus routing using harmony search. In *Genetic and Evolutionary Computation Conference (GECCO 2005)*, Washington, DC.

Glover, F. W., & Kochenberger, G. A. (Eds.). (2006). *Handbook of metaheuristics* (Vol. 57). Springer Science & Business Media.

Golden, B., Bodin, L., Doyle, T., & Stewart, W. Jr. (1980). Approximate traveling salesman algorithms. *Operations research, 28*(3-part-ii), 694-711.

Golden, B. L., Magnanti, T. L., & Nguyan, H. Q. (1972). Implementing vehicle routing algorithms. *Networks, 7*(2), 113–148.

Hendalianpour, A., Razmi, J., & Gheitasi, M. (2017). Comparing clustering models in bank customers: Based on Fuzzy relational clustering approach. *Accounting, 3*(2), 81–94. doi:10.5267/j.ac.2016.8.003

Holland, J. H. (1975). *Adaptation in natural and artificial systems.* Ann Arbor, MI: University of Michigan Press.

Huang, Z. (1998). Extensions to the k-means algorithm for clustering large data sets with categorical values. *Data Mining and Knowledge Discovery, 2*(3), 283–304.

Johnson, D. S. & McGeoch, L. A. (1997). The traveling salesman problem: A case study in local optimization. *Local search in combinatorial optimization, 1*(1), 215-310.

Kang, M., Kim, S. K., Felan, J. T., Choi, H. R., & Cho, M. (2015). Development of a genetic algorithm for the school bus routing problem. *International Journal of Software Engineering and Its Applications, 9*(5), 107–126.

Kim, T., & Park, B. J. (2013). Model and algorithm for solving school bus problem. *Journal of Emerging Trends in Computing and Information Sciences, 4*(8), 596–600.

Kinable, J., Spieksma, F. C., & Vanden Berghe, G. (2014). School bus routing—a column generation approach. *International Transactions in Operational Research, 21*(3), 453–478.

Kiriş, S. B., & Tüysüz, F. (2017). Performance comparison of different clustering methods for manufacturing cell formation. *Sakarya University Journal of Science, 21*(5), 1031–1044.

Kırkpatrick, S., Gelatt, C. D., & Vecchi, M. P. (1983). Optimization by simulated annealing. *Science, 220*(4598), 671–680. PMID:17813860

Koziel, S. & Yang, X.-S. (Eds.). (2011). Computational Optimization, Methods and Algorithms, Berlin, Germany: Springer, Verlag.

Laporte, G. (1992). The vehicle routing problem: An overview of exact and approximate algorithms. *European Journal of Operational Research, 59*(3), 345–358.

Li, L. Y. O., & Fu, Z. (2002). The school bus routing problem: A case study. *The Journal of the Operational Research Society, 53*(5), 552–558.

Liu, J., Zhao, X. D., & Xu, Z. H. (2017). Identification of rock discontinuity sets based on a modified affinity propagation algorithm. *International Journal of Rock Mechanics and Mining Sciences, 94,* 32–42. doi:10.1016/j.ijrmms.2017.02.012

Liu, Y., Özyer, T., Alhajj, R., & Barker, K. (2005). Integrating multi-objective genetic algorithm and validity analysis for locating and ranking alternative clustering. *Informatica, 29*(1).

MacQueen, J. (1967). Some methods for classification and analysis of multivariate observations. In *Proceedings of the fifth Berkeley symposium on mathematical statistics and probability* (Vol. 1, No. 14, pp. 281-297).

Metropolis, N., Rosenbluth, A. W., Rosenbluth, M. N., Teller, A. H., & Teller, E. (1953). Equation of state calculations by fast computing machines. *The Journal of Chemical Physics, 21*(6), 1087–1092.

Minocha, B. & Tripathi, S. (2014). Solving school bus routing problem using hybrid genetic algorithm: a case study. In *Proceedings of the Second International Conference on Soft Computing for Problem Solving (SocProS 2012), Dec. 28-30, 2012* (pp. 93-103). New Delhi, India: Springer. 10.1007/978-81-322-1602-5_11

Mokhtari, N. A., & Ghezavati, V. (2018). Integration of efficient multi-objective ant-colony and a heuristic method to solve a novel multi-objective mixed load school bus routing model. *Applied Soft Computing, 68,* 92–109. doi:10.1016/j.asoc.2018.03.049

Newton, R. M., & Thomas, W. H. (1974). Bus routing in a multi-school system. *Computers & Operations Research*, *1*(2), 213–222.

Oluwadare, S. A., Oguntuyi, I. P., & Nwaiwu, J. C. (2018). Solving school bus routing problem using genetic algorithm-based model. *International Journal of Intelligent Systems and Applications*, *11*(3), 50.

Park, J., & Kim, B. I. (2010). The school bus routing problem: A review. *European Journal of Operational Research*, *202*(2), 311–319.

Park, J., Tae, H., & Kim, B. I. (2012). A post-improvement procedure for the mixed load school bus routing problem. *European Journal of Operational Research*, *217*(1), 204–213.

Rashidi, T. H., Zokaei-Aashtiani, H., & Mohammadian, A. (2009). School bus routing problem in large-scale networks: new approach utilizing tabu search on a case study in developing countries. *Transportation Research Record: Journal of the Transportation Research Board*, *2137*(1), 140–147. doi:10.3141/2137-15

Ray, S., & Turi, R. H. (1999). Determination of number of clusters in k-means clustering and application in colour image segmentation. In *Proceedings of the 4th international conference on advances in pattern recognition and digital techniques* (pp. 137-143).

Riera-Ledesma, J., & Salazar-González, J. J. (2012). Solving school bus routing using the multiple vehicle traveling purchaser problem: A branch-and-cut approach. *Computers & Operations Research*, *39*(2), 391–404. doi:10.1016/j.cor.2011.04.015

Riera-Ledesma, J., & Salazar-González, J. J. (2013). A column generation approach for a school bus routing problem with resource constraints. *Computers & Operations Research*, *40*(2), 566–583. doi:10.1016/j.cor.2012.08.011

Russell, R. A., & Morrel, R. B. (1986). Routing special-education school buses. *Interfaces*, *16*(5), 56–64. doi:10.1287/inte.16.5.56

Samadi-Dana, S., Paydar, M. M., & Jouzdani, J. (2017). A simulated annealing solution method for robust school bus routing. *International Journal of Operational Research*, *28*(3), 307–326. doi:10.1504/IJOR.2017.081908

Schittekat, P., Kinable, J., Sörensen, K., Sevaux, M., Spieksma, F., & Springael, J. (2013). A metaheuristic for the school bus routing problem with bus stop selection. *European Journal of Operational Research*, *229*(2), 518–528.

Schittekat, P., Sevaux, M., & Sorensen, K. (2006). A mathematical formulation for a school bus routing problem. In *2006 International Conference on Service Systems and Service Management,* (Vol. 2, pp. 1552-1557). IEEE.

Schrage, L. E. & LINDO Systems, Inc. (1997). *Optimization modeling with LINGO*, CA: Duxbury Press.

Shafahi, A., Wang, Z., & Haghani, A. (2018). SpeedRoute: Fast, efficient solutions for school bus routing problems. *Transportation Research Part B: Methodological*, *117*, 473–493. doi:10.1016/j.trb.2018.09.004

Spada, M., Bierlaire, M., & Liebling, T. M. (2005). Decision-aiding methodology for the school bus routing and scheduling problem. *Transportation Science*, *39*(4), 477–490. doi:10.1287/trsc.1040.0096

Sugar, C. A., & James, G. M. (2003). Finding the number of clusters in a dataset: An information-theoretic approach. *Journal of the American Statistical Association, 98*(463), 750–763. doi:10.1198/016214503000000666

Sule, D. R. (2001). Logistics of facility location and allocation (p. 240). Boca Raton, FL: CRC Press. doi:10.1201/9780203910405

Swersey, A. J., & Ballard, W. (1984). Scheduling school buses. *Management Science, 30*(7), 844–853. doi:10.1287/mnsc.30.7.844

Thangiah, S. R., & Nygard, K. E. (1992, March). School bus routing using genetic algorithms. In *Applications of Artificial Intelligence X* [International Society for Optics and Photonics.]. *Knowledge-Based Systems, 1707*, 387–399.

Tibshirani, R., Walther, G., & Hastie, T. (2001). Estimating the number of clusters in a data set via the gap statistic. *Journal of the Royal Statistical Society. Series B, Statistical Methodology, 63*(2), 411–423.

Toth, P., & Vigo, D. (Eds.). (2002). *The vehicle routing problem.* Society for Industrial and Applied Mathematics. doi:10.1137/1.9780898718515

Ünsal, Ö., & Yiğit, T. (2018). Using the genetic algorithm for the optimization of dynamic school bus routing problem. *BRAIN. Broad Research in Artificial Intelligence and Neuroscience, 9*(2), 6–21.

Witten, I. H., Frank, E., Hall, M. A., & Pal, C. J. (2016). The WEKA Workbench. In *Data Mining: Practical machine learning tools and techniques* (2nd ed.). Morgan Kaufmann.

Yang, X. S. (2010). *Nature-inspired metaheuristic algorithms* (2nd ed.). Luniver Press.

Yigit, T., & Unsal, O. (2016). Using the ant colony algorithm for real-time automatic route of school buses. [IAJIT]. *The International Arab Journal of Information Technology, 13*(5).

ADDITIONAL READING

Braekers, K., Ramaekers, K., & Van Nieuwenhuyse, I. (2016). The vehicle routing problem: State of the art classification and review. *Computers & Industrial Engineering, 99*, 300–313. doi:10.1016/j.cie.2015.12.007

Corberán, A., Fernández, E., Laguna, M., & Marti, R. A. F. A. E. L. (2002). Heuristic solutions to the problem of routing school buses with multiple objectives. *The Journal of the Operational Research Society, 53*(4), 427–435. doi:10.1057/palgrave.jors.2601324

Eksioglu, B., Vural, A. V., & Reisman, A. (2009). The vehicle routing problem: A taxonomic review. *Computers & Industrial Engineering, 57*(4), 1472–1483. doi:10.1016/j.cie.2009.05.009

Ellegood, W. A., Solomon, S., North, J., & Campbell, J. F. (2019). School bus Routing Problem: Contemporary Trends and Research Directions. *Omega*, 102056. doi:10.1016/j.omega.2019.03.014

Golden, B. L., Raghavan, S., & Wasil, E. A. (Eds.). (2008). *The vehicle routing problem: latest advances and new challenges* (Vol. 43). Springer Science & Business Media. doi:10.1007/978-0-387-77778-8

Koç, Ç., Bektaş, T., Jabali, O., & Laporte, G. (2016). Thirty years of heterogeneous vehicle routing. *European Journal of Operational Research*, *249*(1), 1–21. doi:10.1016/j.ejor.2015.07.020

Laporte, G. (1992). The vehicle routing problem: An overview of exact and approximate algorithms. *European Journal of Operational Research*, *59*(3), 345–358.

Laporte, G., Gendreau, M., Potvin, J. Y., & Semet, F. (2000). Classical and modern heuristics for the vehicle routing problem. *International Transactions in Operational Research*, *7*(4-5), 285–300. doi:10.1111/j.1475-3995.2000.tb00200.x

Park, J., & Kim, B. I. (2010). The school bus routing problem: A review. *European Journal of Operational Research*, *202*(2), 311–319. doi:10.1016/j.ejor.2009.05.017

KEY TERMS AND DEFINITIONS

Genetic Algorithm: A metaheuristic, which is inspired by the theory of Darvin on natural selection and evaluation.

Heuristic: An optimization method that tries to exploit problem-specific knowledge and for which we have no guarantee that it finds the optimal solution.

K-Means Clustering: One of the simplest and popular unsupervised machine learning algorithms, which group similar data points together.

Metaheuristic: A master strategy that guides and modifies other heuristics to produce solutions beyond those that are normally generated in a quest for local optimality.

Saving Algorithm: One of the most known heuristics for vehicle routing problem.

Simulated Annealing: An optimization method, which mimics the slow cooling of metals.

Vehicle Routing Problem: A classical problem in operation research, which is one of the most challenging combinatorial optimization tasks.

Chapter 13

An Assessment of Imbalanced Control Chart Pattern Recognition by Artificial Neural Networks

Ramazan Ünlü

https://orcid.org/0000-0002-1201-195X

Gumushane University, Turkey

ABSTRACT

Manual detection of abnormality in control data is an annoying work which requires a specialized person. Automatic detection might be simpler and effective. Various methodologies such as ANN, SVM, Fuzzy Logic, etc. have been implemented into the control chart patterns to detect abnormal patterns in real time. In general, control chart data is imbalanced, meaning the rate of minority class (abnormal pattern) is much lower than the rate of normal class (normal pattern). To take this fact into consideration, authors implemented a weighting strategy in conjunction with ANN and investigated the performance of weighted ANN for several abnormal patterns, then compared its performance with regular ANN. This comparison is also made under different conditions, for example, abnormal and normal patterns are separable, partially separable, inseparable and the length of data is fixed as being 10,20, and 30 for each. Based on numerical results, weighting policy can better predict in some of the cases in terms of classifying samples belonging to minority class to the correct class.

INTRODUCTION

Quality control engineering provides some strategies to ensure a product is satisfied with some predetermined quality standards before market release. It provides the necessary mathematical and statistical tools to improve a process, to assure safety, and to analyze reliability (Montgomery, 2007). Quality control process can also help to detect a failure in the production systems such as machine failure. Sequential production of an item that does not satisfy the quality standards can be a sign of a machine

DOI: 10.4018/978-1-7998-0301-0.ch013

failure (Panagiotidou & Tagaras, 2012; Paté-Cornell, Lee, & Tagaras, 1987). Early detection of a machine failure can help to avoid expensive equipment and reducing repair cost. Over the years, various rules are implemented such as zone tests or run tests (Jill A Swift & Mize, 1995). Manual quality control process can be a tedious task and highly relies on human skills and experience. For this reason, automated systems to detect abnormal behavior in a control chart is developed by researchers (Hachicha & Ghorbel, 2012). Automated methods provide sophisticated techniques to distinguish abnormal and normal pattern during the production process. Over the years, various normal and abnormal patterns reported in real production systems. In an early study of Western Electric Company, seven abnormal patterns are identified and formulized which are named as uptrend (UT), downtrend (DT), upshift (US), downshift (DS), cyclic (C), systematic (S), stratification (F) patterns are shown in Figure 1, also the mathematical formulations of all these abnormal patterns are given in APPENDIX-A.

Each pattern is associated with a particular failure type. In the crankcase manufacturing operations, tool wear and malfunction problems yield Uptrend and downtrend abnormality (Hachicha & Ghorbel, 2012). They are also observed in stamping tonnage and paper making and viscosity process (Chinnam, 2002; Cook & Chiu, 1998; Jin & Shi, 2001). Upshift and downshift patterns might be related to the variation of operator, material or machine instrument (Davy, Desobry, Gretton, & Doncarli, 2006; El-Midany, El-Baz, & Abd-Elwahed, 2010). Variability of the power supply voltage is usually pinpointed by cyclic abnormal patterns (Kawamura, Chuarayapratip, & Haneyoshi, 1988). Uptrend, downtrend, cyclic, and systematic patterns are also observed in the automotive industry (Jang, Yang, & Kang, 2003). Since each pattern is associated with some certain type of malfunctions, the best possible method has to be implemented with the purpose of efficient identification of abnormal patterns. By doing this, system

Figure 1. Example of six abnormal patterns vs normal pattern

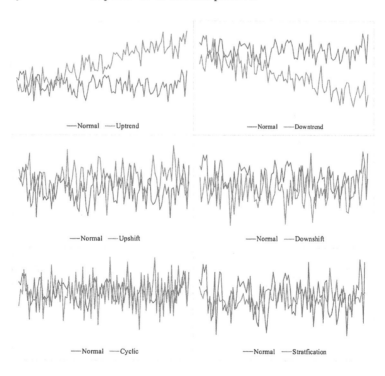

sustainability can be provided, an extensive cost can be avoided, and maintenance can be lowered. Early control chart pattern recognition studies focused on basic statistical approaches to identify changes in mean and variance with regards to normal pattern (Jill Anne Swift, 1987). Knowledge-based expert systems and artificial intelligence-based methods are employed in several studies. (H Brian Hwarng & Hubele, 1992) implemented Boltzmann machines (BM), a special type of neural network, to detect cyclical and stratification pattern. Based on their numerical results, BM is a quite powerful control chart pattern recognition tool. In another study of (H. B. Hwarng, 1995), multilayer perceptron is used to detect a cyclic pattern in which each perceptron deals with the cycles of a certain period. The proposed methodology performs superiorly in detecting higher noise and cycles of higher amplitudes. Backpropagation approach is also used for control chart pattern recognition. (H. B. Hwarng & Hubele, 1993) implemented backpropagation algorithm, suitable for real-time statistical process control, to recognize cyclic or trend patterns and they have evaluated the results based on Type 1 and Type 2 errors. Artificial neural networks is used in various control chart pattern recognition studies. (Pugh, 1989) used ANN with backpropagation algorithm and they alleged that ANN performs reasonably well under most conditions. (Cheng & Cheng, 2008) implemented ANN and SVM which are used to detect shift pattern and classify them. In addition to regular SVM, weighted SVM (WSVM) is used by (Xanthopoulos & Razzaghi, 2014) in which they compared SVM and WSVM in terms of detecting seven abnormal patterns and comparison is made based on data mining-based evaluation metrics and ARDLIX. One can see some other control chart pattern recognition methods in the literature such as principal component analysis (PCA) (Aparisi, 1996), time series analyzing (Alwan & Roberts, 1988), regression(Mandel, 1969), and correlation analysis technique (Al-Ghanimai, Amjed M Ludeman, 1997; Yang & Yang, 2005). Soft computing and data mining approaches are also utilized to detect abnormality during the production process. (Ghazanfari, Alaeddini, Niaki, & Aryanezhad, 2008) proposed a clustering methodology to estimate Shewhart control chart change points. As different from classical methods, the proposed methodology does not depend on known values of parameters and even the distribution of the process variables. Despite the capability in monitoring the variability of the process, control charts can fail to identify a real-time of changes in a process. In order to handle this, (Zarandi & Alaeddini, 2010) proposed a fuzzy clustering approach for estimating real-time change in a variable. Another data mining-based approach called decision trees used by (C.-H. Wang, Guo, Chiang, & Wong, 2008). They have used decision trees approach to identify six anomaly types including systematic pattern, cyclic pattern, upshift pattern, downshift pattern, uptrend pattern, and downtrend pattern. In terms of classification accuracy and computational cost, decision trees produce reasonably good results based on given numerical results. In addition to these methods, Support Vector Machines is another one of the commonly used data mining based methods. It has some advantages over other methods such as its performance is not affected when control chart data does not fit to a normal distribution (Sun & Tsung, 2003). Rather than other statistical as well as machine learning methods, SVM might give a better result especially is the control chart data from facilities are often autocorrelated (Chinnam, 2002).

Since control chart pattern recognition can be considered as a classification problem, Artificial neural networks (ANNs) might be a powerful tool to detect an abnormal pattern. Over the last decades, with the advancement of computational power, deep ANNs architectures have been started to use to reveal hidden patterns from an extremely complex dataset. To date, to the best of author knowledge, despite ANNs is used in control chart pattern recognition, size of normal and abnormal classes are not taking into consideration. If we consider a production dataset, we can expect that the rate of abnormal pattern will be much lower than the rate of normal class. Thus, the equal contribution of samples from normal class and abnormal class has no bases and might get misguiding classification accuracy. In order to handle with this problem, different weights can be assigned to each class based on their size. Thanks to the simplicity of high-level ANNs libraries, assigning different weights to each class is a straightforward process but can yield much better results. Thus, in this study, our core objective is explaining how class weighting strategy can be utilized in conjunction with ANNs and show how the algorithm's performance differ from no weighted case. For this objective, we have used simulated abnormal and normal patterns with fixed size and different parameters. The remainder of the study is organized as follows. In the Materials and Methods section, we have described the dataset, gave the details of the design of the experiment, algorithms evaluation metrics. In section Computational Results, we have given the detailed numerical results, the graphical illustration of the study and discussed the performance of the methods. Finally, we have completed the study with the Conclusion section.

MATERIALS AND METHODS

Dataset

In this section we have described the dataset and mathematical background of used strategies through the experiment. As we mentioned above, there available seven basic abnormal pattern, namely (1) normal (N) (2) uptrend (Ut) (3) downtrend (Dt) (4) upshifts (Us) (5) downshift (Ds) (6) cyclic (C) (7) systematic (S) (8) stratification (F).

As we can see in Figure 1, each abnormal pattern has deviated from the normal pattern based on its own parameters. Based on the literature, these parameters lie within the range of the certain values. The magnitude of these parameters (see Table 1) play the main role in how well an abnormal pattern distinguished from the normal pattern.

Among given six abnormal patterns we have chosen four of them UT, DT, US, and DS. The mathematical formulation of all abnormal patterns is given in APPENDIX-A. Because the rate of samples from abnormal class is much lower than the normal pattern during the entire process, we have focused on weighting policies and convert them balanced in the context of the classification problem. To do this we have implemented two main strategies as follows 1) Simple Artificial Neural Network 2) Weighted Artificial Neural network.

Table 1. Summary of parameters for abnormal patterns

Name	Symbol	Range
Mean of process (all pattern)	μ	0
Standard deviation of normal process	σ	1
Slope (UT and DT)	λ	[0.005 σ, 0.605 σ]
Shift (US and DS)	ω	[0.005 σ, 1.805 σ]
Standard deviation (F)	ε_t	[0.005 σ, 0.8 σ]
Cyclic parameter (C)	α	[0.005 σ, 1.805 σ]
Systematic parameter (S)	k	[0.005 σ, 1.805 σ]

Based on abnormal parameters, it might very difficult to distinguish the normal pattern from an abnormal pattern while its parameter value is too low, and it might be a very easy task while its parameter value is high. In real-world application, it is possible to see any kind of difficulty level. That's why we have set parameter values as being inseparable, partial separable, and separable.

Artificial Neural Network (ANNs)

The main idea of ANNs comes from the neural structure of the brain (Anderson & McNeill, 1992). Since long, it has been widely used to reveal hidden nonlinear relationship in many different problems such as image processing (N. Wang & Yeung, 2013), text mining (Berry & Castellanos, 2004), natural language processing (Manning, Manning, & Schütze, 1999), clustering (Jain, Murty, & Flynn, 1999; Ünlü & Xanthopoulos, 2017, 2019) etc.. A simple ANN architecture consists of some connected layers formed by neurons. The first layer in an ANN architecture is called the input layer in which the number of neurons is equal to the number of data features plus one (bias value). The next layer is named as the hidden layer which plays the main role to uncover the nonlinearity (Onur, İmrak, & Onur, 2017). There is no explicit rule of determining the number of neurons in this layer. However, some heuristics and statistical approaches are proposed in different studies (Fujita, 1998; Hagiwara, 1994; Islam & Murase, 2001; Keeni, Nakayama, & Shimodaira, 1999; Onoda, 1995; Tamura & Tateishi, 1997). The last layer of the ANNs architecture is called as the output layer in which the number of neurons is equal to the number of class size (i.e. the number of neurons in the last layer is two for a binary classification problem). And, each layer is connected by some weights $W \in \mathbb{R}^{m+n}$.

The most well-known systematic of an ANNs is known as a feed-forward neural network. The main idea of the feed-forward system is that each layer (or neurons) is fed by the previous layers' outcomes until the output layer. Then, the overall cost of the system is minimized by an algorithm which is called backpropagation algorithm. To make it more concrete assume we have a given dataset $X \in \mathbb{R}^{n+1}$, where n is the number of features and ground true outputs $y \in \mathbb{R}^m$ where m is the number of samples. In this case, the input layer is formed by n+1 neurons. The output of each neuron is equal to the value of each feature and the output of the bias neuron is equal to a constant bias term. Every single neuron except the bias one in the first hidden layers will be fed by the output of each neuron in input layer in a way that the input of the neurons will be calculated as in Equation 1.

$$h_w(x) = w^T X \tag{1}$$

The next layer, which can be another hidden layer or the output layer, is fed by the outcome of an activation function which is a function of $h_w(x)$. There are numerous activation functions in the literature and we have used the rectified linear unit (ReLU). So, inputs of each neuron in the next layer will be calculated based on the following rule shown in Equation 2 . The function of ReLu is shown in Figure 2

$$R(h_w(x)) = Max(0, h_w(x)) \tag{2}$$

The outcome of the activation function will be the input of each neuron in the next layer including the output layer. The outcome of the activation function in the output layer is the prediction of the systems $\hat{y} \in \mathbb{R}^m$. Finally, one can easily calculate the average error of the system based on Equation 3.

Figure 2. The function of rectified linear unit (ReLU)

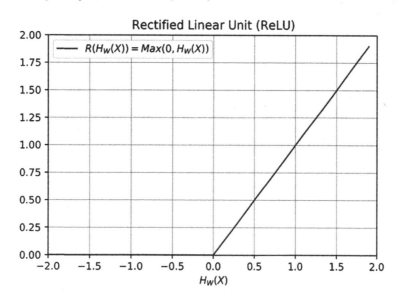

$$J\left(W\right) = -\frac{1}{m}\left[\sum_{i=1}^{m}\sum_{k=1}^{K}y_k^{(i)}\log\left(h_W\left(x^{(i)}\right)\right)_k + \left(1-y_k^{(i)}\right)\log(1-\left(h_W\left(x^{(i)}\right)\right)_k\right] + \frac{\lambda}{2m}\sum_{l=1}^{L-1}\sum_{i=1}^{s_l}\sum_{j=1}^{s_{l+1}}\left(W_{ji}^{(l)}\right)^2 \quad (3)$$

where $h_W\left(x\right) \in R^K$, L is the total number of layers in the network, is the number of non-bias neurons in layer , K is the total number of classes (in other words number of neurons in output layer), is the regularization term. After completing the first iteration, the system produces the initial cost of . The next step is being to utilize the back-propagation algorithm to optimize every single with the purpose of minimizing overall error. In other words, ANNs will learn through the iterations what each weight should be to get the minimum average error. The detailed representation of the ANNs for a binary classification problem (i.e. the number of neurons in the output layer is equal to 2) is shown in Figure 3.

Described ANN model is not a cost-sensitive model. So, it assigns equal weights to each sample regardless of their class size. In order to handle with this problem, it can be easily assigned different

Figure 3. A) Representation of the ANN model B) The process of a single neuron

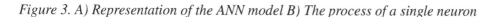

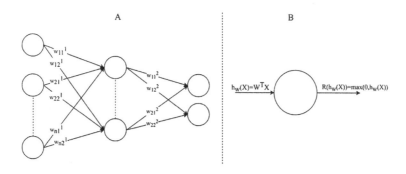

weights to each class based on their class size. By doing this, we can force samples to equally contributes to the training process.

Weighting the Samples

In real-world applications, having an imbalanced data (i.e. machine failure) is a commonly seen fact and it is not as easy as analyzing a balanced dataset. In order to solve this problem, we can transform the learning strategy by giving different weights to each class depend on their class size. These weights are simply calculated as shown in Equation 4.

$$classweights = \frac{n}{K * n^{K_i}} \tag{4}$$

where n is the total number of samples, K is the number of classes, n^{K_i} is the number of samples in the ith class.

Performance Measures: Various performance measures can be used to appraise the performance of a machine learning method for an imbalanced dataset. Some of the evaluation measurements can be calculated directly or indirectly based on confusion matrix which is a square matrix $C \in R^{KxK}$ with each entry of the matrix represent the number of samples assigned to the class K_i while belongs to the class K_j. For binary classification problem, for example, the confusion matrix is structured as shown in Table 2 in which TP, TF, FP, and FN stand for True positive, True False, False Positive, and False Negative. Obviously, the higher values of the diagonal refer a better classification result. A commonly used performance measure is so-called accuracy which can be calculated as TP+TN / (TP+TN+FP+FN). For an imbalanced problem, the accuracy can be a valid performance measure. However, it can possibly be a misleading measurement for the imbalanced problem because majority class dominates the behavior of this metric. For example, assigning all data samples to the class of y=1 will yield 98% accuracy in an imbalanced data where 98% of the data belongs to the class of y=1 and 2% of the class of y=0. To avoid this problem, some other metrics can be used based on the given formulations in Table 3.

COMPUTATIONAL RESULTS AND DISCUSSIONS

In this section, we present the experimental results of applied ANNs strategies. The model is created in Python version of 3.5 by using Keras version of 2.2.1 and run on an Intel Core i5, 2.6 GHz with 16gb Ram. It is because we have focused on detecting tool wear and malfunction and some kind of variations (i.e. variation of a machine instrument), we have used associated patterns called Uptrend (UT)

Table 2. Confusion matrix for a binary classification problem

	Predicted Positive Class	**Predicted Negative Class**
Actual Positive Class	TP	FP
Actual Negative Class	FN	TN

Table 3. The formulations of the evaluation metrics

Evaluation Metrics	Formulations
Sensitivity/Recall	TP/Actual Positive
Specificity	TN/Actual Negative
Precision	TP/Predicted Positive
Prevalence	Actual Positive/Total number of samples
F1 score	2 x (Precision x Recall)/(Precision + Recall)
G-Mean	$\sqrt{Sensitivity \times Specificity}$

and Downtrend (DT). The synthetic data is created based on Equation 4 and Equation 5 for UT and DT patterns, respectively.

$$\zeta(i) = \mu - \varepsilon_i - \lambda i \tag{4}$$

$$\zeta(i) = \mu - \varepsilon_i + \omega \tag{5}$$

where $\zeta(i)$ represents the simulated control chart pattern, μ is a constant term process mean, ε_i is a random and normally distributed term, λ is the trend slope in terms of σ_e, and i is the time unit. One needs to note that the parameter $\lambda > 0$ should be chosen for UT and $\lambda < 0$ for DT, ω is the shift magnitude. $\Omega > 0$ for US pattern and $\omega < 0$ for DS pattern. Without loss of generality, we use $\mu = 0$ and $\varepsilon_i \sim N(0,1)$. All parameters are given in Table 4.

Through the experiment, we have determined sample size as 1000, window lengths are 10,20, and 30 for both trend and shift pattern. UT and DT patterns categorized as separable, partially separable, and inseparable by choosing the value of λ as 0.6, 0.06, 0.006 respectively. In the same way, US and DS are categorized as separable, partially separable, and inseparable by choosing the value of ω as 1.8, 1, 0.006 respectively. Each pattern versus the normal pattern is illustrated in Figure 4 and Figure 5.

At first glance, we can see that inseparable patterns are not distinguishable from normal patterns. So that it is not an easy task to classify them correctly especially with having limited data (i.e w=10). On the other hand, separable cases are rather different from the normal pattern. In this case, important point

Table 4. Summary of parameters for chosen abnormal patterns

Name	Symbol	Range
Length of the data	w	[10,30]
Mean of the process	μ	0
Standard deviation of normal process	σ	1
Slope (UT and DT)	λ	0.006 σ, 0.06 σ, 0.6 σ
Shift (US and DS)	ω	0.006 σ, 1 σ, 1.8 σ

Figure 4. Downtrend and uptrend patterns vs. normal pattern along with different parameters

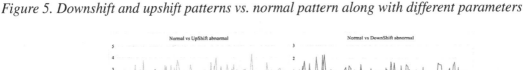

is to detect failure earlier such as while w<=10. The critical case is partially separable cases. It might be expected to get different performance measures from WANN and ANN in this case since it can be said neither partially separable cases are very similar to normal pattern nor totally different.

Through experiment the rate of abnormal samples is determined as 5% and all data samples are normalized with z-score before training phase. After completing mentioned data preprocessing, we have compared regular ANN and weighted ANN (WANN) for the separable problem ($\lambda=0.6$, $\omega=1.8$) for both trend and shift patterns for which window lengths are assigned as 10, 20, and 30. The following Table 5 shows the assignment results of ANN and WANN for the separable problem. Regardless of the window length, ANN and WANN are a powerful prediction model for the separable problem. They successfully assign all points to the correct class for both trend and shift patterns. Obviously, both methodologies give the best possible results in terms of the evaluation metrics as shown in Table 6. As we pointed out before, this is an expected outcome from both method due to the high difference between normal patterns and separable uptrend, downtrend, upshift, and downshift patterns. It can be visually seen in that there is no difference between the performance of both methodologies in Figure 6. Although both approaches perform well for separable problems, they might not be practical in real-world application. Separable problems can be defined as abnormality becomes a serious problem and it detected just before the machine failure. That's why the purpose of protecting overall systems by detecting might be unsuccessful due to late detection.

For the separable problem, it can be expected to get similar results because of the simplicity of the problem. However, solving the classification problem for partially separable or inseparable cases might be more difficult. To compare the performance of the chosen methodology, we set the value of λ as 0.06 and $\omega=1$ to create a partially separable dataset for downtrend, uptrend, downshift and upshift patterns. Same training and test set is used for both ANN and WANN. The following Table 7 shows the

Figure 5. Downshift and upshift patterns vs. normal pattern along with different parameters

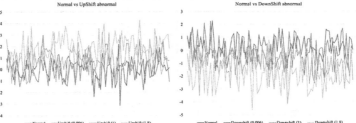

Table 5. Confusion matrices for the separable problem ($\lambda=0.6$, $\omega=1.8$) with different window lengths

			w=10		w=20		w=30	
			PC	NC	PC	NC	PC	NC
Uptrend	WANN	PC	285	0	283	0	283	0
		NC	0	15	0	17	0	17
	ANN	PC	285	0	283	0	283	0
		NC	0	15	0	17	0	17
Downtrend	WANN	PC	283	0	283	0	283	0
		NC	0	17	0	17	0	17
	ANN	PC	283	0	283	0	283	0
		NC	0	17	0	17	0	17
Upshift	WANN	PC	285	0	283	0	283	0
		NC	0	15	0	17	0	17
	ANN	PC	285	0	283	0	283	0
		NC	0	15	0	17	0	17
Downshift	WANN	PC	283	0	283	0	283	0
		NC	0	17	0	17	0	17
	ANN	PC	283	0	283	0	283	0
		NC	0	17	0	17	0	17

classification performance of the methods and Table 8 illustrated the comparison of both methods in terms of evaluation metrics.

As we pointed out above, the performance of ANN and WANN differs to solve the partially separable problem. In the case of feeding neural network with enough historical data (in other words having enough attributes) (i.e. w=30), ANN and WANN can produce similar results. This fact can be seen in the values of evaluation metrics for w=30 and in Table 8. In this case, regardless of the class size, both ANN and WANN behaves similarly. On the other hand, if limited information is given (i.e. w=20), WANN outperforms ANN much better with higher G-mean for downtrend, uptrend, downshift, and upshift patterns. However, the biggest problem for the partially separable cases is that both methods fail to classify samples if there available very limited information such as w=10. This can be commented as that neural networks might likely yields poor performance in terms of early detection of the abnormal patterns no matter what weighting strategy used. But in general, WANN can be preferred rather than regular ANN

Table 6. Performance of ANN and WANN for the separable UT, DT, DS, and US patterns

		w=10		w=20		w=30	
		WANN	ANN	WANN	ANN	WANN	ANN
Uptrend	Sensitivity/Recall	1.00	1.00	1.00	1.00	1.00	1.00
	Specificity	1.00	1.00	1.00	1.00	1.00	1.00
	Precision	1.00	1.00	1.00	1.00	1.00	1.00
	Prevalence	0.95	0.95	0.94	0.94	0.94	0.94
	F1 score	1.00	1.00	1.00	1.00	1.00	1.00
	G-Mean	1.00	1.00	1.00	1.00	1.00	1.00
Downtrend	Sensitivity/Recall	1.00	1.00	1.00	1.00	1.00	1.00
	Specificity	1.00	1.00	1.00	1.00	1.00	1.00
	Precision	1.00	1.00	1.00	1.00	1.00	1.00
	Prevalence	0.94	0.94	0.94	0.94	0.94	0.94
	F1 score	1.00	1.00	1.00	1.00	1.00	1.00
	G-Mean	1.00	1.00	1.00	1.00	1.00	1.00
Upshift	Sensitivity/Recall	1.00	1.00	1.00	1.00	1.00	1.00
	Specificity	1.00	1.00	1.00	1.00	1.00	1.00
	Precision	1.00	1.00	1.00	1.00	1.00	1.00
	Prevalence	0.95	0.95	0.94	0.94	0.94	0.94
	F1 score	1.00	1.00	1.00	1.00	1.00	1.00
	G-Mean	1.00	1.00	1.00	1.00	1.00	1.00
Downshift	Sensitivity/Recall	1.00	1.00	1.00	1.00	1.00	1.00
	Specificity	1.00	1.00	1.00	1.00	1.00	1.00
	Precision	1.00	1.00	1.00	1.00	1.00	1.00
	Prevalence	0.94	0.94	0.94	0.94	0.94	0.94
	F1 score	1.00	1.00	1.00	1.00	1.00	1.00
	G-Mean	1.00	1.00	1.00	1.00	1.00	1.00

Figure 6. Representation of the performance of methods for separable problem

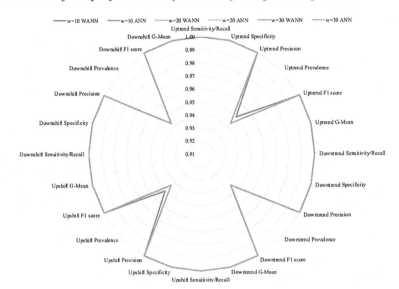

based on especially G-mean score. In Figure 7, we can also visually see how ANN and WANN differ based on given parameters and set window lengths in terms of given performance measures.

For the cases from the real-world production dataset, it is possible to have an inseparable normal and abnormal pattern. For this reason, we have compared the chosen methods for inseparable control chart patterns with the value of λ=0.006, ω=0.006. The following Table 9 and Table 10 shows the comparison of the ANN and WANN for the inseparable problem for downtrend, uptrend, downshift and upshift abnormal patterns.

As shown in Table 10, ANN or WANN cannot classify samples in the minority class if given data is very limited (w=10). In other words, it cannot differentiate the abnormal patterns from the normal patterns. On the other hand, WANN gives better results if there available limited or enough information before the abnormal pattern occurred.

Table 7. Confusion matrices for the separable problem (λ=0.6, ω=1) with different window lengths

			w=10		w=20		w=30	
			PC	NC	PC	NC	PC	NC
Uptrend	WANN	PC	288	0	280	4	281	0
		NC	12	0	6	10	0	19
	ANN	PC	288	0	280	3	281	0
		NC	12	0	12	5	0	19
Downtrend	WANN	PC	281	0	285	5	290	0
		NC	19	0	2	8	0	10
	ANN	PC	281	0	285	5	289	1
		NC	19	0	4	6	0	10
Upshift	WANN	PC	285	0	281	2	283	0
		NC	15	0	2	15	0	17
	ANN	PC	285	0	279	4	283	0
		NC	15	0	7	10	0	17
Downshift	WANN	PC	283	0	280	3	283	0
		NC	17	0	3	14	0	17
	ANN	PC	283	0	280	3	283	0
		NC	17	0	4	13	0	17

Table 8. Performance of ANN and WANN in terms of the evaluation metrics

		w=10		w=20		w=30	
		WANN	ANN	WANN	ANN	WANN	ANN
Uptrend	Sensitivity/Recall	1.00	1.00	0.99	0.99	1.00	1.00
	Specificity	0.00	0.00	0.63	0.29	1.00	1.00
	Precision	0.96	0.96	0.98	0.96	1.00	1.00
	Prevalence	0.96	0.96	0.95	0.94	0.94	0.94
	F1 score	0.98	0.98	0.98	0.97	1.00	1.00
	G-Mean	0.00	0.00	0.78	0.54	1.00	1.00
Downtrend	Sensitivity/Recall	1.00	1.00	0.98	0.98	1.00	1.00
	Specificity	0.00	0.00	0.80	0.60	1.00	1.00
	Precision	0.94	0.94	0.99	0.99	1.00	1.00
	Prevalence	0.94	0.94	0.97	0.97	0.97	0.97
	F1 score	0.97	0.97	0.99	0.98	1.00	1.00
	G-Mean	0.00	0.00	0.89	0.77	1.00	1.00
Upshift	Sensitivity/Recall	1.00	1.00	0.99	0.99	1.00	1.00
	Specificity	0.00	0.00	0.88	0.59	1.00	1.00
	Precision	0.95	0.95	0.99	0.98	1.00	1.00
	Prevalence	0.95	0.95	0.94	0.94	0.94	0.94
	F1 score	0.97	0.97	0.99	0.98	1.00	1.00
	G-Mean	0.00	0.00	0.94	0.76	1.00	1.00
Downshift	Sensitivity/Recall	1.00	1.00	0.99	0.99	1.00	1.00
	Specificity	0.00	0.00	0.82	0.76	1.00	1.00
	Precision	0.94	0.94	0.99	0.99	1.00	1.00
	Prevalence	0.94	0.94	0.94	0.94	0.94	0.94
	F1 score	0.97	0.97	0.99	0.99	1.00	1.00
	G-Mean	0.00	0.00	0.90	0.87	1.00	1.00

If we take the hardness of the problem into the consideration, performance of the WANN is respectively good even though it correctly classifies a few samples from minority class for both trends and shift abnormal patterns. In the case of feeding neural network with enough information (i.e. w=30), both ANN and WANN gives good results in terms of evaluation metrics. The overall poor performance of the methods can be visually seen in Figure 8.

Based on the result for 3 different scenarios, there are some main facts should be taken into consideration.

Figure 7. Representation of the performance of methods for partially separable problem

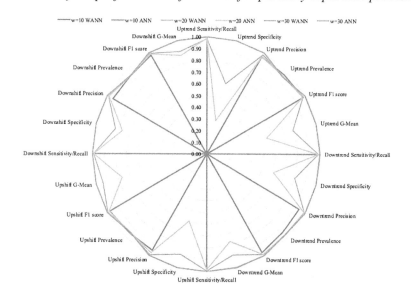

Table 9. Confusion matrices for the inseparable problem ($\lambda=0.006$, $\omega=0.006$) with different window lengths

			w=10		w=20		w=30	
			PC	NC	PC	NC	PC	NC
Uptrend	WANN	PC	280	0	288	0	288	0
		NC	20	0	9	3	0	12
	ANN	PC	280	0	288	0	288	0
		NC	20	0	12	0	0	12
Downtrend	WANN	PC	283	0	283	0	283	0
		NC	17	0	13	4	0	17
	ANN	PC	283	0	283	0	283	0
		NC	17	0	17	0	0	17
Upshift	WANN	PC	285	0	286	1	283	0
		NC	15	0	8	5	0	17
	ANN	PC	285	0	287	0	283	0
		NC	15	0	13	0	0	17
Downshift	WANN	PC	283	0	287	0	283	0
		NC	17	0	8	5	0	17
	ANN	PC	283	0	287	0	283	0
		NC	17	0	13	0	0	17

- ANN framework can be used with the purpose of early detection of the abnormal patterns during a control chart pattern process. However, the success of the methods highly depends on the parameter of the abnormality. Although giving extra weight to the abnormal pattern increase the performance of the ANN, it might fail in the case of having very limited information before an abnormal pattern occurs. This can cause late failure detection in a system.

- By having limited information such as w=20 WANN can predict better than regular ANN if the pattern is partially separable even though its performance is not perfect. Thus, we can say that weighting policy can help to earlier detection.

- In the case of the inseparable problem, feeding neural networks with enough data (i.e. w=100) might help to detect an abnormal pattern. However, this can cause late detection of the tool wear and malfunctions (uptrend and downtrend patterns) and variation of material, operator, or machine (downshift and upshift pattern), so serious problem can occur.

Table 10. Performance of ANN and WANN in terms of the evaluation metrics

		w=10		w=20		w=30	
		WANN	ANN	WANN	ANN	WANN	ANN
Uptrend	Sensitivity/Recall	1.00	1.00	1.00	1.00	1.00	1.00
	Specificity	0.00	0.00	0.25	0.00	1.00	1.00
	Precision	0.93	0.93	0.97	0.96	1.00	1.00
	Prevalence	0.93	0.93	0.96	0.96	0.96	0.96
	F1 score	0.97	0.97	0.98	0.98	1.00	1.00
	G-Mean	0.00	0.00	0.50	0.00	1.00	1.00
Downtrend	Sensitivity/Recall	1.00	1.00	1.00	1.00	1.00	1.00
	Specificity	0.00	0.00	0.24	0.00	1.00	1.00
	Precision	0.94	0.94	0.96	0.94	1.00	1.00
	Prevalence	0.94	0.94	0.94	0.94	0.94	0.94
	F1 score	0.97	0.97	0.98	0.97	1.00	1.00
	G-Mean	0.00	0.00	0.49	0.00	1.00	1.00
Upshift	Sensitivity/Recall	1.00	1.00	1.00	1.00	1.00	1.00
	Specificity	0.00	0.00	0.38	0.00	1.00	1.00
	Precision	0.95	0.95	0.97	0.96	1.00	1.00
	Prevalence	0.95	0.95	0.96	0.96	0.94	0.94
	F1 score	0.97	0.97	0.98	0.98	1.00	1.00
	G-Mean	0.00	0.00	0.62	0.00	1.00	1.00
Downshift	Sensitivity/Recall	1.00	1.00	1.00	1.00	1.00	1.00
	Specificity	0.00	0.00	0.38	0.00	1.00	1.00
	Precision	0.94	0.94	0.97	0.96	1.00	1.00
	Prevalence	0.94	0.94	0.96	0.96	0.94	0.94
	F1 score	0.97	0.97	0.99	0.98	1.00	1.00
	G-Mean	0.00	0.00	0.62	0.00	1.00	1.00

Figure 8. Representation of the performance of methods for partially inseparable problem

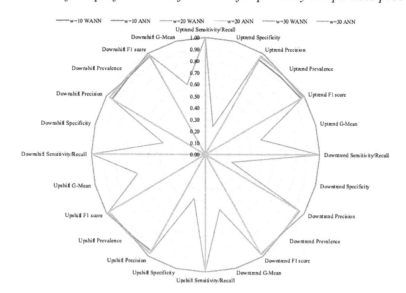

- In general, no matter which methods are used a classical evaluation metrics (i.e. accuracy) can misguide the users in terms of detecting abnormal control chart patterns. Instead, some other metrics given in Table 3 should be used to get the realistic measurements regarding the performance of methods.

CONCLUSION

In this study, we have investigated a weighting strategy in conjunction with artificial neural networks to detect abnormal patterns during a production system. Because the rate of abnormal sample size is much lower than the rate of normal sample size, giving different weights to each class based on their sizes might help to create a more powerful classification model. With the purpose of explaining how to apply weighting strategy for control chart pattern recognition, we have chosen four different abnormal patterns namely uptrend, downtrend, upshift, and downshift. For each pattern, we have created a synthetic dataset with different parameters in order to create separable, partial separable, and inseparable problems. And, each problem is analyzed with different window length (different size of input data). Based on numerical results, both methodologies fail if input data is very limited (w=10) regardless of how well separable data is.

On the other hand, if there exist limited data, WANN outperforms ANN for each inseparable, partially separable, and separable problems in terms of all evaluation metrics. In the case of having enough data, both methodologies perform reasonably well.

One of the weaknesses of giving different weights to each class is that can reduce the sensitivity.

Because our main objective is giving a detailed explanation of how to use weighting strategy we have used limited window lengths and just three different parameters. For the extension of the study, more different parameter values and window lengths should be examined. Another comparison can be made in computational time. In real-world applications, this is a crucial metric since data will be in the online form and algorithm have to deal with the ongoing data in real time. Another approach can be analyzing online data in which multiple abnormal patterns can occur. In this case, multiclassification problem needs to be solved by adjusting weights in the same way.

To sum up, we have investigated how different weights can be given to abnormal class based on class size and how the performance of the artificial neural network can be affected. It can be said that with enough input data artificial neural networks gives better results when classes are weighted.

REFERENCES

Al-Ghanim, A. M. & Ludeman, L. C. (1997). Automated unnatural pattern recognition on control charts using correlation analysis techniques. *Computers & Industrial Engineering, 32*(3), 679-690.

Alwan, L. C., & Roberts, H. V. (1988). Time-series modeling for statistical process control. *Journal of Business & Economic Statistics, 6*(1), 87–95.

Anderson, D., & McNeill, G. (1992). Artificial neural networks technology. *Kaman Sciences Corporation, 258*(6), 1–83.

Aparisi, F. (1996). Hotelling's T2 control chart with adaptive sample sizes. *International Journal of Production Research, 34*(10), 2853–2862.

Berry, M. W., & Castellanos, M. (2004). Survey of text mining. *Computer Review, 45*(9), 548.

Cheng, C.-S., & Cheng, H.-P. (2008). Identifying the source of variance shifts in the multivariate process using neural networks and support vector machines. *Expert Systems with Applications, 35*(1–2), 198–206. doi:10.1016/j.eswa.2007.06.002

Chinnam, R. B. (2002). Support vector machines for recognizing shifts in correlated and other manufacturing processes. *International Journal of Production Research, 40*(17), 4449–4466. doi:10.1080/00207540210152920

Cook, D. F., & Chiu, C.-C. (1998). Using radial basis function neural networks to recognize shifts in correlated manufacturing process parameters. *IIE Transactions, 30*(3), 227–234. doi:10.1080/07408179808966453

Davy, M., Desobry, F., Gretton, A., & Doncarli, C. (2006). An online support vector machine for abnormal events detection. *Signal Processing, 86*(8), 2009–2025. doi:10.1016/j.sigpro.2005.09.027

El-Midany, T. T., El-Baz, M. A., & Abd-Elwahed, M. S. (2010). A proposed framework for control chart pattern recognition in multivariate process using artificial neural networks. *Expert Systems with Applications, 37*(2), 1035–1042. doi:10.1016/j.eswa.2009.05.092

Fujita, O. (1998). Statistical estimation of the number of hidden units for feedforward neural networks. *Neural Networks, 11*(5), 851–859. doi:10.1016/S0893-6080(98)00043-4 PMID:12662787

Ghazanfari, M., Alaeddini, A., Niaki, S. T. A., & Aryanezhad, M.-B. (2008). A clustering approach to identify the time of a step change in Shewhart control charts. *Quality and Reliability Engineering International, 24*(7), 765–778. doi:10.1002/qre.925

Hachicha, W., & Ghorbel, A. (2012). A survey of control-chart pattern-recognition literature (1991–2010) based on a new conceptual classification scheme. *Computers & Industrial Engineering, 63*(1), 204–222. doi:10.1016/j.cie.2012.03.002

Hagiwara, M. (1994). A simple and effective method for removal of hidden units and weights. *Neuro-computing, 6*(2), 207–218. doi:10.1016/0925-2312(94)90055-8

Hwarng, H. B. (1995). Multilayer perceptions for detecting cyclic data on control charts. *International Journal of Production Research, 33*(11), 3101–3117. doi:10.1080/00207549508904863

Hwarng, H. B., & Hubele, N. F. (1992). Boltzmann machines that learn to recognize patterns on control charts. *Statistics and Computing, 2*(4), 191–202. doi:10.1007/BF01889679

Hwarng, H. B., & Hubele, N. F. (1993). Back-propagation pattern recognizers for X control charts: Methodology and performance. *Computers & Industrial Engineering, 24*(2), 219–235.

Islam, M. M., & Murase, K. (2001). A new algorithm to design compact two-hidden-layer artificial neural networks. *Neural Networks, 14*(9), 1265–1278. doi:10.1016/S0893-6080(01)00075-2 PMID:11718425

Jain, A. K., Murty, M. N., & Flynn, P. J. (1999). Data clustering: A review. [CSUR]. *ACM Computing Surveys, 31*(3), 264–323. doi:10.1145/331499.331504

Jang, K.-Y., Yang, K., & Kang, C. (2003). Application of artificial neural network to identify non-random variation patterns on the run chart in automotive assembly process. *International Journal of Production Research, 41*(6), 1239–1254. doi:10.1080/0020754021000042409

Jin, J., & Shi, J. (2001). Automatic feature extraction of waveform signals for in-process diagnostic performance improvement. *Journal of Intelligent Manufacturing, 12*(3), 257–268. doi:10.1023/A:1011248925750

Kawamura, A., Chuarayapratip, R., & Haneyoshi, T. (1988). Deadbeat control of PWM inverter with modified pulse patterns for uninterruptible power supply. *IEEE Transactions on Industrial Electronics, 35*(2), 295–300.

Keeni, K., Nakayama, K., & Shimodaira, H. (1999). Estimation of initial weights and hidden units for fast learning of multilayer neural networks for pattern classification. In *Proceedings IJCNN'99-International Joint Conference on Neural Networks*. (Vol. 3, pp. 1652–1656). IEEE. 10.1109/IJCNN.1999.832621

Mandel, B. J. (1969). The regression control chart. *Journal of Quality Technology, 1*(1), 1–9. doi:10.1 080/00224065.1969.11980341

Manning, C. D., Manning, C. D., & Schütze, H. (1999). *Foundations of statistical natural language processing*. Cambridge, MA: MIT Press.

Montgomery, D. C. (2007). *Introduction to statistical quality control*. Hoboken, NJ: John Wiley & Sons.

Onoda, T. (1995). Neural network information criterion for the optimal number of hidden units. In *Proceedings of ICNN'95-International Conference on Neural Networks* (Vol. 1, pp. 275–280). IEEE. 10.1109/ICNN.1995.488108

Onur, Y. A., İmrak, C. E., & Onur, T. Ö. (2017). Investigation on bending over sheave fatigue life determination of rotation resistant steel wire rope. *Experimental Techniques*, *41*(5), 475–482. doi:10.100740799-017-0188-z

Panagiotidou, S., & Tagaras, G. (2012). Optimal integrated process control and maintenance under general deterioration. *Reliability Engineering & System Safety*, *104*, 58–70. doi:10.1016/j.ress.2012.03.019

Paté-Cornell, M. E., Lee, H. L., & Tagaras, G. (1987). Warnings of malfunction: The decision to inspect and maintain production processes on schedule or on demand. *Management Science*, *33*(10), 1277–1290. doi:10.1287/mnsc.33.10.1277

Pugh, G. A. (1989). Synthetic neural networks for process control. *Computers & Industrial Engineering*, *17*(1–4), 24–26. doi:10.1016/0360-8352(89)90030-2

Sun, R., & Tsung, F. (2003). A kernel-distance-based multivariate control chart using support vector methods. *International Journal of Production Research*, *41*(13), 2975–2989. doi:10.1080/13528160310000075224

Swift, J. A. (1987). *Development of a knowledge based expert system for control chart pattern recognition and analysis*. Oklahoma State University.

Swift, J. A., & Mize, J. H. (1995). Out-of-control pattern recognition and analysis for quality control charts using LISP-based systems. *Computers & Industrial Engineering*, *28*(1), 81–91. doi:10.1016/0360-8352(94)00028-L

Tamura, S., & Tateishi, M. (1997). Capabilities of a four-layered feedforward neural network: Four layers versus three. *IEEE Transactions on Neural Networks*, *8*(2), 251–255. PMID:18255629

Ünlü, R., & Xanthopoulos, P. (2017). A weighted framework for unsupervised ensemble learning based on internal quality measures. *Annals of Operations Research*, 1–19.

Ünlü, R., & Xanthopoulos, P. (2019). Estimating the number of clusters in a dataset via consensus clustering. *Expert Systems with Applications*.

Wang, C.-H., Guo, R.-S., Chiang, M.-H., & Wong, J.-Y. (2008). Decision tree-based control chart pattern recognition. *International Journal of Production Research*, *46*(17), 4889–4901. doi:10.1080/00207540701294619

Wang, N. & Yeung, D.-Y. (2013). Learning a deep compact image representation for visual tracking. In Advances in neural information processing systems (pp. 809–817).

Xanthopoulos, P., & Razzaghi, T. (2014). A weighted support vector machine method for control chart pattern recognition. *Computers & Industrial Engineering*, *70*, 134–149. doi:10.1016/j.cie.2014.01.014

Yang, J.-H., & Yang, M.-S. (2005). A control chart pattern recognition system using a statistical correlation coefficient method. *Computers & Industrial Engineering*, *48*(2), 205–221. doi:10.1016/j.cie.2005.01.008

Zarandi, M. H. F., & Alaeddini, A. (2010). A general fuzzy-statistical clustering approach for estimating the time of change in variable sampling control charts. *Information Sciences*, *180*(16), 3033–3044. doi:10.1016/j.ins.2010.04.017

APPENDIX

The western electric company first proposed several abnormal patterns and those abnormal patterns are used in different studies over the years. The synthetically created control charts data $\zeta(i)$ have three main parameters 1) mean of the process μ 2) a random and normally distributed term ε_i 3) $\Delta(i)$ that is the function of abnormal patterns. Thus, the general formulation of the abnormal patterns can be written as:

$$\zeta(i) = \mu - \varepsilon_i + \Delta(i) \tag{6}$$

The formulation of each abnormal patterns can be written as follows:

1. Up/Downtrends

$$\zeta(i) = \mu - \varepsilon_i + \lambda t \tag{7}$$

where λ is the trend slope in terms of σ_e. The parameter $\lambda > 0$ for uptrend pattern and $\lambda < 0$ for the downtrend pattern.

2. Up and Downshift

$$\zeta(i) = \mu - \varepsilon_i + \omega \tag{8}$$

where ω is the shift magnitude. $\omega > 0$ for upshift pattern and $\omega < 0$ for downshift pattern.

3. Cyclic pattern

$$\zeta(i) = \mu - \varepsilon_i + \alpha \sin\left(\frac{2\pi t}{\Omega}\right) \tag{9}$$

where α is the amplitude of the cyclic pattern, and Ω is the cyclic pattern period.

4. Systematic pattern

$$\zeta(i) = \mu - \varepsilon_i + k(-1)^t \tag{10}$$

where k is the magnitude of the systematic pattern.

5. Stratification trend

$$\zeta(i) = \mu - \varepsilon_i + \dot{\varepsilon}_t \qquad\qquad (10)$$

where $\dot{\varepsilon}_t$ is a fraction of normal process standard deviation

Chapter 14
An Exploration of Machine Learning Methods for Biometric Identification Based on Keystroke Dynamics

Didem Filiz
Ankara University, Turkey

Ömer Özgür Tanrıöver
Ankara University, Turkey

ABSTRACT

In this chapter, the authors explore keystroke dynamics as behavioral biometrics and the effectiveness of state-of-the-art machine learning algorithms for identifying and authenticating users based on keystroke data. One of the motivations of this study is to explore the use of classifiers to the field of keystroke dynamics. In different settings, recent machine learning models have been relatively inexpensive and effective with limited data. Therefore, the authors conducted experiments with two different keystroke dynamics datasets with limited data. They demonstrated the effectiveness of the models using a dataset obtained from touch screen devices (mobile phones) and also on normal keyboards. Although there are similar recent studies that explore different classification algorithms, their main aim has been anomaly detection. However, the authors experimented with classification methods for binary and multiclass user identification and authentication using two different keystroke datasets from touchscreens and keyboards.

INTRODUCTION

Information technology development over the past decades has contributed to increase of access and storage need of confidential information. Users are usually required to sign up with passwords for services offered by online service providers. Keeping track of numerous accounts may make the users' lives

DOI: 10.4018/978-1-7998-0301-0.ch014

difficult as one may forget his/her password or, worse, some malicious party can steal it (Giot, El-Abed, Hemery, & Rosenberger, 2011).

Using passwords for authentication alone is not considered reliable. Therefore, various approaches of authentication systems rely on using humans' biometric information, such as fingerprint, retina, voice, and so on. Another approach is the use of human behavior characteristics for strengthening the user identification process. One of the behavioral approaches is based on "Keystroke Dynamics" (Xi, Tang, & Hu, 2011). Keystroke Dynamics is considered a behavioral biometric that can identify a user uniquely. It is important to note that keystroke dynamics is known with a few different names in the literature: keyboard dynamics, keystroke analysis, typing biometrics, and typing rhythms.

Researchers consider a user's typing pattern to be unique to the person because of differences in neuro-physiological based, character-based, and physical factors. In other words, researchers analyze not what is typed, but how it is typed. The characteristics of typing patterns are considered to be a good sign of identity, and therefore, they may be used as biometrics without the need of other information about the user and without any extra hardware. Keystroke dynamics may be based on various different measurements with keyboard keys or touch screen (Revett, 2009).

In the last two decades, researchers have conducted numerous studies on data acquisition methods, feature representations, classification methods, experimental protocols, and evaluations. In this chapter, the authors focus on the effectiveness of recent classification methods. First, the authors review the work related to keystroke dynamics for classification. Then, the authors investigate the accuracy of different algorithms on two different known datasets collected from mobile device screens (Hwang, Cho, & Park, 2009). and keyboards, namely "The Mobikey Keystroke Dynamics Password Database" (Antal, & Nemes, 2016) and "Benchmark Data Set for Anomaly-Detection with Keystroke Dynamics" (Killourhy, & Maxion,2009). By using multi-class approach, the authors approach the authentication problem from a different perspective. There is one class for every user, and each training point belongs to one of these different classes. Additionally, binary class classification approach is also tested using positive and negative samples. In these tests, positive samples are genuine users and negative samples are assumed as impostors. The authors used two different datasets to assess the reliability of the applied methods.

Therefore, the authors explore the influence and results of different classification algorithms on these datasets. This study is done by using a set of keystroke datasets and performing classification through different methods. In this way, the authors could analyze the experimentation to show the best data-classification algorithm for keystroke dynamics classification. The conclusions that arise from the experimentation can be a useful contribution to studies in this field. In the next sections, Section 2 presents a review of the related literature and the difference of this study from other studies. Sections 3 describes how the authors perform their experiments. Finally, Section 4 discusses the conclusions of the work.

RECENT RESEARCH STUDIES BASED ON KEYSTORKE DYNAMICS

Studies about keystroke dynamics have been continued for a long time since 1980. The first study originated from the observation of individual rhythms when typing telegraphs (Abualgasim, & Osman, 2011). The authors assumed that the same behavior was exhibited by people using computer keyboards. The authors used a digraph, a pair of two keystrokes and the time elapsed between the typing of the first and second keystroke, as a sequence for identification. The authors took mean, kurtosis, and variance into account. For classifying a user, the authors used t-tests. The authors of the Rand report were able to achieve a 100% success rate in classification, but many researchers argue that this is insignificant because only seven test subjects were involved in the study. Yet this is not enough subjects to ensure the validity of the study. Then, many the authors carried out different studies in this area. However, the authors were able to mention some of them that are remarkable and important studies (Zhou, 2008).

Kevin Killourhy and Roy Maxion conducted a study to compare anomaly-detection algorithms for keystroke dynamics in 2009. Their objective was similar to the objective of the authors of this paper, but they used the same data for anomaly detection rather than identification. They collected a keystroke dynamics dataset, measured the performance of anomaly detection algorithms, and then compared the results on the error rates. They collected data from 51 subjects typing 400 passwords each. They implemented and evaluated 14 detectors from the keystroke dynamics and pattern-recognition literature. The best equal-error rate was 0.096, as obtained by the Manhattan (scaled) detector. The Nearest Neighbor (Mahalanobis) detector, the Outlier Count (z-score) detector, and the SVM (one-class) detector were the other top-performing detectors.

After a couple of years from the above work, in 2016, Antal and Nemes (2016) studied touch screen keystroke dynamics. The "Mobikey Keystroke Dynamics Password Database" was formed. To collect data, an android application was implemented. Data was collected from different people using several types of passwords. The authors also used anomaly detection algorithms to classify the dataset and to decide the success of their works. The authors obtained the best equal error rates by the Random Forests classifier. There were not significant differences between different types of password.

In 2017, Avar Pentel studied a different area using keystroke dynamics (Pentel, 2017), in his article entitled "Predicting Age and Gender by Keystroke Dynamics and Mouse Pattern." In this study, the authors used mouse and keyboard data together. They collected 1519 subjects between 2011 and 2017. Two of the sources were real life working systems: the intranet and the feedback questionnaire of the Estonian K-12 school of Toila. They tested five popular machine-learning algorithms for binary classification: Logistic regression, Support Vector Machine, Nearest Neighbor, C4.5, and Random Forest. Next, they used Java implementations of listed algorithms that are available in Weka. They also compared age groups and mean typing speed between male and female subjects. The results of this study are still important, and they still try to get a better understanding of the general links between age, gender and keyboard and mouse usage.

In the same year, Mindaugas Ulinskas, Marcin Woźniak, Robertas Damaševičius studied keystroke dynamics for fatigue recognition (Ulinskas, & Wozniak, 2017). Their aim was to analyze the problem of fatigue recognition using keystroke dynamics data. They could reach up to 91% accuracy (using key release in recognizing the relatively higher state of fatigue in a binary classification setting). However, they used passwords that were short for evaluation. Additionally, lack of sufficient measurements presented a problem because of the small amount of information used to capture behavior of an individual. Thus, researchers need to conduct more extensive experiments to collect more keystroke dynamics data with different people over an extended period under different stress conditions to achieve higher levels of fatigue identification accuracy. Also, the authors could increase the data set size for specific individuals to develop user-specific models of fatigue.

Stress detection based on keystroke dynamics was the focus of interest in a recent study (Lau, & Shing-hon, 2018). The authors of this study collected five types of data from 116 subjects: demographic data, psychological data, physiological data, self-report data, and typing data. The authors identified markers for stress (e.g., a 10% increase in typing speed) that were similar to the antigens used as markers for blood type. To search specific markers for stress, the authors investigated differences in typing rhythms between neutral and stressed states. The authors used a variety of statistical and machine learning techniques with typing data. They employed clustering (e.g., K-means) to detect groups of users whose responses to stress are similar. They were able to identify markers for stress within each subject separately. Accordingly, they were able to discriminate between neutral and stressed typing when examining subjects individually. However, despite their best attempts and the use of state-of-the-art machine learning techniques, they were not able to identify universal markers for stress across subjects. They were also unable to identify clusters of subjects whose stress responses were similar.

On the other hand, in the same year, Jay Richards Young from Brigham Young University, conducted a study on utilizing key print biometrics to identify users in online courses (Brigham Young University, 2018). This study was set up to determine how well key prints are able to identify individuals when typing under various treatment conditions, such as free typing, copy typing, and typing with mild. The authors collected the data from undergraduate students using the Google Chrome browser. They embedded JavaScript into HTML pages to track dwell and transition times for specific keys and key combinations. The results of this study indicated that when determining an individual may not be who he or she is supposed to be, key prints can be utilized effectively for user identification purposes. However, this study had some limitations. An individual's typing ability would change over time with practice. On the other hand, to require typing to be done in a system environment so that keystroke tracking can take place may be difficult to enforce. For future work, researchers can investigate other conditions that are likely to be encountered in an online course or assessment situation.

While the recent research on keystroke dynamics focuses on anomaly detection, the authors of this study use the same datasets for authentication by classification. Classification is simply to distribute data between the various classes defined on a data set. The classification algorithms learn this distribution from the given training set and then try to classify them correctly when the data which has not defined its class. However, in anomaly detection, "normal" and "anomalous" observations are distinguished. Anomalous observations do not conform to the expected pattern of other observations in a data set. Because anomalous events are, by definition, rare events can imply that data sets are imbalanced.

Also, in the case of classification for authentication purposes, in a two-class classification model, the positive samples belong to the genuine user and negatives from the others are selected for both training and test. By contrast, the authors of this study employed a multiclass classification method, in which there is one class for every user and each training point belongs to one of these different classes. The goal is to construct a function which, given a new user data, will correctly predict the class to which the user belongs. The authors take every user as a class and use other attributes for predictors.

The authors also use different classification algorithms to contribute to previous studies and show the reliability of keystroke authentication, which algorithms is best for this. For decision, the authors used correctly classified values and EER results using classification methods in Weka.

APPLICATION OF CLASSIFICATION MODELS FOR IDENTIFICATION WITH KEYSTROKE DYNAMICS

In this section, the authors present the application results of classification models for identification with keystroke patterns. As an experimental platform, the authors used Weka (Shah, 2017) and its classification algorithms. Weka is an open source Java-based platform that implements several machine learning algorithms and data preprocessing tools that can be applied to a data set directly.

In this study, the authors have experimented with two keystroke datasets. The first data set had been collected by Margit Antal and Lehel Nemes in a study for anomaly detection (Antal, & Lehel, 2016). The data were collected using 13 identical Nexus 7 tablets in three sessions one week apart. In each session, users typed at least 60 passwords and at least 20 passwords from each type. Fifty-four volunteers took part in the experiment – five women and 49 men – with an average age of 20.61 years (range: 19-26). At the registration stage, they stated their experience with touch screen devices as follows: two inexperienced, six beginners, 17 intermediate and 29 advanced touch screen users. At the end of data collection, each user had provided at least 60 samples from each type of password (easy: 3,323 samples, strong: 3,303, logical strong: 3,308). The authors used three main types of features in the evaluation: time-based, touch-based, and accelerometer-based. In this dataset, there are 54 users, and every user represent only one class. In Table 1, the authors present selected measurements for keyboard and touch screen keyboard. Also, the dataset included the attributes as summarized in Table 2. Except for the user_id attribute, all other attributes are numeric. The authors classified the attributes using user_id.

The second dataset was collected by Kevin Killourhy and Roy Maxion. The authors refer to this dataset as DSL-strong dataset. This dataset consists of keystroke-timing information of 51 subjects (typists) within their university. Subjects completed eight data-collection sessions (of 50 passwords each) for a total of 400 password-typing samples. The total instance is 20,400. The dataset consists of 30 males and 21 females with an average age of 31–40. To make a password that is representative of typical, strong passwords, they employed a publicly available password generator and password-strength checker. Participants generated a 10-character password containing letters, numbers, and punctuation, and then modified it slightly, interchanging some punctuation and casing to better conform with the general perception of a strong password. The result of this procedure was the following password: ".tie5Roanl". For data collection, the researcher used a laptop with an external keyboard. The researcher ran a Windows application that prompts a subject to type the password. The password was displayed in a screen with a text-entry field by the application. The subject should have typed the password correctly 50 times to be accepted by the program and to have completed a data-collection session. Whenever the subject pressed or released a key, the application recorded the event, the name of the key involved, and at what time the event occurred.

Table 1. Possible keystroke dynamics keyboard measurements

Possible measurements for keyboard	Possible measurements of touchscreen (in addition keyboard metrics)
Latency between consecutive keystrokes	Key press pressure for each key
Duration of the keystroke, hold-time	Mean velocity
Overall typing speed	Mean x acceleration
Frequency of errors (how often the user has to use backspace)	Mean z acceleration
The habit of using additional keys in the keyboard, for example writing numbers with the num pad	Mean hold time
In what order does the user press keys when writing capital letters, is shift or the letter key released first	Mean finger area
Total time for writing something	Total distance
Down-down time for each key	Key hold time for each key

Table 2. Attributes for mobikey dataset

Attribute Name	Numberof Data Points	Attribute Name	Number of Data Points
holdtime	12	Meanholdtime	1
downdown	12	meanxaccelaration	1
updown	12	meanyaccelaration	1
pressure	13	meanzaccelaration	1
fingerarea	13	Velocity	1
meanholdtime	1	Totaltime	1
Meanpressure	1	Totaldistance	1
meanfingerarea	1	user_id	1
meanpressure	1		

Attributes in the data set are presented in Table 3. Except the subject attribute, all other attributes are numeric. However, the classification column (subject) is nominal. These attributes correspond to the timing information for a single repetition of the password by a single subject. The first attribute "subject" is a unique identifier for each subject. Actually, the data set contains 51 subjects, but the identifiers do not range from s001 to s051; subjects have been assigned unique IDs, and not every subject participated in every experiment. For instance, Subject 1 did not perform the password typing task, and so s001 does not appear in the data set. In the actual dataset, the second column, "sessionIndex", is the session in which the password was typed (ranging from one to eight). The authors do not use session index in the study presented in this chapter. Because it is not a discriminative property, the authors eliminated this attribute from the dataset. The third attribute "rep" is the repetition of the password within the session (ranging from one to 50). The remaining attributes represent the timing information for the password, in milliseconds. Attribute H.key designates hold time for the key (i.e., the time from when key $w$as pressed to when it was released). Attribute names of the form DD.key1.key2 designate a keydown-keydown time (i.e., the time from when key1 was pressed to when key2 was pressed). Attribute names of the form UD.key1.key2 designate a keyup-keydown time (i.e., the time from when key1 was released to when key2 was pressed).

In the experiments, six methods are used namely; SVM, KNN (K nearest neighbor), Random Forest, Naive Bayes, Decision Tree, Decision Tables. Taking into account that datasets included numeric attributes as predictors and the class is nominal, the authors used corresponding classifiers. Bayesian

Table 3. Attributes for DSL-Strong dataset as an example

Attributes Name	Number of Data Points
Subject	1
Rep	1
H.period	11
DD.period.	10
UD.period	10

Network, Support Vector Machine (SVM), K nearest neighbour (KNN), and Random forest classifiers are known to be effective for numeric features and nominal classification tasks (Weka 2019). Firstly, the authors used Bayesian Network, which is based on random variables and conditional dependencies using a directed acyclic graph for classification. Nodes of the graph represent random variables. By using Bayesian network classifier, one can make a more detailed (true) model of the problem using several layers of dependencies. Naive Bayes is an extension that produces good results with high dimensional data. The implementation is also fairly simple and it does not require high computation time.

Thirdly, SVM is another popular and successful classification algorithm. It is known to be better from the computation speed and memory usage perspectives. It also works well in both linear separation and non-linear boundary testing tasks. Depending on the number of cores used in the implementation of this algorithm, the algorithm handles high dimensional data in an efficient way. Finally, KNN is a statistical method that uses sample-based learning. It is a classifier that produces good results when there is no preliminary information about data distributions. Furthermore, sometimes, increasing the K value (number of clusters) to a certain point can increase the effectiveness of the method. KNN method may be based on different metrics. The authors used both Manhattan Distance and Euclidean Distance metrics in this study. Before continuing with the presentations of experiments following, the authors give the two metrics that one uses to evaluate the experiments:

Equal Error Rate (EER): In a biometric security system, researchers use EER to determine the threshold values for its false acceptance rate and its false rejection rate. When the rates are equal, the common value is referred to as the equal error rate. The lower the equal error rate value, the higher the accuracy of the biometric system. Thus, EER is a measure to be used when deciding on a classification algorithm with lowest EER value.

Kappa statistic (KS): Researchers use KS to test interrater reliability, which refers to statistical measurements that determine how similar the data collected by different raters/individuals are. KS measures the extent to which the data collected in the study is a correct representation of the variables measured ("Biochemia Medica", McHugh, 2012).

Having described the datasets and algorithms and metrics for comparison, now the authors will describe how they use these datasets in their work. The authors have conducted experiments with easy, strong, and logically strong password of Mobikey dataset. Also, for the strong password data of password. As a

Table 4. Easy data classification results(percentage split %20) for the mobikey keystroke dynamics password database

Easy Data	Correctly Classified	EER Results	Time Taken to Build Model	Kappa Statistic
Bayesian Network	%65.7637	0.0004	0.85 seconds	0.6512
Decision Table	%31.3017	0.0216	13.09 seconds	0.3005
Decision Tree	%46.8397	0.0056	0.08 seconds	0.4584
Naive Bayes	%63.3559	0.1111	0.17 seconds	0.8187
Random Forest	%82.2047	0.0004	0.36 seconds	0.6267
SVM	%68.3597	0.0097	0.84 seconds	0.6777
KNN(Euclidian)	%63.0173	0.0027	0.92 seconds	0.6232
KNN(Manhattan)	%73.702	0.0004	0 seconds	0.7321

classical approach, the authors use 80% of data for training and 20% of data for testing. To validate the classification model, the authors use 10-fold cross validation method supporter as an option in Weka. In this method, the dataset is randomly partitioned into 10 subsamples. The cross-validation process is repeated 10 times. The authors use each of the 10 subsamples only once. Next, the authors average the results from the 10 repetitions (folds) to produce the result. In this way, the authors ensured that each fold contains approximately the correct proportion of the class values. It can be observed from the results shown in Table 4 that when the data for training is low, the number of correctly classified instances are also low. Having more than 10 exposures created some problems. First, it required more computation time. Second, the dataset was small, so when exposure number increased, the instance number decreased, and target classes with few instances were tested and but not used for training. Then, the authors have conducted the same experiment with a higher training rate. As evident in Table 5, the training rate must be high to obtain better classification performance.

Having performed the experiments with the methods described with default hyperparameters in Weka, the authors obtained the results in Tables 5 and Table 6.

During the experimentation, first, the authors conducted a classification experiment by taking multiple classes based on user_id. As evident from Table 5, Random Forest Algorithm has the highest accuracy for three datasets (easy, strong, logical-strong) in the Mobikey dataset. Its correctly classified instance rate is around 94% for the three datasets. The Bayesian networks result, which has 89% correctly classified instances, is also acceptable. Additionally, SVM follows the Bayesian Network with 86% of instances correctly classified. According to these results, one can use Random Forest, Bayesian Network, and SVM classification techniques for identification. However, these results are not conclusive as the size of the datasets in the experiments were relatively small.

Then, the authors conducted the experiments with the second dataset, namely DSL-Strong, which was collected using a regular keyboard. Subject attribute in the data set is taken as the class to be predicted for each individual. Random Forest again was found to be the best classifier with an accuracy of 951373. 2550% instance. The authors used Random Forest for this classification, and 2426 of these instances were correctly classified.

Bayesian Network classifier correctly classified instances with a rate of 89.5245%. Then, K-Nearest Neighbor (Manhattan) correctly classified instances with a rate of 85.0294%, which may be considered

Table 5. Easy data classification results (cross-validation 10) for the mobikey keystroke dynamics password database

Easy Data	Correctly Classified	EER Results	Time Taken to Build Model	Kappa Statistic
Bayesian Network	%86.24	0.0004	0.69 seconds	0.8599
Decision Table	%47.30	0.126	9.98 seconds	0.4629
Decision Tree	%69.69	0.079	0.93 seconds	0.6912
Naive Bayes	% 73	0.095	0.09 seconds	0.9411
Random Forest	% 94	0.0004	4.61 seconds	0.7351
Support Vector Machine	%84.29	0.0061	5.06 seconds	0.8399
KNN(Euclidian)	%72.16	0.0754	0.01 seconds	0.7164
KNN (Manhattan)	%83.65	0.0159	0 second	0.8335

Table 6. Strong data classification results(cross-validation 10) for the mobikey keystroke dynamics password database

Strong Data	Correctly Classified	EER Results	Time Taken to Build Model	Kappa Statistic
Bayesian Network	%88.85	0.0004	0.67 seconds	0.8865
Decision Table	%50.31	0.093	8.89 seconds	0.4935
Decision Tree	% 69	0.066	0.81 seconds	0.6863
Naive Bayes	%74.26	0.066	0.11 seconds	0.9395
Random Forest	% 94	0.0004	4.95 seconds	0.7378
SupportVector Machine	%86.98	0.0004	7.19 seconds	0.8674
KNN (Euclidian)	% 75	0.057	0 second	0.7498
KNN (Manhattan)	%85	0.0017	0 second	0.8473

Table 7. Logicaly-strong data classification results (cross-validation 10) for the mobikey keystroke dynamics password database

Logical Strong Data	Correctly Classified	EER Results	Time Taken to Build Model	Kappa Statistics
Bayesian Network	%88.72	0.0004	0.46 seconds	0.8851
Decision Table	%49.21	0.126	10.65 seconds	0.4823
Decision Tree	%69.37	0.116	1.27 seconds	0.688
Naive Bayes	%75.81	0.05	0.13 seconds	0.939
Random Forest	%94.014	0.0004	6.24 seconds	0.7536
Support Vector Machine	%86.85	0.0004	6.73 seconds	0.866
KNN(Euclidian)	%73.87	0.001	0 seconds	0.7333
KNN(Manhattan)	%83.43	0.0003	0 seconds	0.8312

Table 8. DSL_Strong data classification results (cross-validation 10) for keystroke dynamics - benchmark data set

DSL-Strong Data	Correctly Classified	EER Results	Time Taken to Build Model	Kappa Statistic
Bayesian Network	89.5245%	0.025	1.47 seconds	0.8931
Decision Table	42.098%	0.2625	34.9 seconds	0.4094
Decision Tree	74.3235%	0.1432	4.59 second	0.7381
Naive Bayes	67.0294%	0.0975	0.2 seconds	0.9504
Random Forest	95.1373%	0.0088	3.44 seconds	0.6637
SVM	78.4902%	0.0654	12.93 seconds	0.7806
KNN (Euclidian)	74.1667%	0.0059	0 second	0.7365
KNN (Manhattan)	85.0294%	0.0029	0 second	0.8473

suitable. Next, in this experiment, SVM's accuracy was 78.4902%. EER results are comparable to other algorithms.

One of the essential observations is that the authors obtained different results for two datasets. This may be attributed to the fact that the number of attributes, the number of instances, and the values are different for two datasets. For example, the first dataset, which was obtained thorough a mobile phone tablet (Medium, 2016), contains three different subsets with around 3000 instances. However, the second dataset, which was obtained through the keyboard, has around 20400 instances. Additionally, this difference leads to very different classification results. For instance, when testing fewer instances algorithms, x can be trained quickly; if the number of samples increases, the algorithms can be trained in an exponentially longer time.

Finally, from a completely different perspective, the authors have conducted experiments with Mobikey-easy dataset, but this time, instead of multiple classifications, the authors opted for an application situation where binary classification may be sufficient. This is a typical case for authentication applications where the application will try to associate a known "user id" with the user biometric. If the known association is satisfied, the authentication is granted. Thus, a binary classification is sufficient for this type of application. For this purpose, first, the authors selected two positive and negative samples for a given user. Positive samples are actual people, and negative samples are two randomly selected samples from each of the other users. Then, the authors applied this method for every user in the dataset, and they obtained the average of the classification results of all the users in the data set. Two-class classification results for the Easy dataset are given in Table 12. As it can be seen for this type application, again, on average, Random Forest, Bayesian Network, and SVM provide the most correctly classified instances.

DISCUSSION AND CONCLUSION

Classification is one of the fundamental techniques for the keystroke dynamics evaluation area. An effective estimation of classification will contribute to the development of studies in this field. In this chapter, the authors have applied different classification models to estimate the reliability of classification techniques on keystroke dynamic, using two different datasets. The objective in this work was to evaluate classification algorithms for deciding the best one to use for this type of dataset by applying different methods. The authors investigated multi-class classification and, from another perspective,

Table 9. Binary (Two-class) classification results for Easy dataset

Easy Data	Correctly Classified
Bayesian Network	92.67%
Decision Table	92.51%
Decision Tree	88.81%
Naive Bayes	89.66%
Random Forest	96.7%
Support Vector Machine	92.48%
K-Nearest Neighbor(Euclidian)	87.56%

other than binary classification. The authors also compared and correctly classified results, EER results, and execution speed.

In particular, the authors concluded that the random forest algorithm is the best model for classification in this setting. Following the Random Forest algorithm, Bayesnet and SVM algorithms also give high accuracy rates. However, in two datasets, the success rates for the algorithms are dispersed. For further development of the work and to achieve more reliable results, authors are planning to use larger datasets, different classification techniques, and different detection parameters or features (Chang, Tsai, Lin, 2012).

REFERENCES

Chang, T. Y., Tsai, C. J., & Lin, J. H. (2012). A graphical-based password keystroke dynamic authentication system for touch screen handheld mobile devices. *Journal of Systems and Software, 85*(5), 1157–1165. doi:10.1016/j.jss.2011.12.044

Giot, R., El-Abed, M., Hemery, B., & Rosenberger, C. (2011). Unconstrained keystroke dynamics authentication with shared secret. *Computers & Security, 30*(6-7), 427–445. doi:10.1016/j.cose.2011.03.004

Hwang, S. S., Cho, S., & Park, S. (2009). Keystroke dynamics-based authentication for mobile devices. *Computers & Security, 28*(1-2), 85–93. doi:10.1016/j.cose.2008.10.002

Kevin, S., & Killourhy, R. A. (2009). Comparing anomaly-detection algorithms for keystroke dynamics. *2009 IEEE/IFIP International Conference on Dependable Systems & Networks* (p. 125-134). Lisbon, Portugal: IEEE.

Lau, S.-H. (2018): Stress detection for keystroke dynamics. (Doctoral thesis), Carnegie Mellon University. doi:10.1184/R1/6723227.v3

Antal, & M.,, Lehel, N. (2016). *The MOBIKEY keystroke dynamics password database: benchmark results., software engineering perspectives and application in intelligent systems* (pp. 35–46). Cham, Switzerland: Springer. doi:10.1007/978-3-319-33622-0_4

McHugh, M. (2012). Interrater reliability: The kappa statistic. *Biochemia Medica, 22*(3), 276–282. doi:10.11613/BM.2012.031 PMID:23092060

Medium. (2016). MAE and RMSE — Which metric is better? – Human in a machine world – Medium. [online] Available at https://medium.com/human-in-a-machine-world/mae-and-rmse-which-metric-is-better-e60ac3bde13d

Pentel, A. (2017). Predicting age and gender by keystroke dynamics and mouse patterns. In *Proceedings 25th Conference on User Modeling, Adaptation and Personalization (UMAP'2017)*, Bratislava, Slovakia. 10.1145/3099023.3099105

Revett, K. (2009). A bioinformatics-based approach to user authentication via keystroke dynamics. *International Journal of Control, Automation, and Systems, 7*(1), 7–15. doi:10.100712555-009-0102-2

Sally Dafaallah Abualgasim, I. O. (2011). *An application of the keystroke dynamics biometric for securing pins and passwords. World of Computer Science and Information Technology Journal* (pp. 398–404). WCSIT.

Shah, P. (2017). An introduction to Weka - open source for you. [online] Open Source For You. Available at https://opensourceforu.com/2017/01/an-introduction-to-weka/

Ulinskas, M., Woźniak, M., & Damaševičius, R. (2017). Analysis of keystroke dynamics for fatigue recognition. In International Conference on Computational Science and Its Applications (pp. 235-247). Cham, Switzerland: Springer.

Weka (2019). Capabilities of Weka. [online] Available at http://weka.sourceforge.net/doc.stable/weka/core/Capabilities.html

Xi, K., Tang, Y., & Hu, J. (2011). Correlation keystroke verification scheme for user access control in cloud computing environment. *The Computer Journal*, *54*(10), 1632–1644. doi:10.1093/comjnl/bxr064

Young, J. R. (2018). *Keystroke Dynamics: Utilizing Keyprint Biometrics to Identify Users in Online Courses*. Brigham Young University.

Zhou, C. (2008). A study of keystroke dynamics as a practical form of authentication. [ebook] Available at https://www.semanticscholar.org/paper/A-Study-of-Keystroke-Dynamics-as-a-Practical-Form-May/39b6cfab3f58231df47d0604c196387356d503bc

Compilation of References

Abdullah, A. S., Selvakumar, S., & Abirami, A. M. (2017). An Introduction to Data Analytics: Its Types and Its Applications. In Handbook of Research on Advanced Data Mining Techniques and Applications for Business Intelligence. Hershey, PA: IGI Global. doi:10.4018/978-1-5225-2031-3.ch001

Abdullah, S. N. & Zeng, X. (2010, July). Machine learning approach for crude oil price prediction with Artificial Neural Networks-Quantitative (ANN-Q) model. In *The 2010 International Joint Conference on Neural Networks* (IJCNN) (pp. 1-8). IEEE.

Adamowski, J., & Karapataki, C. (2010). Comparison of multivariate regression and artificial neural networks for peak urban water-demand forecasting: Evaluation of different ANN learning algorithms. *Journal of Hydrologic Engineering*, *15*(10), 729–743. doi:10.1061/(ASCE)HE.1943-5584.0000245

Adebowale, A., Idowu, S. A., & Amarachi, A. (2013). Comparative study of selected data mining algorithms used for intrusion detection. [IJSCE]. *International Journal of Soft Computing and Engineering*, *3*(3), 237–241.

Adıvar çelik halat. (2010). *Adıvar çelik halat kataloğu*. Retrieved from http://www.adivarcelikhalat.com

Adusumilli, S., Bhatt, D., Wang, H., Bhattacharya, P., & Devabhaktuni, V. (2013). A low-cost INS/GPS integration methodology based on random forest regression. *Expert Systems with Applications*, *40*(11), 4653–4659. doi:10.1016/j.eswa.2013.02.002

Agamennoni, G., Nieto, J., & Nebot, E. (2009). Mining GPS data for extracting significant places. In *Proceedings IEEE International Conference on Robotics and Automation, 2009. ICRA'09*. IEEE. 10.1109/ROBOT.2009.5152475

Ahmadi-Nedushan, B. (2012). An optimized instance-based learning algorithm for estimation of compressive strength of concrete. *Engineering Applications of Artificial Intelligence*, *25*(5), 1073–1081. doi:10.1016/j.engappai.2012.01.012

Aichouri, I., Hani, A., Bougherira, N., Djabri, L., Chaffai, H., & Lallahem, S. (2015). River flow model using artificial neural networks. *Energy Procedia*, *74*, 1007–1014. doi:10.1016/j.egypro.2015.07.832

Akande, K. O., Owolabi, T. O., Twaha, S., & Olatunji, S. O. (2014). Performance comparison of SVM and ANN in predicting compressive strength of concrete. *IOSR Journal of Computer Engineering*, *16*(5), 88–94. doi:10.9790/0661-16518894

Akpinar, P., & Khashman, A. (2017). Intelligent classification system for concrete compressive strength. *Procedia Computer Science*, *120*, 712–718. doi:10.1016/j.procs.2017.11.300

Aksu, M. Ç. & Karaman, E. (n.d.). Link analysis and detection in a web site with decision trees. *Acta INFOLOGICA, 1*(2), 84-91.

Al-Ghanim, A. M. & Ludeman, L. C. (1997). Automated unnatural pattern recognition on control charts using correlation analysis techniques. *Computers & Industrial Engineering, 32*(3), 679-690.

Alliance, S. C. (2011). *Transit and contactless open payments: an emerging approach for fare collection.* White paper, Princeton: Smart Card Alliance.

Alsger, A. A., Mesbah, M., Ferreira, L., & Safi, H. (2015). Use of smart card fare data to estimate public transport origin–destination matrix. *Transportation Research Record: Journal of the Transportation Research Board, 2535*(1), 88–96. doi:10.3141/2535-10

Al-Shamiri, A. K., Kim, J. H., Yuan, T. F., & Yoon, Y. S. (2019). Modeling the compressive strength of high-strength concrete: An extreme learning approach. *Construction & Building Materials, 208*, 204–219. doi:10.1016/j.conbuildmat.2019.02.165

Alwan, L. C., & Roberts, H. V. (1988). Time-series modeling for statistical process control. *Journal of Business & Economic Statistics, 6*(1), 87–95.

American Concrete Institute. (2005). *Building code requirements for structural concrete and commentary* (ACI 318M-05).

Anderson, D., & McNeill, G. (1992). Artificial neural networks technology. *Kaman Sciences Corporation, 258*(6), 1–83.

Anderson, J. A. (1983). Cognitive and psychological computation with neural models. *IEEE Transactions on Systems, Man, and Cybernetics, SMC-13*(5), 799–814. doi:10.1109/TSMC.1983.6313074

Andrienko, G., Andrienko, N., Hurter, C., Rinzivillo, S., & Wrobel, S. (2011). From movement tracks through events to places: Extracting and characterizing significant places from mobility data. In *Proceedings 2011 IEEE Conference on Visual Analytics Science and Technology (VAST)*. IEEE. 10.1109/VAST.2011.6102454

Angel, R. D., Caudle, W. L., Noonan, R., & Whinston, A. N. D. A. (1972). Computer-assisted school bus scheduling. *Management Science, 18*(6), B-279–B-288. doi:10.1287/mnsc.18.6.B279

Antal, & M.,, Lehel, N. (2016). *The MOBIKEY keystroke dynamics password database: benchmark results., software engineering perspectives and application in intelligent systems* (pp. 35–46). Cham, Switzerland: Springer. doi:10.1007/978-3-319-33622-0_4

Anwar, K. M. (2014). Analysis of groundwater quality using statistical techniques : a case study of Aligarh city (India), *2*(5), 100–106.

Aparisi, F. (1996). Hotelling's T2 control chart with adaptive sample sizes. *International Journal of Production Research, 34*(10), 2853–2862.

Arias-Rojas, J. S., Jiménez, J. F., & Montoya-Torres, J. R. (2012). Solving of school bus routing problem by ant colony optimization. *Revista EIA*, (17), 193-208.

Ashbrook, D., & Starner, T. (2003). Using GPS to learn significant locations and predict movement across multiple users. *Personal and Ubiquitous Computing, 7*(5), 275–286. doi:10.100700779-003-0240-0

Atici, U. (2011). Prediction of the strength of mineral admixture concrete using multivariable regression analysis and an artificial neural network. *Expert Systems with Applications, 38*(8), 9609–9618. doi:10.1016/j.eswa.2011.01.156

Atkeson, C. G., Moore, A. W., & Schaal, S. (1997). Locally weighted learning. In *Lazy learning* (pp. 11–73). Dordrecht, The Netherlands: Springer. doi:10.1007/978-94-017-2053-3_2

Awoyera, P. O. (2018). Predictive models for determination of compressive and split-tensile strengths of steel slag aggregate concrete. *Materials Research Innovations, 22*(5), 287–293. doi:10.1080/14328917.2017.1317394

Aydogmus, H. Y., Erdal, H. İ., Karakurt, O., Namli, E., Turkan, Y. S., & Erdal, H. (2015). A comparative assessment of bagging ensemble models for modeling concrete slump flow. *Computers and Concrete, 16*(5), 741–757. doi:10.12989/cac.2015.16.5.741

Barceló, J., Montero, L., Marqués, L., & Carmona, C. (2010). Travel time forecasting and dynamic origin-destination estimation for freeways based on Bluetooth traffic monitoring. *Transportation Research Record: Journal of the Transportation Research Board, 2175*(1), 19–27. doi:10.3141/2175-03

Bar-Gera, H. (2007). Evaluation of a cellular phone-based system for measurements of traffic speeds and travel times: A case study from Israel. *Transportation Research Part C: Emerging Technologies, 15*(6), 380-391.

Bar-Gera, H., & Boyce, D. (2003). Origin-based algorithms for combined travel forecasting models. *Transportation Research Part B: Methodological, 37*(5), 405–422. doi:10.1016/S0191-2615(02)00020-6

Barry, J., Newhouser, R., Rahbee, A., & Sayeda, S. (2002). Origin and destination estimation in New York City with automated fare system data. *Transportation Research Record: Journal of the Transportation Research Board, 1817*(1), 183-187.

Bartels, J. R., McKewan, W. M., & Miscoe, A. J. (1992). *Bending fatigue tests 2 and 3 on 2-Inch 6 x 25 fiber core wire rope.* Retrieved from https://www.cdc.gov/niosh/mining/UserFiles/works/pdfs/ri9429.pdf

Bas, E., Tekalp, A. M., & Salman, F. S. (2007). Automatic vehicle counting from video for traffic flow analysis. In *2007 IEEE Intelligent Vehicles Symposium* (pp. 392–397). 10.1109/IVS.2007.4290146

Bastian, M., Heymann, S., & Jacomy, M. (2009). Gephi: an open source software for exploring and manipulating networks. Icwsm, *8*(2009), 361–362. doi:10.1016/j.aei.2011.01.003

Batty, M., Axhausen, K. W., Giannotti, F., Pozdnoukhov, A., Bazzani, A., Wachowicz, M., ... Portugali, Y. (2012). Smart cities of the future. *The European Physical Journal. Special Topics, 214*(1), 481–518. doi:10.1140/epjst/e2012-01703-3

Beam, A. L., & Kohane, I. S. (2018). Big data and machine learning in health care. *Journal of the American Medical Association, 319*(13), 1317–1318. doi:10.1001/jama.2017.18391 PMID:29532063

Becker, R. A., Caceres, R., Hanson, K., Loh, J. M., Urbanek, S., Varshavsky, A., &Volinsky, C. (2011). A tale of one city: Using cellular network data for urban planning. *IEEE Pervasive Computing, 10*(4), 18-26.

Behnood, A., Behnood, V., Gharehveran, M. M., & Alyamac, K. E. (2017). Prediction of the compressive strength of normal and high-performance concretes using M5P model tree algorithm. *Construction & Building Materials, 142*, 199–207. doi:10.1016/j.conbuildmat.2017.03.061

Behnood, A., & Golafshani, E. M. (2018). Predicting the compressive strength of silica fume concrete using hybrid artificial neural network with multi-objective grey wolves. *Journal of Cleaner Production, 202*, 54–64.

Bekdaş, G., Nigdeli, S. M., & Yang, X. S. (2015). Sizing optimization of truss structures using flower pollination algorithm. *Applied Soft Computing, 37*, 322–331. doi:10.1016/j.asoc.2015.08.037

Bektaş, T., & Elmastaş, S. (2007). Solving school bus routing problems through integer programming. *The Journal of the Operational Research Society, 58*(12), 1599–1604. doi:10.1057/palgrave.jors.2602305

Bellmore, M., & Nemhauser, G. L. (1968). The traveling salesman problem: A survey. *Operations Research, 16*(3), 538–582. doi:10.1287/opre.16.3.538

Benes, P. & Bukovsky, I. (2016). On the intrinsic relation between linear dynamical systems and higher order neural units. In R. Silhavy, R. Senkerik, Z. K. Oplatkova, Z. Prokopova, & P. Silhavy (Eds.), *Intelligent Systems in Cybernetics and Automation Theory*. doi:10.1007/978-3-319-18503-3_27

Benes, P. M. & Bukovsky, I. (2014). Neural network approach to hoist deceleration control. In *Proceedings 2014 International Joint Conference on Neural Networks (IJCNN)*, 1864–1869. IEEE. Retrieved from http://ieeexplore.ieee.org/xpls/abs_all.jsp?arnumber=6889831

Benes, P. M., Bukovsky, I., Cejnek, M., & Kalivoda, J. (2014). Neural network approach to railway stand lateral skew control. Computer Science & Information Technology (CS& IT), 4, 327–339. Sydney, Australia: AIRCC.

Benes, P. M., Erben, M., Vesely, M., Liska, O., & Bukovsky, I. (2016). HONU and supervised learning algorithms in adaptive feedback control. In Applied Artificial Higher Order Neural Networks for Control and Recognition (pp. 35–60). Hershey, PA: IGI Global.

Benes, P., & Bukovsky, I. (2018). An input to state stability approach for evaluation of nonlinear control loops with linear plant model. In R. Silhavy, R. Senkerik, Z. K. Oplatkova, Z. Prokopova, & P. Silhavy (Eds.), *Cybernetics and Algorithms in Intelligent Systems* (pp. 144–154). Springer International Publishing.

Bennett, B. T. & Gazis, D. C. (1972). School bus routing by computer. *Transportation Research/UK/, 6*(4).

Berry, M. W., & Castellanos, M. (2004). Survey of text mining. *Computer Review, 45*(9), 548.

Bhargava, N., Sharma, G., Bhargava, R., & Mathuria, M. (2013). Decision tree analysis on J48 algorithm for data mining. In Proceedings of International Journal of Advanced Research in Computer Science and Software Engineering, 3(6).

Bilgehan, M., & Turgut, P. (2010). Artificial neural network approach to predict compressive strength of concrete through ultrasonic pulse velocity. *Research in Nondestructive Evaluation, 21*(1), 1–17. doi:10.1080/09349840903122042

Biswas, J. (2018). *9 Complex machine learning applications that even a beginner can build today* Retrieved from https://analyticsindiamag.com/machine-learning-applications-beginners/

Blainey, S. P., & Preston, J. M. (2010). Modelling local rail demand in South Wales. *Transportation Planning and Technology, 33*(1), 55–73. doi:10.1080/03081060903429363

Bodin, L. D., & Berman, L. (1979). Routing and scheduling of school buses by computer. *Transportation Science, 13*(2), 113–129. doi:10.1287/trsc.13.2.113

Bowden, G. J., Maier, H. R., & Dandy, G. C. (2002). Optimal division of data for neural network models in water resources applications. *Water Resources Research, 38*(2), 2–1. doi:10.1029/2001WR000266

Bowerman, R., Hall, B., & Calamai, P. (1995). A multi-objective optimization approach to urban school bus routing: Formulation and solution method. *Transportation Research Part A, Policy and Practice, 29*(2), 107–123. doi:10.1016/0965-8564(94)E0006-U

Boyle, D. K. (1998). Passenger counting technologies and procedures.

Bozkır, A. S., Sezer, E., & Bilge, G. (2019). *Determination of the factors influencing student's success in student selection examination (OSS) via data mining techniques. IATS'09.* Karabük Turkiye.

Braca, J., Bramel, J., Posner, B., & Simchi-Levi, D. (1997). A computerized approach to the New York City school bus routing problem. *IIE Transactions, 29*(8), 693–702. doi:10.1080/07408179708966379

Braekers, K., Ramaekers, K., & Van Nieuwenhuyse, I. (2016). The vehicle routing problem: State of the art classification and review. *Computers & Industrial Engineering, 99*, 300–313. doi:10.1016/j.cie.2015.12.007

Braga, P. L., Oliveira, A. L., & Meira, S. R. (2007, September). Software effort estimation using machine learning techniques with robust confidence intervals. In *7th International Conference on Hybrid Intelligent Systems (HIS 2007)* (pp. 352-357). IEEE. 10.1109/HIS.2007.56

Breiman, L. (1996). Bagging predictors. *Machine Learning, 24*(2), 123–140. doi:10.1007/BF00058655

Breiman, L. (2001). Random forests. *Machine Learning, 45*(1), 5–32. doi:10.1023/A:1010933404324

Buchanan, B. G. (2006). *Brief history.* Retrieved from https://aitopics.org/misc/brief-history

Buchanan, B. G. (2005). A (very) brief history of artificial intelligence. *AI Magazine, 26*(4), 53–53.

Bühlmann, P., & Yu, B. (2002). Analyzing bagging. *Annals of Statistics, 30*(4), 927–961. doi:10.1214/aos/1031689014

Bui, D. K., Nguyen, T., Chou, J. S., Nguyen-Xuan, H., & Ngo, T. D. (2018). A modified firefly algorithm-artificial neural network expert system for predicting compressive and tensile strength of high-performance concrete. *Construction & Building Materials, 180*, 320–333. doi:10.1016/j.conbuildmat.2018.05.201

Bukovsky, I., Benes, P., & Slama, M. (2015). Laboratory systems control with adaptively tuned higher order neural units. In R. Silhavy, R. Senkerik, Z. K. Oplatkova, Z. Prokopova, & P. Silhavy (Eds.), *Intelligent Systems in Cybernetics and Automation Theory* (pp. 275–284). doi:10.1007/978-3-319-18503-3_27

Bukovsky, I., Homma, N., Smetana, L., Rodriguez, R., Mironovova, M., & Vrana, S. (2010, July). Quadratic neural unit is a good compromise between linear models and neural networks for industrial applications. 556–560. doi:10.1109/COGINF.2010.5599677

Bukovsky, I., *Voracek, J., Ichiji, K., & Noriyasu, H.* (2017). Higher order neural units for efficient adaptive control of weakly nonlinear systems. (pp. 149–157). doi:10.5220/0006557301490157

Bukovsky, I., & Homma, N. (2017). An approach to stable gradient-descent adaptation of higher order neural units. *IEEE Transactions on Neural Networks and Learning Systems, 28*(9), 2022–2034. doi:10.1109/TNNLS.2016.2572310 PMID:27295693

Bukovsky, I., Hou, Z.-G., Bila, J., & Gupta, M. M. (2008). Foundations of nonconventional neural units and their classification. *International Journal of Cognitive Informatics and Natural Intelligence, 2*(4), 29–43. doi:10.4018/jcini.2008100103

Bukovsky, I., Redlapalli, S., & Gupta, M. M. (2003). Quadratic and cubic neural units for identification and fast state feedback control of unknown nonlinear dynamic systems. In *Proceedings Fourth International Symposium on Uncertainty Modeling and Analysis, 2003. ISUMA 2003*, 330–334. 10.1109/ISUMA.2003.1236182

Burton, R. E., & Kebler, R. W. (1960). The "half-life" of some scientific and technical literatures. *American Documentation, 11*(1), 18–22.

Businessinsider. (2016, July). Social media engagement: the surprising facts about how much time people spend on the major social networks. Retrieved from http://www.businessinsider.com/social-media-engagement-statistics-2013-12

Caccetta, L., Alameen, M., & Abdul-Niby, M. (2013). An improved Clarke and Wright algorithm to solve the capacitated vehicle routing problem. *Engineering, Technology & Applied Scientific Research, 3*(2), 413–415.

Caceres, N., Wideberg, J., & Benitez, F. (2007). Deriving origin destination data from a mobile phone network. *IET Intelligent Transport Systems, 1*(1), 15–26. doi:10.1049/iet-its:20060020

Caceres, N., Wideberg, J., & Benitez, F. (2008). Review of traffic data estimations extracted from cellular networks. *IET Intelligent Transport Systems, 2*(3), 179–192. doi:10.1049/iet-its:20080003

Calabrese, F., Colonna, M., Lovisolo, P., Parata, D., & Ratti, C. (2011). Real-time urban monitoring using cell phones: A case study in Rome. *IEEE Transactions on Intelligent Transportation Systems, 12*(1), 141-151.

Calabrese, F., Pereira, F. C., Di Lorenzo, G., Liu, L., & Ratti, C. (2010). The geography of taste: analyzing cell-phone mobility and social events. In *Proceedings International Conference on Pervasive Computing,* Springer. 10.1007/978-3-642-12654-3_2

Cao, H., Mamoulis, N., & Cheung, D. W. (2005). Mining frequent spatio-temporal sequential patterns. In *Proceedings Fifth IEEE International Conference on Data Mining (ICDM'05)*, pp. 8-pp. IEEE.

Carbonell, J. G., Michalski, R. S., & Mitchell, T. M. (1983). An overview of machine learning. In *Machine learning* (pp. 3–23). Morgan Kaufmann.

Carvalho, S. (2010). Real-time sensing of traffic information in twitter messages.

Casas, I. & Delmelle, E. C. (2017). Tweeting about public transit—Gleaning public perceptions from a social media microblog. *Case Studies on Transport Policy, 5*(4), 634-642.

Cascardi, A., Micelli, F., & Aiello, M. A. (2017). An Artificial Neural Networks model for the prediction of the compressive strength of FRP-confined concrete circular columns. *Engineering Structures, 140*, 199–208. doi:10.1016/j.engstruct.2017.02.047

Castro-Neto, M., Jeong, Y. S., Jeong, M. K., & Han, L. D. (2009). Online-SVR for shortterm traffic flow prediction under typical and atypical traffic conditions. *Expert Systems with Applications, 36*(3), 6164–6173. doi:10.1016/j.eswa.2008.07.069

Caudill, M. (1988). Neural networks primer, part III. *AI Expert, 3*(6), 53–59.

Cebelak, M. K. (2013). *Location-based social networking data: doubly-constrained gravity model origin-destination estimation of the urban travel demand for Austin*, TX.

Çelik Halat ve Tel Sanayi A.Ş. (1999). *Çelik halat ürün kataloğu.* İzmit, Turkey.

Chang, T. Y., Tsai, C. J., & Lin, J. H. (2012). A graphical-based password keystroke dynamic authentication system for touch screen handheld mobile devices. *Journal of Systems and Software, 85*(5), 1157–1165. doi:10.1016/j.jss.2011.12.044

Chaniotakis, E., Antoniou, C., & Pereira, F. C. (2017). Enhancing resilience to disasters using social media. In *Proceedings 2017 5th IEEE International Conference on Models and Technologies for Intelligent Transportation Systems (MT-ITS)*. IEEE. 10.1109/MTITS.2017.8005602

Chaplin, C. R., Ridge, I. M. L., & Zheng, J. (1999, July). *Rope degradation and damage.* Retrieved from http://www.hse.gov.uk/research/otopdf/1999/oto99033.pdf

Chatterjee, S., Sarkar, S., Dey, N., Sen, S., Goto, T., & Debnath, N. C. (2017). Water quality prediction: multi-objective genetic algorithm coupled artificial neural network-based approach. In *July 2017 IEEE 15th International Conference on Industrial Informatics (INDIN)* (pp. 963-968). IEEE. 10.1109/INDIN.2017.8104902

Chatterjee, S., Sarkar, S., Hore, S., Dey, N., Ashour, A. S., & Balas, V. E. (2017). Particle swarm optimization trained neural network for structural failure prediction of multistoried RC buildings. *Neural Computing & Applications, 28*(8), 2005–2016. doi:10.100700521-016-2190-2

Cha, Y.-J., Choi, W., & Buyukozturk, O. (2017). Deep learning-based crack damage detection using convolutional neural networks. *Computer-Aided Civil and Infrastructure Engineering, 32*(5), 361–378. doi:10.1111/mice.12263

Chen, C. (2006). CiteSpace II: Detecting and visualizing emerging trends and transient patterns in scientific literature. *Journal of the American Society for Information Science and Technology, 57*(3), 359–377. doi:10.1002/asi.20317

Chen, D. S., Kallsen, H. A., Chen, H. C., & Tseng, V. C. (1990). A bus routing system for rural school districts. *Computers & Industrial Engineering*, *19*(1-4), 322–325. doi:10.1016/0360-8352(90)90131-5

Cheng, C.-S., & Cheng, H.-P. (2008). Identifying the source of variance shifts in the multivariate process using neural networks and support vector machines. *Expert Systems with Applications*, *35*(1–2), 198–206. doi:10.1016/j.eswa.2007.06.002

Cheng, M. Y., Chou, J. S., Roy, A. F., & Wu, Y. W. (2012). High-performance concrete compressive strength prediction using time-weighted evolutionary fuzzy support vector machines inference model. *Automation in Construction*, *28*, 106–115. doi:10.1016/j.autcon.2012.07.004

Chen, M., Hao, Y., Hwang, K., Wang, L., & Wang, L. (2017). Disease prediction by machine learning over big data from healthcare communities. *IEEE Access: Practical Innovations, Open Solutions*, *5*, 8869–8879. doi:10.1109/ACCESS.2017.2694446

Chen, M., Liu, X. B., Xia, J. X., & Chien, S. I. (2004). A dynamic bus-arrival time prediction model based on APC data. *Computer-Aided Civil and Infrastructure Engineering*, *19*, 364–376.

Chen, M., Mao, S., & Liu, Y. (2014). Big data: A survey. *Mobile Networks and Applications*, *19*(2), 171–209. doi:10.100711036-013-0489-0

Chen, Q., Li, W., & Zhao, J. (2011). The use of LS-SVM for short-term passenger flow prediction. *Transport*, *26*(1), 5–10. doi:10.3846/16484142.2011.555472

Chen, X., Kong, Y., Dang, L., Hou, Y., & Ye, X. (2015). Exact and metaheuristic approaches for a bi-objective school bus scheduling problem. *PLoS One*, *10*(7), e0132600. doi:10.1371/journal.pone.0132600 PMID:26176764

Chiang, M. M. T. & Mirkin, B. (2007). Experiments for the number of clusters in k-means. In *Proceedings Portuguese Conference on Artificial Intelligence* (pp. 395-405). Berlin, Germany: Springer. 10.1007/978-3-540-77002-2_33

Chiang, Y. M., Hao, R. N., Zhang, J. Q., Lin, Y. T., & Tsai, W. P. (2018). Identifying the sensitivity of ensemble streamflow prediction by artificial intelligence. *Water (Basel)*, *10*(10), 1341. doi:10.3390/w10101341

Chien, I.-Jy., Ding, Y., & Wei, C. (2002). Dynamic bus arrival time prediction with artificial neural networks. *Journal of Transportation Engineering*, *128*(5), 429–438.

Chinnam, R. B. (2002). Support vector machines for recognizing shifts in correlated and other manufacturing processes. *International Journal of Production Research*, *40*(17), 4449–4466. doi:10.1080/00207540210152920

Chi, S., & Caldas, C. H. (2011). Automated object identification using optical video cameras on construction sites. *Computer-Aided Civil and Infrastructure Engineering*, *26*(5), 368–380.

Chithra, S., Kumar, S. S., Chinnaraju, K., & Ashmita, F. A. (2016). A comparative study on the compressive strength prediction models for high performance concrete containing nano silica and copper slag using regression analysis and artificial neural networks. *Construction & Building Materials*, *114*, 528–535. doi:10.1016/j.conbuildmat.2016.03.214

Cho, S.-B. J. N. (2016). Exploiting machine learning techniques for location recognition and prediction with smartphone logs. *Neurocomputing*, *176*, 98-106.

Chong, E. K. P., & Zak, S. H. (2001). *An introduction to optimization* (2nd ed.). New York, NY: John Wiley & Sons.

Chopde, N. R., & Nichat, M. (2013). Landmark based shortest path detection by using A* and Haversine formula. *International Journal of Innovative Research in Computer and Communication Engineering*, *1*(2), 298–302.

Chopra, P., Sharma, R. K., & Kumar, M. (2016). Prediction of compressive strength of concrete using artificial neural network and genetic programmings. *Advances in Materials Science and Engineering*, *2016*, 1–10. doi:10.1155/2016/7648467

Choudhry, R., & Garg, K. (2008). A hybrid machine learning system for stock market forecasting. *World Academy of Science, Engineering and Technology, 39*(3), 315–318.

Chou, J. S., Lin, C. W., Pham, A. D., & Shao, J. Y. (2015). Optimized artificial intelligence models for predicting project award price. *Automation in Construction, 54*, 106–115. doi:10.1016/j.autcon.2015.02.006

Chou, J. S., & Tsai, C. F. (2012). Concrete compressive strength analysis using a combined classification and regression technique. *Automation in Construction, 24*, 52–60. doi:10.1016/j.autcon.2012.02.001

Chou, J. S., Tsai, C. F., Pham, A. D., & Lu, Y. H. (2014). Machine learning in concrete strength simulations: Multi-nation data analytics. *Construction & Building Materials, 73*, 771–780. doi:10.1016/j.conbuildmat.2014.09.054

Chourabi, H., Nam, T., Walker, S., Gil-Garcia, J. R., Mellouli, S., Nahon, K., . . . Scholl, H. J. (2012). Understanding smart cities: An integrative framework. In *Proceedings 2012 45th Hawaii International Conference on System Science (HICSS)*. IEEE. 10.1109/HICSS.2012.615

Cichosz, P. (2015). *Data mining algorithms: explained using R*. Chichester, UK: John Wiley & Sons. doi:10.1002/9781118950951

Civalek, Ö., & Çatal, H. H. (2004). Geriye yayılma yapay sinir ağları kullanılarak elastik kirişlerin statik ve dinamik analizi. *DEÜ Mühendislik Fakültesi Fen ve Mühendislik Dergisi, 1*, 1–16.

Clarke, G., & Wright, J. W. (1964). Scheduling of vehicles from a central depot to a number of delivery points. *Operations Research, 12*(4), 568–581. doi:10.1287/opre.12.4.568

Clark, S. (2003). Traffic prediction using multivariate nonparametric regression. *Journal of Transportation Engineering, 129*(2), 161–168.

Clifton, C. & Tassa, T. (2013). On syntactic anonymity and differential privacy. In *Proceedings 2013 IEEE 29th International Conference on Data Engineering Workshops (ICDEW)*. IEEE. 10.1109/ICDEW.2013.6547433

Collins, C., Hasan, S., & Ukkusuri, S. (2013). A novel transit rider satisfaction metric: Rider sentiments measured from online social media data. *Journal of Public Transportation, 16*(2), 2.

Cook, D. F., & Chiu, C.-C. (1998). Using radial basis function neural networks to recognize shifts in correlated manufacturing process parameters. *IIE Transactions, 30*(3), 227–234. doi:10.1080/07408179808966453

Cookes Limited. (2007). *Wire rope handbook*. Auckland, New Zealand: Cookes Limited.

Corberán, A., Fernández, E., Laguna, M., & Marti, R. (2002). Heuristic solutions to the problem of routing school buses with multiple objectives. *The Journal of the Operational Research Society, 53*(4), 427–435. doi:10.1057/palgrave.jors.2601324

Coumou, D., & Rahmstorf, S. (2012). A decade of weather extremes. *Nature Climate Change, 2*(7), 491–496. doi:10.1038/nclimate1452

Council, S. C. (2013). Smart cities readiness guide. The planning manual for building tomorrow's cities today.

Cranshaw, J., Schwartz, R., Hong, J. I., & Sadeh, N. (2012). The livehoods project: Utilizing social media to understand the dynamics of a city. In *Proceedings International AAAI Conference on Weblogs and Social Media*.

Crawford, M., Khoshgoftaar, T. M., Prusa, J. D., Richter, A. N., & Al Najada, H. (2015). Survey of review spam detection using machine learning techniques. *Journal of Big Data, 2*(1), 23. doi:10.118640537-015-0029-9

Çuhadar, M. (2013). Modeling and forecasting inbound tourism demand to Turkey by MLP, RBF and TDNN artificial neural networks: A comparative analysis. *Journal of Yaşar University, 8*(31).

Cürgül. İ. (1995). Materials handling. İzmit, Turkey, Kocaeli University Publications.

Cyril, A., Mulangi, R. H., & George, V. (2018, August). Modelling and forecasting bus passenger demand using time series method. In *2018 7th International Conference on Reliability, Infocom Technologies and Optimization (Trends and Future Directions)(ICRITO)* (pp. 460-466). IEEE.

Cyril, A., Mulangi, R. H., & George, V. (2019). Bus passenger demand modelling using time-series techniques-big data analytics. *The Open Transportation Journal, 13*(1).

D'Andrea, E., Ducange, P., Lazzerini, B., & Marcelloni, F. (2015). Real-time detection of traffic from twitter stream analysis. *IEEE Transactions on Intelligent Transportation Systems, 16*(4), 2269–2283.

Dabiri, S. & Heaslip, K. (2017). Inferring transportation modes from GPS trajectories using a convolutional neural network. *Transportation Research Part C: Emerging Technologies.*

Dabiri, S. & Heaslip, K. (2019). Developing a Twitter-based traffic event detection model using deep learning architectures. *Expert Systems with Applications, 118,* 425-439.

Dabiri, S., & Heaslip, K. (2018). Introducing a Twitter-based traffic information system using deep learning architectures. *IEEE Transactions on Intelligent Transportation Systems.*

Dahou, Z., Sbartaï, Z. M., Castel, A., & Ghomari, F. (2009). Artificial neural network model for steel–concrete bond prediction. *Engineering Structures, 31*(8), 1724–1733. doi:10.1016/j.engstruct.2009.02.010

Dai, Y. H. & Yuan, Y. (1999). *A nonlinear conjugate gradient method with a strong global convergence property. 10(1),* 177–182. doi:10.1137/S1052623497318992

Dantzig, G. B., & Ramser, J. H. (1959). The truck dispatching problem. *Management Science, 6*(1), 80–92. doi:10.1287/mnsc.6.1.80

Dargay, J., Clark, S., Johnson, D., Toner, J., & Wardman, M. (2010). A forecasting model for long distance travel in Great Britain. *12th World Conference on Transport Research*, Lisbon, Portugal.

Dash, M., Koo, K. K., Krishnaswamy, S. P., Jin, Y., & Shi-Nash, A. (2016). Visualize people's mobility-both individually and collectively-using mobile phone cellular data. In *2016 17th IEEE International Conference on Mobile Data Management (MDM)*. IEEE.

Davy, M., Desobry, F., Gretton, A., & Doncarli, C. (2006). An online support vector machine for abnormal events detection. *Signal Processing, 86*(8), 2009–2025. doi:10.1016/j.sigpro.2005.09.027

De Fabritiis, C., Ragona, R., & Valenti, G. (2008). Traffic estimation and prediction based on real time floating car data. In *Proceedings 11th International IEEE Conference on Intelligent Transportation Systems, 2008. ITSC 2008.* IEEE. 10.1109/ITSC.2008.4732534

Demirkoparan, F., Taştan, S., & Kaynar, O. (2010). Crude oil price forecasting with artificial neural networks. *Ege Academic Review, 10*(2), 559–573.

Demirsoy, M. (1991). *Materials Handling* (Vol. I). İstanbul, Turkey: Birsen Publishing House.

Deo, R. C. (2015). Machine learning in medicine. *Circulation, 132*(20), 1920–1930. doi:10.1161/CIRCULATIONAHA.115.001593 PMID:26572668

Deshpande, N., Londhe, S., & Kulkarni, S. (2014). Modeling compressive strength of recycled aggregate concrete by artificial neural network, model tree and non-linear regression. *International Journal of Sustainable Built Environment*, *3*(2), 187–198.

Desrosiers, J., Ferland, J. A., Rousseau, J.-M., Lapalme, G., & Chapleau, L. (1981). An overview of a school busing system. In N. K. Jaiswal (Ed.), *Scientific Management of Transport Systems* (pp. 235–243). Amsterdam, The Netherlands: North-Holland.

Dietterich, T. G. (2000). An experimental comparison of three methods for constructing ensembles of decision trees: Bagging, boosting, and randomization. *Machine Learning*, *40*(2), 139–157. doi:10.1023/A:1007607513941

Ding, Y. & Chien, S. (2000). The prediction of bus arrival times with link-based artificial neural networks. In *Proceedings of the International Conference on Computational Intelligence & Neurosciences (CI&N)—Intelligent Transportation Systems* (pp. 730–733), Atlantic City, NJ.

Ding, L., Fang, W., Luo, H., Love, P. E. D., Zhong, B., & Ouyang, X. (2018). A deep hybrid learning model to detect unsafe behavior: Integrating convolution neural networks and long short-term memory. *Automation in Construction*, *86*, 118–124.

Do, T. M. T., & Gatica-Perez, D. (2012). Contextual conditional models for smartphone-based human mobility prediction. *Proceedings of the 2012 ACM conference on ubiquitous computing*, ACM. 10.1145/2370216.2370242

Dou, Z., & Wang, M. (2012). *Proceedings of the International Conference on Automatic Control and Artificial Intelligence*, 1614-1616.

Du Preez, J., & Witt, S. F. (2003). Univariate versus multivariate time series forecasting: An application to international tourism demand. *International Journal of Forecasting*, *19*(3), 435–451. doi:10.1016/S0169-2070(02)00057-2

Duan, Z. H., Kou, S. C., & Poon, C. S. (2013). Prediction of compressive strength of recycled aggregate concrete using artificial neural networks. *Construction & Building Materials*, *40*, 1200–1206. doi:10.1016/j.conbuildmat.2012.04.063

Dulac, G., Ferland, J. A., & Forgues, P. A. (1980). School bus routes generator in urban surroundings. *Computers & Operations Research*, *7*(3), 199–213. doi:10.1016/0305-0548(80)90006-4

Efendigil, T., & Eminler, Ö. E. (2017). The importance of demand estimation in the aviation sector: A model to estimate airline passenger demand. *Journal of Yaşar University*, *12*, 14–30.

Eksioglu, B., Vural, A. V., & Reisman, A. (2009). The vehicle routing problem: A taxonomic review. *Computers & Industrial Engineering*, *57*(4), 1472–1483. doi:10.1016/j.cie.2009.05.009

El Faouzi, N.-E., Leung, H., & Kurian, A. (2011). Data fusion in intelligent transportation systems: Progress and challenges–A survey. *Information Fusion*, *12*(1), 4–10. doi:10.1016/j.inffus.2010.06.001

Elango, L., & Subramani, T. (2005). Groundwater quality and its suitability for drinking and agricultural use in Chithar River Basin, Tamilnadu, India. *Environ Geol*, *47*, 1099–1110. doi:10.1007/S00254-005-1243-0(2005)

Elbuluk, M. E., Tong, L., & Husain, I. (2002). Neural-network-based model reference adaptive systems for high-performance motor drives and motion controls. *IEEE Transactions on Industry Applications*, *38*(3), 879–886. doi:10.1109/TIA.2002.1003444

Ellegood, W. A., Campbell, J. F., & North, J. (2015). Continuous approximation models for mixed load school bus routing. *Transportation Research Part B: Methodological*, *77*, 182–198. doi:10.1016/j.trb.2015.03.018

Ellegood, W. A., Solomon, S., North, J., & Campbell, J. F. (2019). School bus routing problem: contemporary trends and research directions. *Omega*, 102056. doi:10.1016/j.omega.2019.03.014

El-Midany, T. T., El-Baz, M. A., & Abd-Elwahed, M. S. (2010). A proposed framework for control chart pattern recognition in multivariate process using artificial neural networks. *Expert Systems with Applications*, *37*(2), 1035–1042. doi:10.1016/j.eswa.2009.05.092

El-Nabarawy, I., Abdelbar, A. M., & Wunsch, D. C. (2013, Aug. 4). *Levenberg-Marquardt and Conjugate Gradient methods applied to a high-order neural network*. 1–7. doi:10.1109/IJCNN.2013.6707004

Erdal, H. I., Karakurt, O., & Namli, E. (2013). High performance concrete compressive strength forecasting using ensemble models based on discrete wavelet transform. *Engineering Applications of Artificial Intelligence*, *26*(4), 1246–1254. doi:10.1016/j.engappai.2012.10.014

Etehad Tavakol, M., Sadri, S., & Ng, E. Y. K. (2010). Application of K-and fuzzy c-means for color segmentation of thermal infrared breast images. *Journal of Medical Systems*, *34*(1), 35–42. PMID:20192053

Euchi, J., & Mraihi, R. (2012). The urban bus routing problem in the Tunisian case by the hybrid artificial ant colony algorithm. *Swarm and Evolutionary Computation*, *2*, 15–24. doi:10.1016/j.swevo.2011.10.002

Fang, H., Hsu, W.-J., & Rudolph, L. (2009). Cognitive personal positioning based on activity map and adaptive particle filter. *Proceedings of the 12th ACM International Conference on Modeling, Analysis and Simulation of Wireless and Mobile Systems*. ACM. 10.1145/1641804.1641873

Fang, W., Ding, L., Luo, H., & Love, P. E. D. (2018). Falls from heights: A computer vision-based approach for safety harness detection. *Automation in Construction*, *91*, 53–61. doi:10.1016/j.autcon.2018.02.018

FAO. (2017). *Water for sustainable food and agriculture. a report produced for the G20 Presidency of Germany. Food and Agriculture Organization of United Nations*. Rome, Italy: FAO.

Faraj, M. F., Sarubbi, J. F. M., Silva, C. M., Porto, M. F., & Nunes, N. T. R. (2014, October). A real geographical application for the school bus routing problem. In *17th International IEEE Conference on Intelligent Transportation Systems (ITSC)* (pp. 2762-2767). IEEE.

Farley, B. G., & Clark, W. A. (1954). Simulation of self-organizing systems by digital computer. *Transactions of the IRE Professional Group on Information Theory*, *4*(4), 76-84. 10.1109/TIT.1954.1057468

Farzin, J. M. (2008). Constructing an automated bus origin–destination matrix using farecard and global positioning system data in Sao Paulo, Brazil. *Transportation Research Record*, *2072*(1), 30-37.

Felzenszwalb, P. F., Girshick, R. B., McAllester, D., & Ramanan, D. (2009). Object detection with discriminatively trained part-based models. *IEEE Transactions on Pattern Analysis and Machine Intelligence*, *32*(9), 1627–1645. doi:10.1109/TPAMI.2009.167 PMID:20634557

Feng, D., & Feng, M. Q. (2018). Computer vision for SHM of civil infrastructure: From dynamic response measurement to damage detection – A review. *Engineering Structures*, *156*, 105–117. doi:10.1016/j.engstruct.2017.11.018

Feyrer, K. (2015). *Wire ropes: tension, endurance, reliability*. New York, NY: Springer.

Fischer, U., & Lehner, W. (2013). Transparent forecasting strategies in database management systems. In E. Zimányi (Ed.), *European Business Intelligence Summer School* (pp. 150–181). Cham, Switzerland: Springer. Retrieved from https://books.google.com.tr/

Fitri, S. (2014). Perbandingan Kinerja Algoritma Klasifikasi Naïve Bayesian, Lazy-Ibk, Zero-R, Dan Decision Tree-J48. *Data Manajemen dan Teknologi Informasi (DASI), 15*(1), 33.

Floridi, L. (2017). *If AI is the future, what does it mean for you?* Retrieved from https://www.weforum.org/agenda/2017/01/the-future-ofai-and-the-implications-for-you

Fügenschuh, A. (2009). Solving a school bus scheduling problem with integer programming. *European Journal of Operational Research, 193*(3), 867–884.

Fujita, O. (1998). Statistical estimation of the number of hidden units for feedforward neural networks. *Neural Networks, 11*(5), 851–859. doi:10.1016/S0893-6080(98)00043-4 PMID:12662787

Gal-Tzur, A., Grant-Muller, S. M., Kuflik, T., Minkov, E., Nocera, S., & Shoor, I. (2014). The potential of social media in delivering transport policy goals. *Transport Policy, 32,* 115-123.

Gandomi, A. H., Yang, X. S., Talatahari, S., & Alavi, A. H. (2013). Metaheuristic algorithms in modeling and optimization. In A. H. Gandomi, X. S. Yang, S. Talatahari, & A. H. Alavi (Eds.), *Metaheuristic applications in structures and infrastructures* (pp. 1–24). Elsevier. doi:10.1016/B978-0-12-398364-0.00001-2

Garcia, C. E., Prett, D. M., & Morani, M. (1989). Model predictive control: Theory and practice—. *Survey (London, England), 25*(3), 335–348. doi:10.1016/0005-1098(89)90002-2

Geem, Z. W. (2005). School bus routing using harmony search. In *Genetic and Evolutionary Computation Conference (GECCO 2005)*, Washington, DC.

Ghaffari, M., Ghadiri, N., Manshaei, M. H., & Lahijani, M. S. (2016). P4QS: a peer to peer privacy preserving query service for location-based mobile applications. *arXiv preprint arXiv:1606.02373.*

Ghazanfari, M., Alaeddini, A., Niaki, S. T. A., & Aryanezhad, M.-B. (2008). A clustering approach to identify the time of a step change in Shewhart control charts. *Quality and Reliability Engineering International, 24*(7), 765–778. doi:10.1002/qre.925

Gholizadeh, S. (2015). Performance-based optimum seismic design of steel structures by a modified firefly algorithm and a new neural network. *Advances in Engineering Software, 81,* 50–65. doi:10.1016/j.advengsoft.2014.11.003

Gibson, P. T., White, F. G., Schalit, L. A., Thomas, R. E., Cote, R. W., & Cress, H. A. (1974, Oct. 31). *A study of parameters that influence wire rope fatigue life.* Retrieved from https://apps.dtic.mil/dtic/tr/fulltext/u2/a001673.pdf

Gibson, P. T. (2001). Operational characteristics of ropes and cables. In J. F. Bash (Ed.), *Handbook of oceanographic winch, wire and cable technology* (pp. 8-3–8-50). USA.

Giot, R., El-Abed, M., Hemery, B., & Rosenberger, C. (2011). Unconstrained keystroke dynamics authentication with shared secret. *Computers & Security, 30*(6-7), 427–445. doi:10.1016/j.cose.2011.03.004

Girardin, F., Calabrese, F., Fiore, F. D., Ratti, C., & Blat, J. (2008). *Digital footprinting: Uncovering tourists with user-generated content.* Institute of Electrical and Electronics Engineers.

Glorot, X., Bordes, A., & Bengio, Y. (2011). Deep sparse rectifier neural networks. In G. Gordon, D. Dunson, & M. Dudík (Eds.), *Proceedings of the Fourteenth International Conference on Artificial Intelligence and Statistics* (pp. 315–323). Retrieved from http://proceedings.mlr.press/v15/glorot11a.html

Glover, F. W., & Kochenberger, G. A. (Eds.). (2006). *Handbook of metaheuristics* (Vol. 57). Springer Science & Business Media.

Golden, B., Bodin, L., Doyle, T., & Stewart, W. Jr. (1980). Approximate traveling salesman algorithms. *Operations research, 28*(3-part-ii), 694-711.

Golden, B. L., Magnanti, T. L., & Nguyan, H. Q. (1972). Implementing vehicle routing algorithms. *Networks, 7*(2), 113–148.

Goodfellow, I., Bengio, Y., & Courville, A. (2016). *Deep learning*. Retrieved from http://www.deeplearningbook.org/

Gorbatov, E. K., Klekovkina, N. A., Saltuk, V. N., Fogel, V., Barsukov, V. K., Barsukov, E. V., ... Kurashov, D. A. (2007). Steel rope with longer service life and improved quality. *Journal of Metallurgist, 51*(5-6), 279–283. doi:10.100711015-007-0052-y

Goulet-Langlois, G., Koutsopoulos, H. N., & Zhao, J. (2016). Inferring patterns in the multi-week activity sequences of public transport users. *Transportation Research Part C: Emerging Technologies, 64*, 1-16.

Grundmann, J., Schütze, N., & Lennartz, F. (2013). Sustainable management of a coupled groundwater–agriculture hydrosystem using multi-criteria simulation-based optimization. *Water Science and Technology, 67*(3), 689–698. doi:10.2166/wst.2012.602 PMID:23202577

Gühnemann, A., Schäfer, R.-P., Thiessenhusen, K.-U., & Wagner, P. (2004). Monitoring traffic and emissions by floating car data.

Gulenko, A., Wallschläger, M., Schmidt, F., Kao, O., & Liu, F. (2016, December). Evaluating machine learning algorithms for anomaly detection in clouds. *In 2016 IEEE International Conference on Big Data (Big Data)* (pp. 2716-2721). IEEE.

Gulez, K. (2008). Yapay sinir ağlarının kontrol mühendisliğindeki uygulamaları (Applications of artificial neural networks in control engineering). (Lecture notes, 2008) Yıldız Teknik University. Retrieved from http://www.yildiz.edu.tr/~gulez/3k1n.pdf

Güllü, H., Pala, M., & Iyisan, R. (2007). Yapay sinir ağları ile en büyük yer ivmesinin tahmin edilmesi (Estimating the largest ground acceleration with artificial neural networks). *Sixth National Conference on Earthquake Engineering*, Istanbul, Turkey.

Gupta, M. M., Homma, N., Hou, Z.-G., Solo, M., & Bukovsky, I. (2010). Higher order neural networks: Fundamental theory and applications. *Artificial Higher Order Neural Networks for Computer Science and Engineering: Trends for Emerging Applications*, 397–422.

Gupta, M. M., Bukovsky, I., Homma, N., Solo, A. M. G., & Hou, Z.-G. (2013). Fundamentals of higher order neural networks for modeling and simulation. In M. Zhang (Ed.), *Artificial Higher Order Neural Networks for Modeling and Simulation* (pp. 103–133). doi:10.4018/978-1-4666-2175-6.ch006

Gu, Y., Qian, Z., & Chen, F. (2016). From Twitter to detector: Real-time traffic incident detection using social media data. *Transportation Research Part C, Emerging Technologies, 67*, 321–342. doi:10.1016/j.trc.2016.02.011

Hachicha, W., & Ghorbel, A. (2012). A survey of control-chart pattern-recognition literature (1991–2010) based on a new conceptual classification scheme. *Computers & Industrial Engineering, 63*(1), 204–222. doi:10.1016/j.cie.2012.03.002

Hagan, M. T., Demuth, H. B., & Beale, M. (1996). *Neural network design*. Boston, MA: PWS.

Haghani, A., Hamedi, M., Sadabadi, K., Young, S., & Tarnoff, P. (2010). Data collection of freeway travel time ground truth with bluetooth sensors. *Transportation Research Record: Journal of the Transportation Research Board, 2160*(1), 60–68. doi:10.3141/2160-07

Haghiabi, A. H. (2017). Estimation of scour downstream of a ski-jump bucket using the multivariate adaptive regression splines. *Scientia Iranica, 24*(4), 1789–1801. doi:10.24200ci.2017.4270

Haghiabi, A. H., Nasrolahi, A. H., & Parsaie, A. (2018). Water quality prediction using machine learning methods. *Water Quality Research Journal*, *53*(1), 3–13. doi:10.2166/wqrj.2018.025

Hagiwara, M. (1994). A simple and effective method for removal of hidden units and weights. *Neurocomputing*, *6*(2), 207–218. doi:10.1016/0925-2312(94)90055-8

Hainen, A., Wasson, J., Hubbard, S., Remias, S., Farnsworth, G., & Bullock, D. (2011). Estimating route choice and travel time reliability with field observations of Bluetooth probe vehicles. *Transportation Research Record: Journal of the Transportation Research Board*, *2256*(1), 43–50. doi:10.3141/2256-06

Hakimpoor, H., Arshad, K. A. B., Tat, H. H., Khani, N., & Rahmandoust, M. (2011). Artificial neural networks applications in management. *World Applied Sciences Journal*, *14*(7), 1008–1019.

Hall, M. A. (1999). *Correlation-based feature selection for machine learning* (Doctoral dissertation). Retrieved from https://www.cs.waikato.ac.nz/~mhall/thesis.pdf

Han, J. & Kamber, M. (2006). Data mining: concepts and techniques (2nd ed.). CL: Morgan Kaufmann.

Han, J., & Gao, J. (2008). Data mining in e-science and engineering. In H. Kargupta, J. Han, S. Y. Philip, R. Motwani, & V. Kumar (Eds.), *Next generation of data mining* (pp. 1–114). Boca Raton, FL: CRC Press. doi:10.1201/9781420085877.pt1

Han, J., & Kamber, M. (2006). *Data mining: concepts and techniques* (2nd ed.). San Francisco, CA: Morgan Kaufmann.

Haq, N. F., Onik, A. R., Hridoy, M. A. K., Rafni, M., Shah, F. M., & Farid, D. M. (2015). Application of machine learning approaches in intrusion detection system: A survey. *IJARAI-International Journal of Advanced Research in Artificial Intelligence*, *4*(3), 9–18.

Hasan, S., & Ukkusuri, S. V. (2014). Urban activity pattern classification using topic models from online geo-location data. *Transportation Research Part C, Emerging Technologies*, *44*, 363–381. doi:10.1016/j.trc.2014.04.003

Haykin, S. (1998). *Neural networks: a comprehensive foundation*. New York, NY: Prentice Hall.

Hebb, D. (1949). *The organization of behavior: a neuropsychological theory*. New York, NY: Wiley.

Hendalianpour, A., Razmi, J., & Gheitasi, M. (2017). Comparing clustering models in bank customers: Based on Fuzzy relational clustering approach. *Accounting*, *3*(2), 81–94. doi:10.5267/j.ac.2016.8.003

Hill, C. M., Malone, L. C., & Trocine, L. (2004). Data mining and traditional regression. In H. Bozdogan (Ed.), *Statistical data mining and knowledge discovery* (p. 242). Boca Raton, FL: CRC Press.

Hochreiter, S., & Schmidhuber, J. (1997). Long short-term memory. *Neural Computation*, *9*(8), 1735–1780. doi:10.1162/neco.1997.9.8.1735 PMID:9377276

Hofmann, M., Wilson, S. P., & White, P. (2009). Automated identification of linked trips at trip level using electronic fare collection data. Transportation Research Board 88th Annual Meeting.

Hoła, J., & Schabowicz, K. (2005). Application of artificial neural networks to determine concrete compressive strength based on non-destructive tests. *Journal of Civil Engineering and Management*, *11*(1), 23–32. doi:10.3846/13923730.2005.9636329

Holland, J. H. (1975). *Adaptation in natural and artificial systems*. Ann Arbor, MI: University of Michigan Press.

Holmes, G., Hall, M., & Prank, E. (1999, December). Generating rule sets from model trees. *In Australasian Joint Conference on Artificial Intelligence* (pp. 1-12). Berlin, Germany: Springer.

Hopfield, J. J. (1982). Neural networks and physical systems with emergent collective computational abilities. *Proceedings of the National Academy of Sciences of the United States of America*, *79*(8), 2554–2558. doi:10.1073/pnas.79.8.2554 PMID:6953413

Hornik, K., Stinchcombe, M., & White, H. (1989). Multilayer feedforward networks are universal approximators. *Neural Networks*, *2*(5), 359–366. doi:10.1016/0893-6080(89)90020-8

Hsu, S. H., Hsieh, J. J. P.-A., Chih, T. C., & Hsu, K. C. (2009). A two-stage architecture for stock price forecasting by integrating self-organizing map and support vector regression. *Expert Systems with Applications*, *36*(4), 7947–7951. doi:10.1016/j.eswa.2008.10.065

Hsu, Y. L., & Liu, T. C. (2007). Developing a fuzzy proportional derivative controller optimization engine for engineering design optimization problems. *Engineering Optimization*, *39*(6), 679–700. doi:10.1080/03052150701252664

Huang, C. L., & Tsai, C. Y. (2009). A hybrid SOFM–SVR with a filter-based feature selection for stock market forecasting. *Expert Systems with Applications*, *36*(2), 1529–1539. doi:10.1016/j.eswa.2007.11.062

Huang, G.-B., Zhu, Q.-Y., & Siew, C.-K. (2006a). Extreme learning machine: Theory and applications. *Neurocomputing*, *70*(1–3), 489–501. doi:10.1016/j.neucom.2005.12.126

Huang, H., Li, Q., & Zhang, D. (2018). Deep learning-based image recognition for crack and leakage defects of metro shield tunnel. *Tunnelling and Underground Space Technology*, *77*, 166–176.

Huang, S., Wang, B., Qiu, J., Yao, J., Wang, G., & Yu, G. (2016). Parallel ensemble of online sequential extreme learning machine based on MapReduce. *Neurocomputing*, *174*, 352–367.

Huang, Y., & Pan, H. (2011, April). Short-term prediction of railway passenger flow based on RBF neural network. In *Proceedings of the 4th International Joint Conference on Computational Sciences and Optimization, CSO 2011* (pp. 594–597). 10.1109/CSO.2011.240

Huang, Z. (1998). Extensions to the k-means algorithm for clustering large data sets with categorical values. *Data Mining and Knowledge Discovery*, *2*(3), 283–304.

Hwang, S. S., Cho, S., & Park, S. (2009). Keystroke dynamics-based authentication for mobile devices. *Computers & Security*, *28*(1-2), 85–93. doi:10.1016/j.cose.2008.10.002

Hwarng, H. B. (1995). Multilayer perceptions for detecting cyclic data on control charts. *International Journal of Production Research*, *33*(11), 3101–3117. doi:10.1080/00207549508904863

Hwarng, H. B., & Hubele, N. F. (1992). Boltzmann machines that learn to recognize patterns on control charts. *Statistics and Computing*, *2*(4), 191–202. doi:10.1007/BF01889679

Hwarng, H. B., & Hubele, N. F. (1993). Back-propagation pattern recognizers for X control charts: Methodology and performance. *Computers & Industrial Engineering*, *24*(2), 219–235.

Imran, M., Castillo, C., Diaz, F., & Vieweg, S. (2015). Processing social media messages in mass emergency: A survey. [CSUR]. *ACM Computing Surveys*, *47*(4), 67. doi:10.1145/2771588

Islam, M. M., & Murase, K. (2001). A new algorithm to design compact two-hidden-layer artificial neural networks. *Neural Networks*, *14*(9), 1265–1278. doi:10.1016/S0893-6080(01)00075-2 PMID:11718425

Ismail, K., Sayed, T., & Saunier, N. (2013). A methodology for precise camera calibration for data collection applications in urban traffic scenes. *Canadian Journal of Civil Engineering*, *40*(1), 57–67. doi:10.1139/cjce-2011-0456

Istanbul Railway Network and Metrobus Line. Retrieved from http://www.metrobusharitasi.com/metrobus-haritasi

Itoh, M., Yokoyama, D., Toyoda, M., Tomita, Y., Kawamura, S., & Kitsuregawa, M. (2014). Visual fusion of mega-city big data: an application to traffic and tweets data analysis of metro passengers. In *Proceedings 2014 IEEE International Conference on Big Data*. IEEE. 10.1109/BigData.2014.7004260

Ivakhnenko, A. G. (1971). Polynomial theory of complex systems. *IEEE Transactions on Systems, Man, and Cybernetics*, *SMC-1*(4), 364–378. doi:10.1109/TSMC.1971.4308320

Jagannathan, G., Pillaipakkamnatt, K., & Wright, R. N. (2009). A practical differentially private random decision tree classifier. In *Proceedings of 2009 IEEE International Conference on Data Mining Workshops*. Miami, FL: IEEE. 10.1109/ICDMW.2009.93

Jain, A. K., Murty, M. N., & Flynn, P. J. (1999). Data clustering: A review. [CSUR]. *ACM Computing Surveys*, *31*(3), 264–323. doi:10.1145/331499.331504

Jamuna, M. (2018). Statistical analysis of ground water quality parameters in Erode District, Taminadu, India, (4), 84–89.

Jang, K.-Y., Yang, K., & Kang, C. (2003). Application of artificial neural network to identify non-random variation patterns on the run chart in automotive assembly process. *International Journal of Production Research*, *41*(6), 1239–1254. doi:10.1080/0020754021000042409

Janovsky, L. (1999). *Elevator mechanical design*. USA: Elevator World Inc.

Jiang, B., Yin, J., & Zhao, S. (2009). Characterizing the human mobility pattern in a large street network. *Physical Review. E*, *80*(2), 021136. doi:10.1103/PhysRevE.80.021136 PMID:19792106

Jiawei, H. & Kamber, M. (2001). Data mining: concepts and techniques. San Francisco, CA: Morgan Kaufmann, 5.

Jin, J., & Shi, J. (2001). Automatic feature extraction of waveform signals for in-process diagnostic performance improvement. *Journal of Intelligent Manufacturing*, *12*(3), 257–268. doi:10.1023/A:1011248925750

Jin, R., Zou, P. X. W., Piroozfar, P., Wood, H., Yang, Y., Yan, L., & Han, Y. (2019). A science mapping approach-based review of construction safety research. *Safety Science*, *113*, 285–297. doi:10.1016/j.ssci.2018.12.006

Johnson, D. S. & McGeoch, L. A. (1997). The traveling salesman problem: A case study in local optimization. *Local search in combinatorial optimization, 1*(1), 215-310.

Jovicic, G., & Hansen, C. O. (2003). A passenger travel demand model for Copenhagen. *Transportation Research Part A, Policy and Practice*, *37*(4), 333–349. doi:10.1016/S0965-8564(02)00019-8

Jurdak, R., Zhao, K., Liu, J., AbouJaoude, M., Cameron, M., & Newth, D. (2015). Understanding human mobility from Twitter. *PLoS One*, *10*(7). doi:10.1371/journal.pone.0131469 PMID:26154597

Kalmegh, S. (2015). Analysis of weka data mining algorithm reptree, simple cart and randomtree for classification of Indian news. *International Journal of Innovative Science, Engineering & Technology*, *2*(2), 438–446.

Kang, M., Kim, S. K., Felan, J. T., Choi, H. R., & Cho, M. (2015). Development of a genetic algorithm for the school bus routing problem. *International Journal of Software Engineering and Its Applications*, *9*(5), 107–126.

Kawamura, A., Chuarayapratip, R., & Haneyoshi, T. (1988). Deadbeat control of PWM inverter with modified pulse patterns for uninterruptible power supply. *IEEE Transactions on Industrial Electronics*, *35*(2), 295–300.

Kayabekir, A. E., Bekdaş, G., Nigdeli, S. M., & Temür, R. (2018). Investigation of cross-sectional dimension on optimum carbon-fiber-reinforced polymer design for shear capacity increase of reinforced concrete beams. In *Proceedings of 7th International Conference on Applied and Computational Mathematics (ICACM '18)*. Rome, Italy: International Journal of Theoretical and Applied Mechanics.

Kayabekir, A. E., Sayin, B., Bekdas, G., & Nigdeli, S. M. (2017). Optimum carbon-fiber-reinforced polymer design for increasing shear capacity of RC beams. In *Proceedings of 3rd International Conference on Engineering and Natural Sciences (ICENS 2017)*. Budapest, Hungary.

Kayabekir, A. E., Sayin, B., Bekdas, G., & Nigdeli, S. M. (2018) The factor of optimum angle of carbon-fiber-reinforced polymers. In *Proceedings of 4th International Conference on Engineering and Natural Sciences (ICENS 2018)*. Kiev, Ukraine.

Kayabekir, A. E., Sayin, B., Nigdeli, S. M., & Bekdaş, G. (2017). Jaya algorithm based optimum carbon-fiber-reinforced polymer design for reinforced concrete beams. In *Proceedings of 15th International Conference of Numerical Analysis and Applied Mathematics*. Thessaloniki, Greece: AIP Conference Proceedings 1978.

Keeni, K., Nakayama, K., & Shimodaira, H. (1999). Estimation of initial weights and hidden units for fast learning of multilayer neural networks for pattern classification. In *Proceedings IJCNN'99-International Joint Conference on Neural Networks*. (Vol. 3, pp. 1652–1656). IEEE. 10.1109/IJCNN.1999.832621

Kerner, B., Demir, C., Herrtwich, R., Klenov, S., Rehborn, H., Aleksic, M., & Haug, A. (2005). *Traffic state detection with floating car data in road networks. In Proceedings 2005 IEEE Intelligent Transportation Systems, 2005*. IEEE.

Kevin, S., & Killourhy, R. A. (2009). Comparing anomaly-detection algorithms for keystroke dynamics. *2009 IEEE/IFIP International Conference on Dependable Systems & Networks* (p. 125-134). Lisbon, Portugal: IEEE.

Kewalramani, M. A., & Gupta, R. (2006). Concrete compressive strength prediction using ultrasonic pulse velocity through artificial neural networks. *Automation in Construction*, *15*(3), 374–379. doi:10.1016/j.autcon.2005.07.003

Khalifa, A., & Nanni, A. (2000). Improving shear capacity of existing RC T-section beams using CFRP composites. *Cement and Concrete Composites*, *22*(3), 165–174. doi:10.1016/S0958-9465(99)00051-7

Khoshgoftaar, T. M., Golawala, M., & Van Hulse, J. (2007). An empirical study of learning from imbalanced data using random forest. In *Proceedings of 19th IEEE International Conference on Tools with Artificial Intelligence (ICTAI 2007)*. Patras, Greece: IEEE. 10.1109/ICTAI.2007.46

Kia, M. B., Pirasteh, S., Pradhan, B., Mahmud, A. R., Sulaiman, W. N. A., & Moradi, A. (2012). An artificial neural network model for flood simulation using GIS: Johor River Basin, Malaysia. *Environmental Earth Sciences*, *67*(1), 251–264.

Kieu, L. M., Bhaskar, A., & Chung, E. (2014). Transit passenger segmentation using travel regularity mined from Smart Card transactions data.

Kılınç, D., Borandağ, E., Yücalar, F., Özçift, A., & Bozyiğit, F. (2015). Yazılım hata kestiriminde kolektif sınıflandırma modellerinin etkisi. In *Proceedings of IX. Ulusal Yazılım Mühendisliği Sempozyumu*. Bornava-İzmir.

Kılınç, D., Borandağ, E., Yücalar, F., Özçift, A., & Bozyiğit, F. (2015). Yazılım hata kestiriminde kolektif sınıflandırma modellerinin etkisi. In *Proceedings of IX. Ulusal Yazılım Mühendisliği Sempozyumu*. İzmir, Turkey: Yaşar Üniversitesi.

Kim, J., Kim, R., Lee, J., Cheong, T., Yum, B., & Chang, H. (2005). Multivariate statistical analysis to identify the major factors governing groundwater quality in the coastal area of Kimje, South Korea Multivariate statistical analysis to identify the major factors governing groundwater quality in the coastal area of Kimje, South Korea, (July 2018). doi:10.1002/hyp.5565

Kim, K. J. (2003). Financial time series forecasting using support vector machines. *Neurocomputing*, *55*(1-2), 307–319. doi:10.1016/S0925-2312(03)00372-2

Kim, T., & Park, B. J. (2013). Model and algorithm for solving school bus problem. *Journal of Emerging Trends in Computing and Information Sciences*, *4*(8), 596–600.

Kinable, J., Spieksma, F. C., & Vanden Berghe, G. (2014). School bus routing—a column generation approach. *International Transactions in Operational Research*, *21*(3), 453–478.

Kiriş, S. B., & Tüysüz, F. (2017). Performance comparison of different clustering methods for manufacturing cell formation. *Sakarya University Journal of Science*, *21*(5), 1031–1044.

Kırkpatrick, S., Gelatt, C. D., & Vecchi, M. P. (1983). Optimization by simulated annealing. *Science*, *220*(4598), 671–680. PMID:17813860

Kisi, O., Shiri, J., Karimi, S., Shamshirband, S., Motamedi, S., Petković, D., & Hashim, R. (2015). A survey of water level fluctuation predicting in Urmia Lake using support vector machine with firefly algorithm. *Applied Mathematics and Computation*, *270*, 731–743. doi:10.1016/j.amc.2015.08.085

Kleinberg, J. (2002). Bursty and hierarchical structure in streams. In *Proceedings of the eighth ACM SIGKDD international conference on Knowledge discovery and data mining* (pp. 91–101). ACM. 10.1145/775047.775061

Kohonen, T. (1982). Self-organized formation of topologically correct feature maps. *Biological Cybernetics*, *43*(1), 59–69. doi:10.1007/BF00337288

Kohonen, T. (1988). An introduction to neural computing. *Neural Networks*, *1*(1), 3–16. doi:10.1016/0893-6080(88)90020-2

Kolari, P., Java, A., Finin, T., Oates, T., & Joshi, A. (2006, July). Detecting spam blogs: A machine learning approach. In *Proceedings of the national conference on artificial intelligence* (Vol. 21, No. 2, p. 1351). Menlo Park, CA: AAAI Press.

Kosmatopoulos, E. B., Polycarpou, M. M., Christodoulou, M. A., & Ioannou, P. A. (1995). High-order neural network structures for identification of dynamical systems. *IEEE Transactions on Neural Networks*, *6*(2), 422–431. doi:10.1109/72.363477 PMID:18263324

Koziel, S. & Yang, X.-S. (Eds.). (2011). Computational Optimization, Methods and Algorithms, Berlin, Germany: Springer, Verlag.

Küçük, B. (2000). Factors providing the strength and durability of concrete. *Pamukkale Üniversitesi Mühendislik Bilimleri Dergisi*, *6*(1), 79–85.

Kulshreshtha, A., & Shanmugam, P. (2018). Assessment of trophic state and water quality of coastal-inland lakes based on fuzzy inference system. *Journal of Great Lakes Research*, *44*(5), 1010–1025. doi:10.1016/j.jglr.2018.07.015

Kuncheva, L. I., & Rodríguez, J. J. (2007). An experimental study on rotation forest ensembles. In M. Haindl, J. Kittler, & F. Roli (Eds.), *Multiple classifier systems* (pp. 459–468). Heidelberg, Germany: Springer-Verlag. doi:10.1007/978-3-540-72523-7_46

Kusakabe, T., & Asakura, Y. (2014). Behavioural data mining of transit smart card data: A data fusion approach. *Transportation Research Part C, Emerging Technologies*, *46*, 179–191. doi:10.1016/j.trc.2014.05.012

Labrinidis, A., & Jagadish, H. V. (2012). Challenges and opportunities with big data. In *Proceedings of the VLDB Endowment International Conference on Very Large Data Bases*, *5*(12), 2032–2033. doi:10.14778/2367502.2367572

Laplante, P. A. (Ed.). (2000). Dictionary of computer science, engineering and technology. Boca Raton, FL: CRC Press.

Laporte, G. (1992). The vehicle routing problem: An overview of exact and approximate algorithms. *European Journal of Operational Research*, *59*(3), 345–358.

Lau, S.-H. (2018): Stress detection for keystroke dynamics. (Doctoral thesis), Carnegie Mellon University. doi:10.1184/R1/6723227.v3

Ławryńczuk, M. (2009). *Neural networks in model predictive control.* 31–63. doi:10.1007/978-3-642-04170-9_2

LeCun, Y., Jackel, L. D., Boser, B., Denker, J. S., Graf, H. P., Guyon, I. ... Hubbard, W. (1989). Handwritten digit recognition: applications of neural net chips and automatic learning. In F. Fogelman, J. Herault, & Y. Burnod (Eds.), Neurocomputing, Algorithms, Architectures and Applications. Les Arcs, France: Springer.

LeCun, Y., Bengio, Y., & Hinton, G. (2015). Deep learning. *Nature, 521*(7553), 436–444. doi:10.1038/nature14539 PMID:26017442

Lee, J. H., Gao, S., Janowicz, K., & Goulias, K. G. (2015). Can Twitter data be used to validate travel demand models? IATBR 2015-WINDSOR.

Lee, J.-G., Han, J., Li, X., & Cheng, H. (2011). Mining discriminative patterns for classifying trajectories on road networks. *IEEE Transactions on Knowledge and Data Engineering, 23*(5), 713–726. doi:10.1109/TKDE.2010.153

Lee, J.-G., Han, J., Li, X., & Gonzalez, H. (2008). TraClass: Trajectory classification using hierarchical region-based and trajectory-based clustering. In *Proceedings of the VLDB Endowment International Conference on Very Large Data Bases, 1*(1), 1081–1094. doi:10.14778/1453856.1453972

Lee, S. G., & Hickman, M. (2014). Trip purpose inference using automated fare collection data. *Public Transport (Berlin), 6*(1), 1–20. doi:10.100712469-013-0077-5

Lee, S., & Hickman, M. (2013). Are transit trips symmetrical in time and space? evidence from the twin cities. *Transportation Research Record: Journal of the Transportation Research Board, 2382*(1), 173–180. doi:10.3141/2382-19

Levy, J. I., Buonocore, J. J., & Von Stackelberg, K. (2010). Evaluation of the public health impacts of traffic congestion: A health risk assessment. *Environmental Health, 9*(1), 1. doi:10.1186/1476-069X-9-65 PMID:20979626

Li, M., Mehrotra, K., Mohan, C. K., & Ranka, S. (1990). Forecasting sunspot numbers using neural networks. *Electrical Engineering and Computer Science Technical Reports,* 67.

Liang, X., Zheng, X., Lv, W., Zhu, T., & Xu, K. (2012). The scaling of human mobility by taxis is exponential. *Physica A, 391*(5), 2135–2144. doi:10.1016/j.physa.2011.11.035

Li, L. Y. O., & Fu, Z. (2002). The school bus routing problem: A case study. *The Journal of the Operational Research Society, 53*(5), 552–558.

Lin, Y. P., Petway, J., Lien, W. Y., & Settele, J. (2018). Blockchain with artificial intelligence to efficiently manage water use under climate change.

Lin, Y., Yang, X., Zou, N., & Jia L. (2013). Real-time bus arrival time prediction: Case study for Jinan, China. *Journal of Transportation Engineering, 139*(11), 1133-1140.

Lin, M., & Hsu, W.-J. (2014). Mining GPS data for mobility patterns: A survey. *Pervasive and Mobile Computing, 12,* 1–16.

Lin, W. H. (2001). A Gaussian maximum likelihood formulation for short-term forecasting of traffic flow. In *Proceedings of the 2001 IEEE Intelligent Transportation Systems* (pp. 150–155), Oakland, CA.

Liu, Y., Özyer, T., Alhajj, R., & Barker, K. (2005). Integrating multi-objective genetic algorithm and validity analysis for locating and ranking alternative clustering. *Informatica, 29*(1).

Liu, D., & Wei, Q. (2014). Policy iteration adaptive dynamic programming algorithm for discrete-time non-linear systems. *IEEE Transactions on Neural Networks and Learning Systems, 25*(3), 621–634. doi:10.1109/TNNLS.2013.2281663 PMID:24807455

Liu, J. H., Shi, W., Elrahman, O. A., Ban, X., & Reilly, J. M. (2016). Understanding social media program usage in public transit agencies. *International Journal of Transportation Science and Technology, 5*(2), 83–92. doi:10.1016/j.ijtst.2016.09.005

Liu, J., Zhao, X. D., & Xu, Z. H. (2017). Identification of rock discontinuity sets based on a modified affinity propagation algorithm. *International Journal of Rock Mechanics and Mining Sciences, 94*, 32–42. doi:10.1016/j.ijrmms.2017.02.012

Liu, S., Ni, L. M., & Krishnan, R. (2014). Fraud detection from taxis' driving behaviors. *IEEE Transactions on Vehicular Technology, 63*(1), 464–472. doi:10.1109/TVT.2013.2272792

Liu, Y., Sui, Z., Kang, C., & Gao, Y. (2014). Uncovering patterns of inter-urban trip and spatial interaction from social media check-in data. *PLoS One, 9*(1). doi:10.1371/journal.pone.0086026 PMID:24465849

Lu, C. J., Lee, T. S., & Chiu, C. C. (2009). Financial time series forecasting using independent component analysis and support vector regression. *Decision Support Systems, 47*(2), 115–125. doi:10.1016/j.dss.2009.02.001

Luger, G. F. (2009). *Artificial intelligence: structures and strategies for complex problem solving* (6th ed.). Boston, MA: Pearson Education.

Lu, X., Wang, C., Yang, J.-M., Pang, Y., & Zhang, L. (2010). Photo2trip: generating travel routes from geo-tagged photos for trip planning. *Proceedings of the 18th ACM international conference on Multimedia*, ACM. 10.1145/1873951.1873972

Ma, X. & Wang, Y. (2014). Development of a data-driven platform for transit performance measures using smart card and GPS data. *Journal of Transport Geography, 140*(12), 04014063.

Ma, X., Liu, C., Wen, H., Wang, Y., & Wu, Y.-J. (2017). Understanding commuting patterns using transit smart card data. *Journal of Transport Geography, 58*, 135-145.

MacQueen, J. (1967). Some methods for classification and analysis of multivariate observations. In *Proceedings of the fifth Berkeley symposium on mathematical statistics and probability* (Vol. 1, No. 14, pp. 281-297).

Madhusudana, C. K., Kumar, H., & Narendranath, S. (2016). Condition monitoring of face milling tool using K-star algorithm and histogram features of vibration signal. *Engineering Science and Technology. International Journal (Toronto, Ont.), 19*(3), 1543–1551.

Maghrebi, M., Abbasi, A., & Waller, S. T. (2016). Transportation application of social media: Travel mode extraction. In *Proceedings 2016 IEEE 19th International Conference on Intelligent Transportation Systems (ITSC)*. IEEE. 10.1109/ITSC.2016.7795779

Maimon, O., & Rokach, L. (2010). *Data mining and knowledge discovery handbook* (2nd ed.). New York, NY: Springer. doi:10.1007/978-0-387-09823-4

Malinov, S., Sha, W., & McKeown, J. J. (2001). Modelling the correlation between processing parameters and properties in titanium alloys using artificial neural network. *Computational Materials Science, 21*(3), 375–394. doi:10.1016/S0927-0256(01)00160-4

Mandel, B. J. (1969). The regression control chart. *Journal of Quality Technology, 1*(1), 1–9. doi:10.1080/00224065.1969.11980341

Mandic, D. P. (2001). *Recurrent neural networks for prediction: Learning algorithms, architectures, and stability*. Chichester, NY: John Wiley. doi:10.1002/047084535X

Mandic, D. P. (2004). A generalized normalized gradient descent algorithm. *IEEE Signal Processing Letters, 11*(2), 115–118. doi:10.1109/LSP.2003.821649

Manning, C. D., Manning, C. D., & Schütze, H. (1999). *Foundations of statistical natural language processing*. Cambridge, MA: MIT Press.

MarketsandMarkets. (2018). *Artificial Intelligence Market worth 190.61 Billion USD by 2025*. Available at https://www.marketsandmarkets.com/PressReleases/artificial-intelligence.asp

MathWorks. (2016). Introducing Machine Learning. Retrieved from https://www.academia.edu/31538731/Introducing_Machine_Learning

Mathworks. (2018). MATLAB R2018a. The MathWorks Inc., Natick, MA.

Maurya, S. P., Ohri, A., & Singh, P. K. (2018). Evaluation of urban water balance using decision support system of Varanasi City, India. *Nature Environment and Pollution Technology*, *17*(4), 1219–1225.

Ma, X., Wu, Y.-J., Wang, Y., Chen, F., & Liu, J. (2013). Mining smart card data for transit riders' travel patterns. *Transportation Research Part C, Emerging Technologies*, *36*, 1–12. doi:10.1016/j.trc.2013.07.010

McCarthy, J. (2007). *What is artificial intelligence?* Retrieved from http://www.formal.stanford. edu/jmc/whatisai.html

McCulloch, W. S., & Pitts, W. (1943). A logical calculus of the ideas immanent in nervous activity. *The Bulletin of Mathematical Biophysics*, *5*(4), 1–19. doi:10.1007/BF02478259

McHugh, M. (2012). Interrater reliability: The kappa statistic. *Biochemia Medica*, *22*(3), 276–282. doi:10.11613/BM.2012.031 PMID:23092060

Medium. (2016). MAE and RMSE — Which metric is better? – Human in a machine world – Medium. [online] Available at https://medium.com/human-in-a-machine-world/mae-and-rmse-which-metric-is-better-e60ac3bde13d

Merton, R. K. (1968). The Matthew effect in science. *Science*, *159*(3810), 56–63. doi:10.1126cience.159.3810.56

Metropolis, N., Rosenbluth, A. W., Rosenbluth, M. N., Teller, A. H., & Teller, E. (1953). Equation of state calculations by fast computing machines. *The Journal of Chemical Physics*, *21*(6), 1087–1092.

Minocha, B. & Tripathi, S. (2014). Solving school bus routing problem using hybrid genetic algorithm: a case study. In *Proceedings of the Second International Conference on Soft Computing for Problem Solving (SocProS 2012), Dec. 28-30, 2012* (pp. 93-103). New Delhi, India: Springer. 10.1007/978-81-322-1602-5_11

Mintsis, G., Basbas, S., Papaioannou, P., Taxiltaris, C., & Tziavos, I. (2004). Applications of GPS technology in the land transportation system. *European Journal of Operational Research*, *152*(2), 399–409. doi:10.1016/S0377-2217(03)00032-8

Mitchell, T. M. (2006). *The discipline of machine learning* (Vol. 9). Pittsburgh, PA: Carnegie Mellon University, School of Computer Science, Machine Learning Department.

Mohamed, K., Côme, E., Baro, J., & Oukhellou, L. (2014). Understanding passenger patterns in public transit through smart card and socioeconomic data.

Mohamed, W. N. H. W., Salleh, M. N. M., & Omar, A. H. (2012, November). A comparative study of reduced error pruning method in decision tree algorithms. In *2012 IEEE International Conference on Control System, Computing and Engineering* (pp. 392-397). IEEE. 10.1109/ICCSCE.2012.6487177

Mohandes, M. A., Halawani, T. O., Rehman, S., & Hussain, A. A. (2004). Support vector machines for wind speed prediction. *Renewable Energy*, *29*(6), 939–947. doi:10.1016/j.renene.2003.11.009

Moher, D., Liberati, A., Tetzlaff, J., & Altman, D. G.Prisma Group. (2009). Preferred reporting items for systematic reviews and meta-analyses: The PRISMA statement. *PLoS Medicine*, *6*(7). PMID:19621072

Mokhtari, N. A., & Ghezavati, V. (2018). Integration of efficient multi-objective ant-colony and a heuristic method to solve a novel multi-objective mixed load school bus routing model. *Applied Soft Computing*, *68*, 92–109. doi:10.1016/j.asoc.2018.03.049

Momeni, E., Armaghani, D. J., & Hajihassani, M. (2015). Prediction of uniaxial compressive strength of rock samples using hybrid particle swarm optimization-based artificial neural networks. *Measurement*, *60*, 50–63. doi:10.1016/j.measurement.2014.09.075

Montgomery, D. C. (2007). *Introduction to statistical quality control*. Hoboken, NJ: John Wiley & Sons.

Morani, M. & Lee, J. H. (1999). *Model predictive control: Past, present and future*. *23*(4–5), 667–682. doi:10.1016/S0098-1354(98)00301-9

Munizaga, M. A., & Palma, C. (2012). Estimation of a disaggregate multimodal public transport Origin–Destination matrix from passive smartcard data from Santiago, Chile. *Transportation Research Part C, Emerging Technologies*, *24*, 9–18. doi:10.1016/j.trc.2012.01.007

Murphy, K., Torralba, A., Eaton, D., & Freeman, W. (2006). Object detection and localization using local and global features. In *Toward Category-Level Object Recognition* (pp. 382–400). Springer. doi:10.1007/11957959_20

Naderpour, H., Kheyroddin, A., & Amiri, G. G. (2010). Prediction of FRP-confined compressive strength of concrete using artificial neural networks. *Composite Structures*, *92*(12), 2817–2829.

Naghibi, S. A., Pourghasemi, H. R., & Dixon, B. (2016). GIS-based groundwater potential mapping using boosted regression tree, classification and regression tree, and random forest machine learning models in Iran. *Environmental Monitoring and Assessment*, *188*(1), 44. doi:10.100710661-015-5049-6 PMID:26687087

Nam, T., & Pardo, T. A. (2011). Conceptualizing smart city with dimensions of technology, people, and institutions. *Proceedings of the 12th Annual International Digital Government Research Conference: Digital Government Innovation in Challenging Times*. ACM. 10.1145/2037556.2037602

Nantes, A., Billot, R., Miska, M., & Chung, E. (2013). Bayesian inference of traffic volumes based on Bluetooth data.

Narendra, K. S., & Valavani, L. S. (1979). Direct and indirect model reference adaptive control. *Automatica*, *15*(6), 653–664. doi:10.1016/0005-1098(79)90033-5

Newton, R. M., & Thomas, W. H. (1974). Bus routing in a multi-school system. *Computers & Operations Research*, *1*(2), 213–222.

Nikolaev, N. Y. & Iba, H. (2006). *Adaptive learning of polynomial networks genetic programming, backpropagation and Bayesian methods*. Retrieved from http://public.eblib.com/choice/publicfullrecord.aspx?p=303002

Nikoo, M., Zarfam, P., & Sayahpour, H. (2015). Determination of compressive strength of concrete using self organization feature map (SOFM). *Engineering with Computers*, *31*(1), 113–121.

Nilsson, N. J. (1998). *Introduction to machine learning an early draft of a proposed textbook*. Retrieved from http://ai.stanford.edu/~nilsson/MLBOOK.pdf

Nishikawa, T., Yoshida, J., Sugiyama, T., & Fujino, Y. (2012). Concrete crack detection by multiple sequential image filtering. *Computer-Aided Civil and Infrastructure Engineering*, *27*(1), 29–47.

Noulas, A., Scellato, S., Lambiotte, R., Pontil, M., & Mascolo, C. (2012). A tale of many cities: Universal patterns in human urban mobility. *PLoS One*, *7*(5), e37027. doi:10.1371/journal.pone.0037027 PMID:22666339

Noulas, A., Scellato, S., Mascolo, C., & Pontil, M. (2011). An empirical study of geographic user activity patterns in foursquare. *ICwSM, 11*, 70–573.

Nunes, A. A., & Dias, T. G. (2016). Passenger journey destination estimation from automated fare collection system data using spatial validation. *IEEE Transactions on Intelligent Transportation Systems, 17*(1), 133–142. doi:10.1109/TITS.2015.2464335

Obermeyer, Z., & Emanuel, E. J. (2016). Predicting the future—Big data, machine learning, and clinical medicine. *The New England Journal of Medicine, 375*(13), 1216–1219. doi:10.1056/NEJMp1606181 PMID:27682033

Office of Naval Research Contract. (1960). *Adaptive switching circuits*. Stanford, CA: B. Widrow and M. E. Hoff.

Olivas, E. S., Guerrero, J. D. M., Sober, M. M., Benedito, J. R. M., & López, A. J. S. (Eds.). (2009). *Handbook of research on machine learning applications and trends: algorithms, methods, and techniques*. Hershey, PA: IGI Global.

Olivas, E. S., Guerrero, J. D. M., Sober, M. M., Benedito, J. R. M., & López, A. J. S. (Eds.). (2009). *Handbook of research on machine learning applications and trends: Algorithms, methods, and techniques*. Hershey, PA: IGI Global.

Oluwadare, S. A., Oguntuyi, I. P., & Nwaiwu, J. C. (2018). Solving school bus routing problem using genetic algorithm-based model. *International Journal of Intelligent Systems and Applications, 11*(3), 50.

Onoda, T. (1995). Neural network information criterion for the optimal number of hidden units. In *Proceedings of ICNN'95-International Conference on Neural Networks* (Vol. 1, pp. 275–280). IEEE. 10.1109/ICNN.1995.488108

Onur, Y. A. (2010). *Theoretical and experimental investigation of parameters affecting to the rope lifetime* (Unpublished doctoral dissertation). Istanbul Technical University Institute of Science, Turkey.

Onur, Y. A. (2012). Condition monitoring of Koepe winder ropes by electromagnetic non-destructive inspection. *Insight (American Society of Ophthalmic Registered Nurses), 54*(3), 144–148.

Onur, Y. A., & İmrak, C. E. (2011). The influence of rotation speed on the bending fatigue lifetime of steel wire ropes. *Proceedings of the Institution of Mechanical Engineers. Part C, Journal of Mechanical Engineering Science, 225*(3), 520–525. doi:10.1243/09544062JMES2275

Onur, Y. A., & İmrak, C. E. (2012). Experimental and theoretical investigation of bending over sheave fatigue life of stranded steel wire rope. *Indian Journal of Engineering and Materials Sciences, 19*, 189–195.

Onur, Y. A., İmrak, C. E., & Onur, T. Ö. (2016). Investigation of bending over sheave fatigue life of stranded steel wire rope by artificial neural networks. *International Journal of Mechanical And Production Engineering, 4*(9), 47–49.

Onur, Y. A., İmrak, C. E., & Onur, T. Ö. (2017). Investigation on bending over sheave fatigue life determination of rotation resistant steel wire rope. *Experimental Techniques, 41*(5), 475–482. doi:10.100740799-017-0188-z

Onwubolu, G. C., & Babu, B. V. (2004). *New optimization techniques in engineering* (Vol. 141). Heidelberg, Germany: Springer-Verlag. doi:10.1007/978-3-540-39930-8

Onyari, E. K., & Ilunga, F. M. (2013). Application of MLP neural network and M5P model tree in predicting streamflow: A case study of Luvuvhu catchment, South Africa. *International Journal of Innovation, Management and Technology, 4*(1), 11.

Ormsbee, L. E. & Reddy, S. L. (1995). Pumping system control using genetic optimization and neural networks. *IFAC Proceedings Volumes, 28*(10), 685-690.

Osburn, P. V. (1961). *New developments in the design of model reference adaptive control systems*. Institute of the Aerospace Sciences.

Owen, A. D., & Philips, G. D. A. (1987). An econometric investigation into the characteristics of railway passenger demand. *Journal of Transport Economics and Policy, 21,* 231–253.

Ozkan, C., & Sunar Erbek, F. (2003). The comparison of activation functions for multispectral landsat TM image classification. *Photogrammetric Engineering and Remote Sensing, 69*(11), 1225–1234. doi:10.14358/PERS.69.11.1225

Ozturk, H., Namli, E., & Erdal, H. I. (2016). Modelling sovereign credit ratings: The accuracy of models in a heterogeneous sample. *Economic Modelling, 54,* 469–478.

Pai, P. F., & Lin, C. S. (2005). A hybrid ARIMA and support vector machines model in stock price forecasting. *Omega, 33*(6), 497–505. doi:10.1016/j.omega.2004.07.024

Pai, P. F., Yang, S. L., & Chang, P. T. (2009). Forecasting output of integrated circuit industry by support vector regression models with marriage honey-bees optimization algorithms. *Expert Systems with Applications, 36*(7), 10746–10751. doi:10.1016/j.eswa.2009.02.035

Pal, U., Mukhopadhyay, G., Sharma, A., & Bhattacharya, S. (2018). Failure analysis of wire rope of ladle crane in steel making shop. *International Journal of Fatigue, 116,* 149–155. doi:10.1016/j.ijfatigue.2018.06.019

Panagiotidou, S., & Tagaras, G. (2012). Optimal integrated process control and maintenance under general deterioration. *Reliability Engineering & System Safety, 104,* 58–70. doi:10.1016/j.ress.2012.03.019

Pan, B., Zheng, Y., Wilkie, D., & Shahabi, C. (2013). Crowd sensing of traffic anomalies based on human mobility and social media. In *Proceedings of the 21st ACM SIGSPATIAL International Conference on Advances in Geographic Information Systems.* ACM. 10.1145/2525314.2525343

Park, J., & Kim, B. I. (2010). The school bus routing problem: A review. *European Journal of Operational Research, 202*(2), 311–319.

Park, J., Tae, H., & Kim, B. I. (2012). A post-improvement procedure for the mixed load school bus routing problem. *European Journal of Operational Research, 217*(1), 204–213.

Park, M.-W., & Brilakis, I. (2012). Construction worker detection in video frames for initializing vision trackers. *Automation in Construction, 28,* 15–25. doi:10.1016/j.autcon.2012.06.001

Parks, P. C. (1966). Liapunov redesign of model reference adaptive control systems. *IEEE Transactions on Automatic Control, 11*(3), 362–367.

Parsaie, A., Azamathulla, H. M., & Haghiabi, A. H. (2017). Physical and numerical modeling of performance of detention dams. *Journal of Hydrology (Amsterdam).* doi:10.1016/j.jhydrol.2017.01.018

Parsaie, A., Ememgholizadeh, S., Haghiabi, A. H., & Moradinejad, A. (2018). Investigation of trap efficiency of retention dams. *Water Science and Technology: Water Supply, 18*(2), 450–459. doi:10.2166/ws.2017.109

Parsaie, A., & Haghiabi, A. H. (2017). Mathematical expression of discharge capacity of compound open channels using MARS technique. *Journal of Earth System Science, 126*(2), 20. doi:10.100712040-017-0807-1

Parsaie, A., Haghiabi, A. H., & Moradinejad, A. (2019). Prediction of scour depth below river pipeline using support vector machine. *KSCE Journal of Civil Engineering, 23*(6), 2503–2513. doi:10.100712205-019-1327-0

Parsaie, A., Yonesi, H., & Najafian, S. (2017). Prediction of flow discharge in compound open channels using adaptive neuro fuzzy inference system method. *Flow Measurement and Instrumentation, 54,* 288–297. doi:10.1016/j.flowmeasinst.2016.08.013

Paté-Cornell, M. E., Lee, H. L., & Tagaras, G. (1987). Warnings of malfunction: The decision to inspect and maintain production processes on schedule or on demand. *Management Science, 33*(10), 1277–1290. doi:10.1287/mnsc.33.10.1277

Patel, J., Shah, S., Thakkar, P., & Kotecha, K. (2015). Predicting stock market index using fusion of machine learning techniques. *Expert Systems with Applications, 42*(4), 2162–2172. doi:10.1016/j.eswa.2014.10.031

Patil, A., Patted, L., Tenagi, M., Jahagirdar, V., Patil, M., & Gautam, R. (2017). Artificial intelligence as a tool in civil engineering – a review. In *Proceedings of National conference on advances in computational biology, communication, and data analytics (ACBCDA 2017).* India: IOSR Journal of Computer Engineering.

Pattara-Atikom, W. & Peachavanish, R. (2007). Estimating road traffic congestion from cell dwell time using neural network. In *Proceedings 2007 7th International Conference on ITS Telecommunications.* IEEE. 10.1109/ITST.2007.4295824

Pelletier, M.-P., Trépanier, M., & Morency, C. (2011). Smart card data use in public transit: A literature review. *Transportation Research Part C, Emerging Technologies, 19*(4), 557–568. doi:10.1016/j.trc.2010.12.003

Pender, B., Currie, G., Delbosc, A., & Shiwakoti, N. (2014). Social media use during unplanned transit network disruptions: A review of literature. *Transport Reviews, 34*(4), 501–521. doi:10.1080/01441647.2014.915442

Peng, Y., Kou, G., Wang, G., Wang, H., & Ko, F. I. (2009). Empirical evaluation of classifiers for software risk management. *International Journal of Information Technology & Decision Making, 8*(04), 749–767. doi:10.1142/S0219622009003715

Pentel, A. (2017). Predicting age and gender by keystroke dynamics and mouse patterns. In *Proceedings 25th Conference on User Modeling, Adaptation and Personalization (UMAP'2017),* Bratislava, Slovakia. 10.1145/3099023.3099105

Pereira, J., Pasquali, A., Saleiro, P., & Rossetti, R. (2017). Transportation in social media: an automatic classifier for travel-related tweets. *Portuguese Conference on Artificial Intelligence.* Springer. 10.1007/978-3-319-65340-2_30

Persson, O. (1994). The intellectual base and research fronts of JASIS 1986–1990. *Journal of the American Society for Information Science, 45*(1), 31–38. doi:10.1002/(SICI)1097-4571(199401)45:1<31::AID-ASI4>3.0.CO;2-G

Preston, J. (1991). Demand forecasting for new local rail stations and services. *Journal of Transport Economics and Policy, 25,* 183–202.

Price, D. J. D. S. (1965). Networks of scientific papers. *Science, 149*(3683), 510–515. doi:10.1126cience.149.3683.510 PMID:14325149

Provost, F., & Kohavi, R. (1998). Glossary of terms. *Journal of Machine Learning, 30*(2-3), 271–274.

Pugh, G. A. (1989). Synthetic neural networks for process control. *Computers & Industrial Engineering, 17*(1–4), 24–26. doi:10.1016/0360-8352(89)90030-2

Raheli, B. A.-S., Aalami, M. T., El-Shafie, A., Ghorbani, M. A., & Deo, R. C. (2017). Uncertainty assessment of the multilayer perceptron (MLP) neural network model with implementation of the novel hybrid MLP-FFA method for prediction of biochemical oxygen demand and dissolved oxygen: A case study of Langat River. *Environmental Earth Sciences, 76*(14), 1–16. doi:10.100712665-017-6842-z

Rahmani, M., Jenelius, E., & Koutsopoulos, H. N. (2015). Non-parametric estimation of route travel time distributions from low-frequency floating car data. *Transportation Research Part C, Emerging Technologies, 58,* 343–362. doi:10.1016/j.trc.2015.01.015

Rahmani, M., & Koutsopoulos, H. N. (2013). Path inference from sparse floating car data for urban networks. *Transportation Research Part C, Emerging Technologies, 30,* 41–54. doi:10.1016/j.trc.2013.02.002

Rao, R. (2016). Jaya: A simple and new optimization algorithm for solving constrained and unconstrained optimization problems. *International Journal of Industrial Engineering Computations, 7*(1), 19–34.

Rao, R. V., Savsani, V. J., & Vakharia, D. P. (2011). Teaching–learning-based optimization: A novel method for constrained mechanical design optimization problems. *Computer Aided Design, 43*(3), 303–315. doi:10.1016/j.cad.2010.12.015

Rashidi, T. H., Abbasi, A., Maghrebi, M., Hasan, S., & Waller, T. S. (2017). Exploring the capacity of social media data for modelling travel behaviour: Opportunities and challenges. *Transportation Research Part C, Emerging Technologies, 75*, 197–211. doi:10.1016/j.trc.2016.12.008

Rashidi, T. H., Zokaei-Aashtiani, H., & Mohammadian, A. (2009). School bus routing problem in large-scale networks: new approach utilizing tabu search on a case study in developing countries. *Transportation Research Record: Journal of the Transportation Research Board, 2137*(1), 140–147. doi:10.3141/2137-15

Ratti, C., Frenchman, D., Pulselli, R. M., & Williams, S. (2006). *Mobile landscapes: using location data from cell phones for urban analysis, 33*(5), 727-748.

Ray, S., & Turi, R. H. (1999). Determination of number of clusters in k-means clustering and application in colour image segmentation. In *Proceedings of the 4th international conference on advances in pattern recognition and digital techniques* (pp. 137-143).

Raza, M. Q., & Khosravi, A. (2015). A review on artificial intelligence-based load demand forecasting techniques for smart grid and buildings. *Renewable & Sustainable Energy Reviews, 50*, 1352–1372. doi:10.1016/j.rser.2015.04.065

Reades, J., Calabrese, F., Sevtsuk, A., & Ratti, C. (2007). Cellular census: Explorations in urban data collection. IEEE Pervasive computing, 6(3), 30-38.

Refaeilzadeh, P., Tang, L., & Liu, H. (2009). Cross-validation. In L. Liu & M. T. Özsu (Eds.), *Encyclopedia of database systems* (pp. 24–154). Boston, MA: Springer.

Revett, K. (2009). A bioinformatics-based approach to user authentication via keystroke dynamics. *International Journal of Control, Automation, and Systems, 7*(1), 7–15. doi:10.100712555-009-0102-2

Rezazadeh Azar, E., Dickinson, S., & McCabe, B. (2012). Server-customer interaction tracker: Computer vision–based system to estimate dirt-loading cycles. *Journal of Construction Engineering and Management, 139*(7), 785–794. doi:10.1061/(ASCE)CO.1943-7862.0000652

Ridge, I. M. L., Chaplin, C. R., & Zheng, J. (2001). Effect of degradation and impaired quality on wire rope bending over sheave fatigue endurance. *Journal of Engineering Failure Analysis, 8*(2), 173–187. doi:10.1016/S1350-6307(99)00051-5

Riera-Ledesma, J., & Salazar-González, J. J. (2012). Solving school bus routing using the multiple vehicle traveling purchaser problem: A branch-and-cut approach. *Computers & Operations Research, 39*(2), 391–404. doi:10.1016/j.cor.2011.04.015

Riera-Ledesma, J., & Salazar-González, J. J. (2013). A column generation approach for a school bus routing problem with resource constraints. *Computers & Operations Research, 40*(2), 566–583. doi:10.1016/j.cor.2012.08.011

Rojas, F., Calabrese, F., Krishnan, S., Ratti, C. J. N., & Studies, C. (2008). *Real Time Rome, 20*(3), 247–258.

Rojas, R. (1996). *Neural networks: a systematic introduction.* Berlin, Germany: Springer-Verlag. doi:10.1007/978-3-642-61068-4

Rokach, L., & Maimon, O. (2015). *Data mining with decision trees theory and applications* (2nd ed., Vol. 81). Singapore: World Scientific.

Rokach, L., & Maimon, O. (2015). *Data mining with decision trees: theory and applications* (2nd ed., Vol. 81). Singapore: World Scientific Publishing.

Rumelhart, D. E., Hinton, G. E., & Williams, R. J. (1986). Learning representations by back propagation error. *Nature*, *32*(6088), 533–536. doi:10.1038/323533a0

Russell, R. A., & Morrel, R. B. (1986). Routing special-education school buses. *Interfaces*, *16*(5), 56–64. doi:10.1287/inte.16.5.56

Russell, S. J., & Norvig, P. (1995). *Artificial intelligence: a modern approach*. Englewood Cliffs, NJ: Prentice Hall.

Russell, S. J., & Norvig, P. (1995). *Artificial intelligence: A modern approach*. Englewood Cliffs, NJ: Prentice Hall.

Saadeh, M., Sleit, A., Sabri, K. E., & Almobaideen, W. (2018). Hierarchical architecture and protocol for mobile object authentication in the context of IoT smart cities. Journal of Network and Computer Applications, 121, 1-19.

Sadat-Noori, S. M., Ebrahimi, K., & Liaghat, A. M. (2013). Groundwater quality assessment using the water quality index and GIS in Saveh-nodaran Aquifer, Iran, Enviro Earth Science. doi:10.1007/S12665-013-2770-8

Sahay, R. R., & Srivastava, A. (2014). Predicting monsoon floods in rivers embedding wavelet transform, genetic algorithm and neural network. *Water Resources Management*, *28*(2), 301–317. doi:10.100711269-013-0446-5

Sahoo, G. B., Ray, C., Mehnert, E., & Keefer, D. A. (2006). Application of artificial neural networks to assess pesticide contamination in shallow groundwater. *The Science of the Total Environment*, *367*(1), 234–251. doi:10.1016/j.scitotenv.2005.12.011 PMID:16460784

Sakaki, T., Okazaki, M., & Matsuo, Y. (2010). Earthquake shakes Twitter users: real-time event detection by social sensors. In *Proceedings of the 19th International Conference on World Wide Web*. Raleigh, NC: ACM. 851-860. 10.1145/1772690.1772777

Salcedo-Sanz, S., Ortiz-Garcı, E. G., Pérez-Bellido, Á. M., Portilla-Figueras, A., & Prieto, L. (2011). Short term wind speed prediction based on evolutionary support vector regression algorithms. *Expert Systems with Applications*, *38*(4), 4052–4057. doi:10.1016/j.eswa.2010.09.067

Sally Dafaallah Abualgasim, I. O. (2011). *An application of the keystroke dynamics biometric for securing pins and passwords. World of Computer Science and Information Technology Journal* (pp. 398–404). WCSIT.

Samadi-Dana, S., Paydar, M. M., & Jouzdani, J. (2017). A simulated annealing solution method for robust school bus routing. *International Journal of Operational Research*, *28*(3), 307–326. doi:10.1504/IJOR.2017.081908

Schittekat, P., Sevaux, M., & Sorensen, K. (2006). A mathematical formulation for a school bus routing problem. In *2006 International Conference on Service Systems and Service Management*, (Vol. 2, pp. 1552-1557). IEEE.

Schittekat, P., Kinable, J., Sörensen, K., Sevaux, M., Spieksma, F., & Springael, J. (2013). A metaheuristic for the school bus routing problem with bus stop selection. *European Journal of Operational Research*, *229*(2), 518–528.

Schrage, L. E. & LINDO Systems, Inc. (1997). *Optimization modeling with LINGO*, CA: Duxbury Press.

Sebaaly, H., Varma, S., & Maina, J. W. (2018). Optimizing asphalt mix design process using artificial neural network and genetic algorithm. *Construction & Building Materials*, *168*, 660–670. doi:10.1016/j.conbuildmat.2018.02.118

Şen, Z. (2004). *Yapay sinir ağları ilkeleri*. Su Vakfı Yayınları.

Seo, J., Han, S., Lee, S., & Kim, H. (2015). Computer vision techniques for construction safety and health monitoring. *Advanced Engineering Informatics*, *29*(2), 239–251. doi:10.1016/j.aei.2015.02.001

Seyhan, A. T., Tayfur, G., Karakurt, M., & Tanoglu, M. (2005). Artificial neural network (ANN) prediction of compressive strength of VARTM processed polymer composites. *Computational Materials Science, 34*(1), 99–105. doi:10.1016/j.commatsci.2004.11.001

Shafahi, A., Wang, Z., & Haghani, A. (2018). SpeedRoute: Fast, efficient solutions for school bus routing problems. *Transportation Research Part B: Methodological, 117*, 473–493. doi:10.1016/j.trb.2018.09.004

Shah, P. (2017). An introduction to Weka - open source for you. [online] Open Source For You. Available at https://opensourceforu.com/2017/01/an-introduction-to-weka/

Shahid, M., & Amba, G. (2018). Groundwater quality assessment of urban Bengaluru using multivariate statistical techniques. *Applied Water Science, 8*(1), 1–15. doi:10.100713201-018-0684-z

Shebani, A., & Iwnicki, S. (2018). Prediction of wheel and rail wear under different contact conditions using artificial neural networks. *Wear, 406*, 173–184. doi:10.1016/j.wear.2018.01.007

Shen, S., Jiang, H., & Zhang, T. (2012). *Stock market forecasting using machine learning algorithms* (pp. 1–5). Stanford, CA: Department of Electrical Engineering, Stanford University.

Sinha, K., Srivastava, D. K., & Bhatnagar, R. (2018). Water quality management through data driven intelligence system in Barmer Region, Rajasthan. *Procedia Computer Science, 132*, 314–322. doi:10.1016/j.procs.2018.05.183

Skurichina, M., & Duin, R. P. (2002). Bagging, boosting and the random subspace method for linear classifiers. *Pattern Analysis & Applications, 5*(2), 121–135. doi:10.1007100440200011

Smith, B. L., Williams, B. M., & Oswald, R. K. (2002). Comparison of parametric and nonparametric models for traffic flow forecasting. *Transportation Research Part C, Emerging Technologies, 10*(4), 303–321. doi:10.1016/S0968-090X(02)00009-8

Smit, R., Brown, A., & Chan, Y. (2008). Do air pollution emissions and fuel consumption models for roadways include the effects of congestion in the roadway traffic flow? *Environmental Modelling & Software, 23*(10), 1262–1270. doi:10.1016/j.envsoft.2008.03.001

Solomatine, D. P., Maskey, M., & Shrestha, D. L. (2006, July). Eager and lazy learning methods in the context of hydrologic forecasting. In *The 2006 IEEE International Joint Conference on Neural Network Proceedings* (pp. 4847-4853). IEEE.

Spada, M., Bierlaire, M., & Liebling, T. M. (2005). Decision-aiding methodology for the school bus routing and scheduling problem. *Transportation Science, 39*(4), 477–490. doi:10.1287/trsc.1040.0096

Stallkamp, J., Schlipsing, M., Salmen, J., & Igel, C. (2012). Man vs. computer: Benchmarking machine learning algorithms for traffic sign recognition. *Neural Networks, 32*, 323–332. PMID:22394690

Steenbruggen, J., Borzacchiello, M. T., Nijkamp, P., & Scholten, H. (2013). Mobile phone data from GSM networks for traffic parameter and urban spatial pattern assessment: A review of applications and opportunities. *GeoJournal, 78*(2), 223–243. doi:10.100710708-011-9413-y

Stefanidis, A., Cotnoir, A., Croitoru, A., Crooks, A., Rice, M., & Radzikowski, J. (2013). Demarcating new boundaries: mapping virtual polycentric communities through social media content. Cartography and Geographic Information Science, 40(2), 116-129.

Sugar, C. A., & James, G. M. (2003). Finding the number of clusters in a dataset: An information-theoretic approach. *Journal of the American Statistical Association, 98*(463), 750–763. doi:10.1198/016214503000000666

Suh, D. Y., & Ryerson, M. S. (2019). Forecast to grow: Aviation demand forecasting in an era of demand uncertainty and optimism bias. *Transportation Research Part E, Logistics and Transportation Review, 128*, 400–416. doi:10.1016/j.tre.2019.06.016

Suh, J. I., & Chang, S. P. (2000). Experimental study on fatigue behaviour of wire ropes. *International Journal of Fatigue, 22*(4), 339–347. doi:10.1016/S0142-1123(00)00003-7

Sujatha, P. & Kumar, G. P. (2011). Prediction of groundwater levels using different artificial neural network architectures and algorithms. In *Proceedings on the International Conference on Artificial Intelligence (ICAI)* (p. 1). The Steering Committee of The World Congress in Computer Science, Computer Engineering and Applied Computing (WorldComp).

Sule, D. R. (2001). Logistics of facility location and allocation (p. 240). Boca Raton, FL: CRC Press. doi:10.1201/9780203910405

Sun, D., Zhang, C., Zhang, L., Chen, F., & Peng, Z.-R. (2014). Urban travel behavior analyses and route prediction based on floating car data. *Transportation Letters, 6*(3), 118–125. doi:10.1179/1942787514Y.0000000017

Suner, F. (1988). *Crane bridges*. Istanbul, Turkey: Egitim Publications.

Sung-Ho, K., Sung-Hoon, H., & Jae-Do, K. (2014). Bending fatigue characteristics of corroded wire ropes. *Journal of Mechanical Science and Technology, 28*(7), 2853–2859. doi:10.100712206-014-0639-8

Sun, R., & Tsung, F. (2003). A kernel-distance-based multivariate control chart using support vector methods. *International Journal of Production Research, 41*(13), 2975–2989. doi:10.1080/1352816031000075224

Sun, Y., Fan, H., Bakillah, M., & Zipf, A. (2015). Road-based travel recommendation using geo-tagged images. *Computers, Environment and Urban Systems, 53*, 110–122. doi:10.1016/j.compenvurbsys.2013.07.006

Svozil, D., Kvasnicka, V., & Pospichal, J. (1997). Introduction to multi-layer feed-forward neural networks. *Chemometrics and Intelligent Laboratory Systems, 39*(1), 43–62. doi:10.1016/S0169-7439(97)00061-0

Sweet, M. (2011). Does traffic congestion slow the economy? *Journal of Planning Literature, 26*(4), 391–404. doi:10.1177/0885412211409754

Swersey, A. J., & Ballard, W. (1984). Scheduling school buses. *Management Science, 30*(7), 844–853. doi:10.1287/mnsc.30.7.844

Swift, J. A. (1987). *Development of a knowledge based expert system for control chart pattern recognition and analysis*. Oklahoma State University.

Swift, J. A., & Mize, J. H. (1995). Out-of-control pattern recognition and analysis for quality control charts using LISP-based systems. *Computers & Industrial Engineering, 28*(1), 81–91. doi:10.1016/0360-8352(94)00028-L

Tabibiazar, A. & Basir, O. (2011). Kernel-based modeling and optimization for density estimation in transportation systems using floating car data. In *Proceedings 2011 14th International IEEE Conference on Intelligent Transportation Systems (ITSC)*. IEEE. 10.1109/ITSC.2011.6083098

Tamura, S., & Tateishi, M. (1997). Capabilities of a four-layered feedforward neural network: Four layers versus three. *IEEE Transactions on Neural Networks, 8*(2), 251–255. PMID:18255629

Tasse, D. & Hong, J. I. (2014). Using social media data to understand cities.

Tay, F. E. H., & Cao, L. J. (2003). Support vector machine with adaptive parameters in financial time series forecasting. *IEEE Transactions on Neural Networks, 14*(6), 1506–1518. doi:10.1109/TNN.2003.820556 PMID:18244595

Taylor, J. G., & Coombes, S. (1993). Learning higher order correlations. *Neural Networks, 6*(3), 423–427. doi:10.1016/0893-6080(93)90009-L

Temür, R., & Bekdaş, G. (2016). Teaching learning-based optimization for design of cantilever retaining walls. *Structural Engineering and Mechanics, 57*(4), 763–783. doi:10.12989em.2016.57.4.763

Thangiah, S. R., & Nygard, K. E. (1992, March). School bus routing using genetic algorithms. In *Applications of Artificial Intelligence X* [International Society for Optics and Photonics.]. *Knowledge-Based Systems, 1707*, 387–399.

Thiessenhusen, K., Schafer, R., & Lang, T. (2003). *Traffic data from cell phones: a comparison with loops and probe vehicle data.* Germany: Institute of Transport Research, German Aerospace Center.

Thissen, U., van Brakel, R., de Weijer, A. P., Melssen, W. J., & Buydens, L. M. C. (2003). Using support vector machines for time series prediction. *Chemometrics and Intelligent Laboratory Systems, 69*(1-2), 35–49. doi:10.1016/S0169-7439(03)00111-4

Thivya, C. & Chidambaram, S. (2013). A study on the significance of lithology in groundwater quality of Madurai district, Tamilnadu (India). *Environment, Development and Sustainability.* doi:10.1007/S10668-013-9439-Z

Thornton, C., Hutter, F., Hoos, H. H., & Leyton-Brown, K. (2013, August). Auto-WEKA: Combined selection and hyperparameter optimization of classification algorithms. In *Proceedings of the 19th ACM SIGKDD international conference on Knowledge discovery and data mining* (pp. 847-855). ACM. 10.1145/2487575.2487629

Tibshirani, R., Walther, G., & Hastie, T. (2001). Estimating the number of clusters in a data set via the gap statistic. *Journal of the Royal Statistical Society. Series B, Statistical Methodology, 63*(2), 411–423.

Timur, M., Aydın, F., & Akıncı, T. Ç. (2011). The prediction of wind speed of Göztepe district of Istanbul via machine learning method. *Electronic Journal of Machine Technologies, 8*(4), 75–80.

Toğan, V. (2012). Design of planar steel frames using teaching–learning based optimization. *Engineering Structures, 34*, 225–232. doi:10.1016/j.engstruct.2011.08.035

Topcu, I. B., & Sarıdemir, M. (2008). Prediction of compressive strength of concrete containing fly ash using artificial neural networks and fuzzy logic. *Computational Materials Science, 41*(3), 305–311.

Topçu, İ. B., & Sarıdemir, M. (2008). Prediction of mechanical properties of recycled aggregate concretes containing silica fume using artificial neural networks and fuzzy logic. *Computational Materials Science, 42*(1), 74–82. doi:10.1016/j.commatsci.2007.06.011

Torkar, M., & Arzensek, B. (2002). Failure of crane wire rope. *Journal of Engineering Failure Analysis, 9*(2), 227–233. doi:10.1016/S1350-6307(00)00047-9

Torkestani, J. A. (2012). Mobility prediction in mobile wireless networks. Journal of Network and Computer Applications, 35(5), 1633-1645.

Toth, P., & Vigo, D. (Eds.). (2002). *The vehicle routing problem.* Society for Industrial and Applied Mathematics. doi:10.1137/1.9780898718515

TransitWiki. (2016, April). Automated Passenger Counter. Retrieved from https://www.transitwiki.org/TransitWiki/index.php/Category:Automated_Passenger_Counter

Trepanier, M., Morency, C., & Blanchette, C. (2009). Enhancing household travel surveys using smart card data. Transportation Research Board 88th Annual Meeting.

Trépanier, M., Morency, C., & Agard, B. (2009). Calculation of transit performance measures using smartcard data. *Journal of Public Transportation, 12*(1), 5. doi:10.5038/2375-0901.12.1.5

Trépanier, M., Tranchant, N., & Chapleau, R. (2007). Individual trip destination estimation in a transit smart card automated fare collection system. *Journal of Intelligent Transport Systems, 11*(1), 1–14. doi:10.1080/15472450601122256

Tripathi, B. K. (2015). Higher-order computational model for novel neurons. In B. K. Tripathi (Ed.), *High Dimensional Neurocomputing* (Vol. 571, pp. 79–103). doi:10.1007/978-81-322-2074-9_4

Trtnik, G., Kavčič, F., & Turk, G. (2009). Prediction of concrete strength using ultrasonic pulse velocity and artificial neural networks. *Ultrasonics, 49*(1), 53–60. doi:10.1016/j.ultras.2008.05.001 PMID:18589471

Tsang, S. W., & Jim, C. Y. (2016). Applying artificial intelligence modeling to optimize green roof irrigation. *Energy and Building, 127*, 360–369.

Tsanis, I. K., Coulibaly, P., & Daliakopoulos, I. N. (2008). Improving groundwater level forecasting with a feed forward neural network and linearly regressed projected precipitation. *Journal of Hydroinformatics, 10*(4), 317–330.

Tsubota, T., Bhaskar, A., Chung, E., & Billot, R. (2011). Arterial traffic congestion analysis using Bluetooth Duration data.

Tsung-Hsien, T., Chi-Kang, L., & Chien-Hung, W. (2005). Design of dynamic neural networks to forecast short-term railway passenger demand. *Journal of the Eastern Asia Society for Transportation Studies, 6*, 1651–1666.

Türk Standardları Enstitüsü. (2005). Çelik tel halatlar-güvenlik-bölüm 2: tarifler, kısa gösteriliş ve sınıflandırma. Ankara, Turkey.

Türkan, Y. S., Erdal, H., Ekinci, A. (2011). *Predictive stock exchange modeling for the developing Balkan countries: An application on Istanbul stock exchange*. Vol. 4, 34-51.

Türkan, Y. S., & Yumurtacı Aydogmuş, H. (2016). Passenger demand prediction for fast ferries based on neural networks and support vector machines. *Journal of Alanya Faculty of Business, 8*(1).

Türkan, Y. S., Yumurtacı Aydogmus, H., & Erdal, H. (2016). The prediction of the wind speed at different heights by machine learning methods. [IJOCTA]. *An International Journal of Optimization and Control: Theories & Applications, 6*(2), 179–187.

Turkish Standards Institute (TSE). (2000). *Requirements for design and construction of reinforced concrete structures (ICS 91.080.40)*. Ankara, Turkey: Turkish Republic Ministry of Public Works and Settlement.

Uçan, O. N., Danacı, E., & Bayrak, M. (2003). *İşaret ve görüntü işlemede yeni yaklaşımlar:yapay sinir ağları*. İstanbul Üniversitesi Mühendislik Fakültesi Yayınları.

Ulinskas, M., Woźniak, M., & Damaševičius, R. (2017). Analysis of keystroke dynamics for fatigue recognition. In International Conference on Computational Science and Its Applications (pp. 235-247). Cham, Switzerland: Springer.

United Nations Department of Economics and Social Affairs. (2015). *World population projected to reach 9.7 billion by 2050*. Available at http://www.un.org/en/development/desa/news/population/2015-report.html

Ünlü, R., & Xanthopoulos, P. (2017). A weighted framework for unsupervised ensemble learning based on internal quality measures. *Annals of Operations Research*, 1–19.

Ünlü, R., & Xanthopoulos, P. (2019). Estimating the number of clusters in a dataset via consensus clustering. *Expert Systems with Applications*.

Ünsal, Ö., & Yiğit, T. (2018). Using the genetic algorithm for the optimization of dynamic school bus routing problem. *BRAIN. Broad Research in Artificial Intelligence and Neuroscience, 9*(2), 6–21.

Urxqgzdwhu, Q. R. I. (n.d.)., qwhusuhwdwlrq ri *urxqgzdwhu 4xdolw\ 8vlqj 6wdwlvwlfdo $qdo\vlv iurp .rsdujdrq 0dkdudvkwud, qgld, *7*. Indian water quality standards code 10500-2012.

Utsunomiya, M., Attanucci, J., & Wilson, N. (2006). Potential uses of transit smart card registration and transaction data to improve transit planning. *Transportation Research Record: Journal of the Transportation Research Board(1971),* 119-126.

Uygunoğlu, T., & Yurtçu, Ş. (2006). Yapay zekâ tekniklerinin inşaat mühendisliği problemlerinde kullanımı. *Yapı Teknolojileri Elektronik Dergisi, 1,* 61–70.

Vadood, M., Semnani, D., & Morshed, M. (2011). Optimization of acrylic dry spinning production line by using artificial neural network and genetic algorithm. *Journal of Applied Polymer Science, 120*(2), 735–744. doi:10.1002/app.33252

Van Eck, N., & Waltman, L. (2009). Software survey: VOSviewer, a computer program for bibliometric mapping. Scientometrics, 84(2), 523-538. PMID:22394690

Veintimilla-Reyes, J., Cisneros, F., & Vanegas, P. (2016). Artificial neural networks applied to flow prediction: A use case for the Tomebamba river. *Procedia Engineering, 162,* 153–161. doi:10.1016/j.proeng.2016.11.031

Veloso, M., Phithakkitnukoon, S., Bento, C., Fonseca, N., & Olivier, P. (2011). Exploratory study of urban flow using taxi traces. *First Workshop on Pervasive Urban Applications (PURBA) in conjunction with Pervasive Computing,* San Francisco, CA.

Verreet, R. (1998). *Calculating the service life of running steel wire ropes.* Retrieved from http://fastlift.co.za/pdf/CASAR%20%20Calculating%20the%20service%20life%20of%20running%20steel%20wire%20ropes.pdf

Visalakshi, S., & Radha, V. (2017). Integrated framework to identify the presence of contamination in drinking water. In *2017 IEEE International Conference on Computational Intelligence and Computing Research (ICCIC)* (pp. 1-5). IEEE.

Wang, N. & Yeung, D.-Y. (2013). Learning a deep compact image representation for visual tracking. In Advances in neural information processing systems (pp. 809–817).

Wang, X., Cheng, Y., & Sun, W. (2007). *A proposal of adaptive PID controller based on reinforcement learning, 17*(1), 40–44. doi:10.1016/S1006-1266(07)60009-1

Wang, C.-H., Guo, R.-S., Chiang, M.-H., & Wong, J.-Y. (2008). Decision tree-based control chart pattern recognition. *International Journal of Production Research, 46*(17), 4889–4901. doi:10.1080/00207540701294619

Wang, C., Quddus, M. A., & Ison, S. G. (2009). Impact of traffic congestion on road accidents: A spatial analysis of the M25 motorway in England. *Accident; Analysis and Prevention, 41*(4), 798–808. doi:10.1016/j.aap.2009.04.002 PMID:19540969

Wang, F. Y., Zhang, H., & Liu, D. (2009, May). Adaptive dynamic programming: an introduction. *IEEE Computational Intelligence Magazine, 4*(2), 39–47. doi:10.1109/MCI.2009.932261

Wang, Q. (2007). Artificial neural networks as cost engineering methods in a collaborative manufacturing environment. *International Journal of Production Economics, 109*(1), 53–64. doi:10.1016/j.ijpe.2006.11.006

Wang, Y., Malinovskiy, Y., Wu, Y.-J., Lee, U. K., & Neeley, M. (2011). *Error modeling and analysis for travel time data obtained from Bluetooth MAC address matching.* Department of Civil and Environmental Engineering, University of Washington.

Wang, Z., & Liu, D. (2014). Stability analysis for a class of systems: from model-based methods to data-driven methods. *IEEE Transactions on Industrial Electronics, 61*(11), 6463–6471. doi:10.1109/TIE.2014.2308146

Wardman, M. (2006). Demand for rail travel and the effects of external factors. *Transportation Research Part E, Logistics and Transportation Review, 42*(3), 129–148. doi:10.1016/j.tre.2004.07.003

Wardman, M., & Tyler, J. (2000). Rail network accessibility and the demand for inter- urban rail travel. *Transport Reviews, 20*(1), 3–24. doi:10.1080/014416400295310

Wei, C., & Wu, K. (1997). Developing intelligent freeway ramp metering control systems. In *Conf. Proc., National Science Council in Taiwan, 7C*(3), 371–389.

Weka (2019). Capabilities of Weka. [online] Available at http://weka.sourceforge.net/doc.stable/weka/core/Capabilities.html

Weka. Retrieved from http://www.cs.waikato.ac.nz/~ml/weka/

Williams, B. M., Durvasula, P. K., & Brown, D. E. (1998). Urban free-way traffic flow prediction: Application of seasonal autoregressive integrated moving average and exponential smoothing models. *Transportation Research Record: Journal of the Transportation Research Board, 1644*(1), 132–141. doi:10.3141/1644-14

Wirerope Works Inc. (2008). *Bethlehem elevator rope technical bulletin 9*. Williamsport, PA.

Witten, I. H., & Frank, E. (2005). *Data mining: Practical machine learning tools and techniques* (2nd ed.). San Francisco, CA: Morgan Kaufmann.

Witten, I. H., & Frank, E. (2005). *Data Mining: Practical machine learning tools and techniques* (2nd ed.). San Francisco, CA: Morgan Kaufmann.

Witten, I. H., Frank, E., Hall, M. A., & Pal, C. J. (2016). The WEKA Workbench. In *Data Mining: Practical machine learning tools and techniques* (2nd ed.). Morgan Kaufmann.

Witten, I. H., Hall, M. A., & Frank, E. (2011). *Data mining: practical machine learning tools and techniques*. USA: Morgan Kaufmann Series.

World Economic Forum. (2018, January). *Harnessing artificial intelligence for the Earth*. Fourth Industrial Revolution for the Earth. In collaboration with PwC and Stanford Woods Institute for the Environment.

Wu, X., Zhu, X., Wu G.-Q., & Ding, W. (2014). Data mining with big data. *IEEE transactions on knowledge and data engineering, 26(1)*, 97-107.

Wu, H.-R., Yeh, M.-Y., & Chen, M.-S. (2013). Profiling moving objects by dividing and clustering trajectories spatio-temporally. *IEEE Transactions on Knowledge and Data Engineering, 25*(11), 2615–2628. doi:10.1109/TKDE.2012.249

Wu, L., Zhi, Y., Sui, Z., & Liu, Y. (2014). Intra-urban human mobility and activity transition: Evidence from social media check-in data. *PLoS One, 9*(5), e97010. doi:10.1371/journal.pone.0097010 PMID:24824892

Xanthopoulos, P., & Razzaghi, T. (2014). A weighted support vector machine method for control chart pattern recognition. *Computers & Industrial Engineering, 70*, 134–149. doi:10.1016/j.cie.2014.01.014

Xi, K., Tang, Y., & Hu, J. (2011). Correlation keystroke verification scheme for user access control in cloud computing environment. *The Computer Journal, 54*(10), 1632–1644. doi:10.1093/comjnl/bxr064

Xu, H., & Cho, S. B. (2014, December). Recognizing semantic locations from smartphone log with combined machine learning techniques. In 2014 IEEE 11th Intl Conf on Ubiquitous Intelligence and Computing and 2014 IEEE 11th Intl Conf on Autonomic and Trusted Computing and 2014 IEEE 14th Intl Conf on Scalable Computing and Communications and Its Associated Workshops (pp. 66-71). IEEE. 10.1109/UIC-ATC-ScalCom.2014.128

Yamaguchi, T., & Hashimoto, S. (2010). Fast crack detection method for large-size concrete surface images using percolation-based image processing. *Machine Vision and Applications, 21*(5), 797–809. doi:10.100700138-009-0189-8

Yang, X. S. (2012). Flower pollination algorithm for global optimization. In Jérôme Durand- Lose & Nataša Jonoska (Eds.), *International conference on unconventional computation and natural computation* (pp. 240-249). Berlin, Germany: Springer.

Yang, X. S. (2012). Flower pollination algorithm for global optimization. In Jérôme Durand- Lose, & Nataša Jonoska (Eds.), *International conference on unconventional computation and natural computation* (pp. 240-249). Berlin, Germany: Springer. 10.1007/978-3-642-32894-7_27

Yang, F., Jin, P. J., Cheng, Y., Zhang, J., & Ran, B. (2015). Origin-destination estimation for non-commuting trips using location-based social networking data. *International Journal of Sustainable Transportation, 9*(8), 551–564. doi:10.108 0/15568318.2013.826312

Yang, J.-H., & Yang, M.-S. (2005). A control chart pattern recognition system using a statistical correlation coefficient method. *Computers & Industrial Engineering, 48*(2), 205–221. doi:10.1016/j.cie.2005.01.008

Yang, X. S. (2010). *Nature-inspired metaheuristic algorithms* (2nd ed.). Luniver Press.

Yang, X. S., Bekdaş, G., & Nigdeli, S. M. (Eds.). (2016). *Metaheuristics and optimization in civil engineering.* Springer International Publishing. doi:10.1007/978-3-319-26245-1

Yaseen, Z. M., Deo, R. C., Hilal, A., Abd, A. M., Bueno, L. C., Salcedo-Sanz, S., & Nehdi, M. L. (2018). Predicting compressive strength of lightweight foamed concrete using extreme learning machine model. *Advances in Engineering Software, 115*, 112–125. doi:10.1016/j.advengsoft.2017.09.004

Yeh, I. C. (1998a). Modeling of strength of high-performance concrete using artificial neural networks. *Cement and Concrete Research, 28*(12), 1797–1808. doi:10.1016/S0008-8846(98)00165-3

Yeh, I. C. (1998b). Modeling concrete strength with augment-neuron networks. *Journal of Materials in Civil Engineering, 10*(4), 263–268. doi:10.1061/(ASCE)0899-1561(1998)10:4(263)

Yesilnacar, M. I., Sahinkaya, E., Naz, M., & Ozkaya, B. (2008). Neural network prediction of nitrate in groundwater of Harran Plain, Turkey. *Environmental Geology, 56*(1), 19–25. doi:10.100700254-007-1136-5

Ye, Z. R., Zhang, Y. L., & Middleton, D. R. (2006). Unscented Kalman filter method for speed estimation using single loop detector data. *Transportation Research Record: Journal of the Transportation Research Board, 1968*(1), 117–125. doi:10.1177/0361198106196800114

Ygnace, J. L. (2001). Travel time/speed estimates on the french rhone corridor network using cellular phones as probes. Final report of the SERTI V program, INRETS, Lyon, France.

Yidana, S. M., Yiran, G. B., Sakyi, P. A., Nude, P. M., & Banoeng-Yakubo, B. (2011). Groundwater evolution in the Voltaian basin, Ghana—an application of multivariate statistical analyses to hydrochemical data. *Nature and Science, 3*(10), 837.

Yigit, T., & Unsal, O. (2016). Using the ant colony algorithm for real-time automatic route of school buses. [IAJIT]. *The International Arab Journal of Information Technology, 13*(5).

Yilmaz, A., Javed, O., & Shah, M. (2006). Object tracking: A survey. *ACM Computing Surveys*, *38*(4), 13, es. doi:10.1145/1177352.1177355

Yoon, H., Zheng, Y., Xie, X., & Woo, W. (2012). Social itinerary recommendation from user-generated digital trails. Personal and Ubiquitous Computing, 16(5), 469-484.

Young, J. R. (2018). *Keystroke Dynamics: Utilizing Keyprint Biometrics to Identify Users in Online Courses*. Brigham Young University.

Yucel, M., Bekdaş, G., Nigdeli, S. M., & Sevgen, S. (2018). Artificial neural network model for optimum design of tubular columns. In *Proceedings of 7th International conference on applied and computational mathematics (ICACM '18)* (pp. 82-86). Rome, Italy: International Journal of Theoretical and Applied Mechanics.

Yucel, M., Bekdaş, G., Nigdeli, S. M., & Sevgen, S. (2019). Estimation of optimum tuned mass damper parameters via machine learning. *Journal of Building Engineering*, 100847.

Yucel, M., Nigdeli, S. M., & Bekdaş, G. (2019, Sept. 22-25). *Generation of an artifical neural network model for optimum design of I-beam with minimum vertical deflection*. Paper presented at 12th HSTAM International Congress on Mechanics. Thessaloniki, Greece.

Yumurtacı Aydogmus, H., Ekinci, A., Erdal, H. İ., & Erdal, H. (2015b). Optimizing the monthly crude oil price forecasting accuracy via bagging ensemble models. *Journal of Economics and International Finance*, *7*(5), 127–136. doi:10.5897/JEIF2014.0629

Yu, Y., Li, W., Li, J., & Nguyen, T. N. (2018). A novel optimised self-learning method for compressive strength prediction of high-performance concrete. *Construction & Building Materials*, *184*, 229–247.

Zarandi, M. H. F., & Alaeddini, A. (2010). A general fuzzy-statistical clustering approach for estimating the time of change in variable sampling control charts. *Information Sciences*, *180*(16), 3033–3044. doi:10.1016/j.ins.2010.04.017

Zhang, D., Feng, C., Chen, K., Wang, D., & Ni, X. (2017). Effect of broken wire on bending fatigue characteristics of wire ropes. *International Journal of Fatigue*, *103*, 456–465.

Zhang, L., & Suganthan, P. N. (2016). A comprehensive evaluation of random vector functional link networks. *Information Sciences*, *367–368*, 1094–1105. doi:10.1016/j.ins.2015.09.025

Zhang, Y., Guo, F., Meng, W., & Wang, X. Q. (2009). Water quality assessment and source identification of Daliao river basin using multivariate statistical methods. *Environmental Monitoring and Assessment*, *152*, 105–121.

Zhang, Y., & Xie, Y. (2007). Forecasting of short-term freeway volume with v-support vector machines. *Transportation Research Record: Journal of the Transportation Research Board*, *2024*(1), 92–99. doi:10.3141/2024-11

Zhao, J., Rahbee, A., & Wilson, N. H. (2007). Estimating a rail passenger trip origin-destination matrix using automatic data collection systems. Computer-Aided Civil and Infrastructure Engineering, 22(5), 376-387.

Zhao, X. (2017). A scientometric review of global BIM research: Analysis and visualization. *Automation in Construction*, *80*, 37–47. doi:10.1016/j.autcon.2017.04.002

Zhao, X., Zuo, J., Wu, G., & Huang, C. (2019). A bibliometric review of green building research 2000–2016. *Architectural Science Review*, *62*(1), 74–88. doi:10.1080/00038628.2018.1485548

Zheng, Y. (2011). Location-based social networks: Users. Computing with spatial trajectories, Springer: 243-276.

Zheng, Y., & Xie, X. (2011). Location-based social networks: Locations. In Computing with spatial trajectories. New York, NY: Springer. 277-308.

Zheng, Y., Capra, L., Wolfson, O., & Yang, H. (2014). Urban computing: Concepts, methodologies, and applications. [TIST]. *ACM Transactions on Intelligent Systems and Technology, 5*(3), 38. doi:10.1145/2629592

Zheng, Y., Li, Q., Chen, Y., Xie, X., & Ma, W.-Y. (2008). Understanding mobility based on GPS data. *Proceedings of the 10th international conference on Ubiquitous computing*, ACM.

Zheng, Y., Liu, L., Wang, L., & Xie, X. (2008). Learning transportation mode from raw gps data for geographic applications on the web. *Proceedings of the 17th international conference on World Wide Web*, ACM. 10.1145/1367497.1367532

Zheng, Y., Xie, X., & Ma, W.-Y. (2010). GeoLife: A collaborative social networking service among user, location and trajectory. *IEEE Data Eng. Bull., 33*(2), 32–39.

Zheng, Y., Zhang, L., Xie, X., & Ma, W.-Y. (2009). Mining interesting locations and travel sequences from GPS trajectories. *Proceedings of the 18th International Conference on World Wide Web*. ACM. 10.1145/1526709.1526816

Zhou, C. (2008). A study of keystroke dynamics as a practical form of authentication. [ebook] Available at https://www.semanticscholar.org/paper/A-Study-of-Keystroke-Dynamics-as-a-Practical-Form-May/39b6cfab3f58231df47d0604c196387356d503bc

Zhou, C., Sun, C., Liu, Z., & Lau, F. (2015). A C-LSTM neural network for text classification. *arXiv preprint arXiv:1511.08630.*

Zhu, T., Yan, Z., & Peng, X. (2017). *A modified nonlinear conjugate gradient method for engineering computation.* 1–11. doi:10.1155/2017/1425857

Zignani, M., & Gaito, S. (2010). *Extracting human mobility patterns from gps-based traces. Wireless Days (WD), 2010 IFIP.* IEEE.

Zou, Q., Cao, Y., Li, Q., Mao, Q., & Wang, S. (2012). CrackTree: Automatic crack detection from pavement images. *Pattern Recognition Letters, 33*(3), 227–238. doi:10.1016/j.patrec.2011.11.004

About the Contributors

Gebrail Bekdaş, Associative Professor, is researcher in Structural Control and Optimization at Istanbul University - Cerrahpaşa. He obtained his DPhil in Structural Engineering from Istanbul University with a thesis subject of design of cylindrical walls. He co-organized 15th EU-ME Workshop: Metaheuristic and Engineering in Istanbul. In optimization, he organized several mini-symposiums or special sections in prestigious international events such as the Biennial International Conference on Engineering Vibration (ICoEV-2015), 3rd International Conference on Multiple Criteria Decision Making (MCDM 2015), 11th. World Congress on Computational Mechanics (WCCM2014), International Conference on Engineering and Applied Sciences Optimization (OPTI 2014), 11th Biennial International Conference on Vibration Problems (ICOVP2013), 3rd European Conference of Civil Engineering and 10th – 11th International Conference of Numerical Analysis and Applied Mathematics (ICNAAM). He co-edited Metaheuristics and Optimization in Civil Engineering published by Springer and he is one of the guest editors in 2017 special issue of KSCE Journal of Civil Engineering. He has authored more than 200 papers for journals and scientific events.

* * *

Sheik Abdullah A. is working as Assistant Professor, Department of Information Technology, Thiagarajar College of Engineering, Madurai, Tamil Nadu, India. He completed his Post Graduate in M.E (Computer Science and Engineering) at Kongu Engineering College under Anna University, Chennai. He is pursuing his Ph.D in the domain of Medical Data Analytics at Anna University Chennai. He has been awarded as gold medalist for his excellence in the degree of Post Graduate, in the discipline of Computer Science and Engineering by Kongu Engineering College. He is an active member of ACM (Association for Computing and Machinery) and being rewarded as ACM Coach Award for ACM World level International Collegiate Programming contest for the year 2013 and 2014. He has handled various E-Governance government projects such as automation system for tracking community certificate, birth and death certificate, DRDA and income tax automation systems. He has published research articles in various reputed journals and International Conferences. He has received the Honorable chief minister award for excellence in E-Governance for the best project in E-Governance for the academic year 2015-16. He has been assigned as a reviewer for various reputed journals such as European Heart Journal, International Journal of Fuzzy Systems, and NASA Springer. He has delivered various guest lecturers in the area of Predictive Analytics, data prediction using R tools and so on. He is working interactively with various industries like CTS, IBM, Xerox Research labs. He has authored various books under IGI Global Publishers, USA.

Ivo Bukovsky graduated from Czech Technical University in Prague (CTU) where he received his Ph.D. in the field of Control and System Engineering in 2007. Ivo is an associate prof. at the Dpt. of Mechanics, Biomechanics and Mechatronics at CTU at the Faculty of Mech. Eng. His interests include neural networks, adaptive novelty detection and computational intelligence methods for industrial systems as well as for biomedical applications.

Sina Dabiri received his MS in Transportation Engineering from Sharif University of Technology, Tehran, Iran in 2012. Then, he simultaneously obtained his Ph.D. in the Department of Civil Engineering and MS in the Department of Computer Science at Virginia Tech in 2018. His research interests include machine learning and deep learning in intelligent transportation systems (ITS).

Didem Filiz has received her Bsc.degree from Computer Engineering at Ankara University, Ankara, Turkey. She has worked in SoundLabAI as a software developer in 2018. Currently, she is an engineer at Software and Design of Avionics System Department of Aselsan, Ankara, Turkey. Her current interests are software engineering and machine learning applications.

Vivek Gundraniya is currently a scholar of Post Graduation in Energy & Environment at Symbiosis Institute of International Business, Symbiosis International University, India. He has received his Bachelor's degree in Environmental Engineering from Gujarat Technological University, India in 2015. His scholarly areas of interest include Energy and Environment domain, Renewable Energy, Waste Management, Sustainability, and Environmental Management Systems.

Kevin Heaslip is an Associate Professor of Civil & Environmental Engineering and Coordinator of the Transportation Infrastructure and Systems Engineering group at Virginia Tech. Dr. Heaslip also serves as the Associate Director of the Electronic Systems Laboratory at the Hume Center for National Security and Technology. Dr. Heaslip is a registered professional engineer in New Hampshire with more than 18 years experience in the transportation field, including consulting firms on transit and highway projects, and university research and education. Dr. Heaslip is the author or co-author of more than 150 publications, including journal articles, conference proceedings, and technical reports. Dr. Heaslip has been the major professor for over 30 graduate students and 25 undergraduate students. Dr. Heaslip has also been a Principal or Co-Principal Investigator of projects funded at over $18 million dollars with his share being over $8,000,000. His research interests lie in the intersections of transportation, smart cities, infrastructure and community resilience, and cybersecurity. He is a member of National Academy of Science's Resilient America Roundtable. Dr. Heaslip earned a B.S. in Civil & Environmental Engineering from the Virginia Tech (with a Public and Urban Affairs Minor); an M.S. in Civil & Environmental Engineering (Transportation & Infrastructure Systems Engineering) from Virginia Tech, and a Ph.D. in Civil & Environmental Engineering (Transportation Engineering) from the University of Massachusetts Amherst.

Keerthy K.'s area of specialization is water resource management and Environmental Engineering. Their research study is water resource management of Madurai city.

Kaveh Bakhsh Kelarestaghi, PhD, is a Senior Transportation Engineer at the ICF Incorporated LLC. Dr. Kelarestaghi is an intelligent transportation systems analyst who is experienced in providing process-oriented data analysis and problem-solving, with in-depth knowledge of statistical modeling, and machine learning to overcome challenges in transportation purview concerning cybersecurity, travel behavior, and safety. Dr. Kelarestaghi has wide-ranging experience with analyzing traveler's behavior related datasets through the means of visual representation of data and conducting machine learning and advanced statistical models. Prior to consulting at ICF, Mr. Kelarestaghi worked at the Federal Highway Administration and Maryland State Highway Administration. He received his M.Sc. in Civil and Environmental Engineering from University of Tehran, Iran in 2013. His research interests are in Intelligent Transportation Systems Security and Safety, transportation resiliency, simulation, and big data analysis.

Sinem Büyüksaatçı Kiriş started her career as a research assistant in 2007 in Industrial Engineering Department of Istanbul University. She received her both BSc and MSc degree on industrial engineering in 2006 and 2009 respectively from the same university. She was awarded with a PhD degree in the same area in 2015 in Istanbul University. She is still a member of the Industrial Engineering Department in Istanbul University - Cerrahpasa, which was born with the separation of Istanbul University according to the law dated on 18 May 2018, enacted by Turkish National Assembly. She is working as an Assistant Professor. Her research area includes optimization, metaheuristic algorithms, simulation, design of experiment and sustainability. She has several national and international research and conference papers published.

Tuğba Özge Onur is an Assistant Professor in the Department of Electrical-Electronics Engineering at Zonguldak Bulent Ecevit University. She received her MSc and PhD degrees, with highest honors, in Electrical-Electronics Engineering from Zonguldak Bulent Ecevit University in 2008 and 2016, respectively. Her doctoral thesis study was partly carried out at Luleå University of Technology (Luleå, Sweden) and University of Wisconsin-Madison (Madison, Wisconsin). Her research interests cover the signal processing, ultrasound signal processing, system identification and nonlinear systems. She has published papers in the field of signal processing.

Yusuf Aytac Onur is an Associate Professor at Zonguldak Bulent Ecevit University, Zonguldak, Turkey. He received the BSc degree from University of Kocaeli, MSc degree from Istanbul Technical University, and PhD degree from Istanbul Technical University in 2003, 2006, and 2010, respectively. He has carried out research into material handling and, in particular, elevator systems.

Tuncay Özcan is an Associate Professor in the Department of Industrial Engineering at the Istanbul University - Cerrahpasa. His research interests include optimization, metaheuristic algorithms, data mining, forecasting. He has several publications in national and international journals including Expert Systems with Applications, International Journal of Computational Intelligence Systems and International Journal of Information Technology & Decision Making.

Chandran S.'s area of Interest is Water Resource Management. They are a secretary of the International Water Association, India Chapter. Their current project works on the reuse of Treated water for wasteland reclamation.

Ravi Sharma is currently working as an Assistant Professor in the Department of Energy and Environment with Symbiosis Institute of International Business (SIIB) under Symbiosis International University, Pune, India. Masters of Science with Environmental Sciences and Toxicology as specialization and Ph.D. in Forestry under discipline Forest Ecology and Environment from Forest Research Institute University (FRIU), Dehradun, India. He was also awarded Junior Research Fellowship (2003-2005) from University Grants Commission (UGC) during his research tenure with Indian Institute of Forest Management (IIFM), Bhopal. Academic area of interest are Environmental Management Systems, Life Cycle Assessment, Sustainability Reporting, Environment Health and Safety, Ecological Risk Assessment and Environmental Monitoring. He is engaged in research and academic activities at University level. IRCA approved Lead Auditor for ISO- 14001:2015 and ISO45001.

Ömer Özgür Tanrıöver has received Science and Technology Policy MSc (2001), Information Systems MSc (2002) and Ph.D. (2008) degrees from the Informatics Institute, Middle East Technical University (METU), Ankara, Turkey. He was a research assistant at the Center for Science and Technology Policy Research — METU between 1998 and 2005. Between 2005 and 2011, he was a certified Information Systems Auditor (CISA) in the Information Management Department at the Banking Regulation Agency of Turkey. Since 2012, he has been with the Computer Engineering Department of Ankara University. His current research interests are Software Quality, Information Technology Management and Machine Learning Applications.

Osman Hürol Türkakın was born in Istanbul and has a Bachelor degree from civil engineering Istanbul Technical University.

Ramazan Ünlü has a Ph.D. in Industrial Engineering from the University of Central Florida with a particular interest in data mining, machine learning, and deep learning. He is currently working in the Department of Management and Information Systems in Gümüşhane University as an Assistant Professor.

Index

Ensure Quality Research is Introduced to the Academic Community

Become an IGI Global Reviewer for Authored Book Projects

Premier Reference Source

Emerging GIS Applications for Emergency and Disaster Management

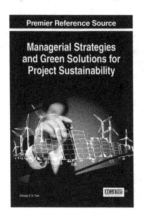

Premier Reference Source

Managerial Strategies and Green Solutions for Project Sustainability

Premier Reference Source

Comparative Approaches to Using R and Python for Statistical Data Analysis

Premier Reference Source

Solutions for High-Touch Communications in a High-Tech World

The overall success of an authored book project is dependent on quality and timely reviews.

In this competitive age of scholarly publishing, constructive and timely feedback significantly expedites the turnaround time of manuscripts from submission to acceptance, allowing the publication and discovery of forward-thinking research at a much more expeditious rate. Several IGI Global authored book projects are currently seeking highly-qualified experts in the field to fill vacancies on their respective editorial review boards:

Applications and Inquiries may be sent to:
development@igi-global.com

Applicants must have a doctorate (or an equivalent degree) as well as publishing and reviewing experience. Reviewers are asked to complete the open-ended evaluation questions with as much detail as possible in a timely, collegial, and constructive manner. All reviewers' tenures run for one-year terms on the editorial review boards and are expected to complete at least three reviews per term. Upon successful completion of this term, reviewers can be considered for an additional term.

If you have a colleague that may be interested in this opportunity, we encourage you to share this information with them.

www.igi-global.com

Publisher of Peer-Reviewed, Timely, and
Innovative Academic Research Since 1988

IGI Global's Transformative Open Access (OA) Model:
How to Turn Your University Library's Database Acquisitions Into a Source of OA Funding

In response to the OA movement and well in advance of Plan S, IGI Global, early last year, unveiled their OA Fee Waiver (Offset Model) Initiative.

Under this initiative, librarians who invest in IGI Global's InfoSci-Books (5,300+ reference books) and/or InfoSci-Journals (185+ scholarly journals) databases will be able to subsidize their patron's OA article processing charges (APC) when their work is submitted and accepted (after the peer review process) into an IGI Global journal.*

How Does it Work?

1. When a library subscribes or perpetually purchases IGI Global's InfoSci-Databases including InfoSci-Books (5,300+ e-books), InfoSci-Journals (185+ e-journals), and/or their discipline/subject-focused subsets, IGI Global will match the library's investment with a fund of equal value to go toward subsidizing the OA article processing charges (APCs) for their patrons.

   Researchers: Be sure to recommend the InfoSci-Books and InfoSci-Journals to take advantage of this initiative.

2. When a student, faculty, or staff member submits a paper and it is accepted (following the peer review) into one of IGI Global's 185+ scholarly journals, the author will have the option to have their paper published under a traditional publishing model or as OA.

3. When the author chooses to have their paper published under OA, IGI Global will notify them of the OA Fee Waiver (Offset Model) Initiative. If the author decides they would like to take advantage of this initiative, IGI Global will deduct the US$ 1,500 APC from the created fund.

4. This fund will be offered on an annual basis and will renew as the subscription is renewed for each year thereafter. IGI Global will manage the fund and award the APC waivers unless the librarian has a preference as to how the funds should be managed.

Hear From the Experts on This Initiative:

"I'm very happy to have been able to make one of my recent research contributions, 'Visualizing the Social Media Conversations of a National Information Technology Professional Association' featured in the *International Journal of Human Capital and Information Technology Professionals*, freely available along with having access to the valuable resources found within IGI Global's InfoSci-Journals database."

– **Prof. Stuart Palmer**,
Deakin University, Australia

For More Information, Visit: www.igi-global.com/publish/contributor-resources/open-access or contact IGI Global's Database Team at eresources@igi-global.com.